Mud Engineering Simplified

MUD ENGINEERING SIMPLIFIED

DR. MILAP GOUD

Become Shakespeare.com

First published in 2017 by

Becomeshakespeare.com
Wordit Content Design & Editing Services Pvt Ltd
Unit - 26, Building A-1, Nr Wadala RTO,
Wadala (East),
Mumbai 400037, India
T:+91 8080226699

ISBN: 978-93-86487-67-4

Contents

ABOUT THE AUTHOR

Dr. Milap, born to a teacher, has inherited the basic traits of learning and sharing as a legacy. The book is an outcome of these values. Having spent 37 years and encountered several situations which warranted an innovative approach towards arriving at solutions, the author has amalgamated his passion of subject and strong desire to give back to mud engineering fraternity whatever he learnt.

The ability of Dr. Milap to adapt and deliver solutions is boosted by his diversity in exposure to all major types of drilling environment as well as drilling fluid systems. A widely travelled professional who has the distinction of working with all types of nationalities without any conflict, is a testimony of the maturity as well as flexibility in personality.

A sports-lover by nature and a marathon runner, has always drawn inspiration to do his bit for the under-privileged sections of the society by raising funds in conjunction with NGOs.

This is an initiative of the firm M/S Knnamp Enterprise, owned and managed by author to provide consultancy services in drilling fluids engineering discipline of oil/gas drilling. The book represents the spirit of sharing, learning and growing together.

ACKNOWLEDGEMENTS

I remain indebted and thankful to all my family members and friends who had been a constant source of inspiration to help me put my thoughts in words. My professional associates also greatly encouraged me to accomplish the task of summing up what all I learnt. All those cups of coffee prepared by my wife to keep me awake and motivated during the writing of book deserve a special mention here, without which, the book would not have reached to the stage of completion. Abhishek Gaur, my brother's son, deserves a pat on his back for his support in drawing some of the graphs in word format. My daughter-in law's unconditional faith in my ability has made the process worth cherishing. My son had been my strongest and most-constructive critic and that has helped me in taking the book to a level where it finally is - simple, yet equally useful to students and fresh engineers. Whenever I got tired and felt like taking long break, my brothers fuelled fire in my intentions to help me bring the book out in planned time. Last but not the least, the innocent smile and playful acts of both my grandsons had been the biggest stress-buster to enable me retain my creativity while writing the book. No amount of thanks will ever be enough to compensate for their love towards me.

DEDICATION

Dedicated to my mother,
Who perpetually kept loving me -
Come what may.

PREFACE

Ever since the dawn of civilization mankind has constantly made conscious efforts to make the life more comfortable by devising various ways of making the technology work for it. This seemingly never-ending race has a long story which cannot be covered in a single book. Although mud engineering is a smaller part of the small energy sector, which works for mankind, but this too has a story of its own which cannot be compiled in a single book. This book also does not aim so and opts only to cover those facets with which the author has spent most of its time, that too focussing only on the basics, leaving apart the advanced concepts to be taken up in volumes to come.

OBJECTIVE

The objective of this book is to provide a platform to new entrants to familiarize themselves with basics of mud engineering in a language with less technical jargon and keep the description simple enough for easy grasping. It also attempts to capture all the author's own learning curves of drilling fluid engineering activities and make the information available along with all other important concepts of drilling fluid engineering at one place for ready reference to help fluids engineers in running the mud system on the rig. It also allows them to innovate their own unique ways of running the mud by explaining the complexities of mud chemistry and amalgamating it with simple concepts of chemistry. Thus, the book aims at encouraging engineers to challenge age old myths and extract the hidden essence out of them to use it in a way, most suited to them in their own prevailing situation. In addition to it, the book also aims at familiarizing the readers with relevant field practices/tips to help the mud engineering fraternity take a leaf out of it for their own benefit, and thus, prove to be an asset to their rig team on-board. For professionals other than drilling fluid engineering in oil field, the book helps them understand better as to what a mud engineer does to achieve and thus narrow down the gap between two entities to help improve the overall progress of the rig by respecting/recognizing and appreciating each-others' capabilities and limitations.

ABOUT THE CONTENT OF THE BOOK:

The book contains technical information on various aspects of mud engineering, where a conscious effort has been made to put proven scientific facts in a way, which makes more sense to its practical application in managing mud on the rig and not only narrate facts of an academic nature that are already well established and known to many, if not all. The interpretations as well as inferences drawn from these academic and theoretical facts, are author's own perception which may / may not match with others. A match in interpretation with others has been just an incident of agreement of professionals on technical aspects and not a case of using the others' information. Many friends have generously shared their views and author would like to acknowledge their contribution in the form of showing gratification to all of them, whether the suggestions find inclusion or not. During the long journey of professional association, the author did get influenced by several illustrious personalities as well as the fellow rig crew who have shaped his own views as well as the outlook towards solving the problems. Author thankfully acknowledges their contributions which may not directly be visible, but has a deep-rooted impact on the way the author has shaped up professionally over last four decades.

Although an attempt has been made to cover all important aspects of drilling fluids, but the subject of mud-engineering is too vast to be accommodated in a single book. Therefore, emphasize is laid on operations which are more relevant to present day trends with extensive coverage on deep-water / ultra-deep-water drilling. Inclusion of all aspects of mud engineering in a book is like trying to cover the entire earth with a single umbrella, which is not realistically possible. Subsequent volumes of the book in future shall cater to the aspects not covered here.

Neither the well-specific program nor the vendor-specific additive description has been included to ensure that the book does not represent any company and is being aimed for projection as a generic book. Since the drilling fluid cannot be discussed in isolation from other aspects of E&P activities in general and drilling in specific, frequent mention of practices relevant to drilling process and other important aspects can be found interspersed in the book. Description about more sophisticated equipment has been excluded in the book since they are more relevant to the research and development type of work and inclusion of research quality of mud engineering related information is beyond the scope of this book. The language of the book is purposely kept simple with less technical jargon, and such framed that the readers can connect with the content of the book at a level where they can use the information to improve their own understanding of the subject and apply the information wherever required and found fit. The book also intends to serve the purpose of a launch pad in terms of information availability for all new entrants, and all the readers in general to acclimatize themselves with the basics of drilling fluid engineering and enable them to understand the relevance of drilling fluid engineering activities in relation to the type of operations for which they have been framed. The book aims at teaching the basics of mud engineering to new entrants and to provoke the middle level mud engineers to devise their own solutions, more than simply offering a solution. In a way, it is a compilation of facts/features from which the experienced engineer can extract their solution in their own way. It aims to trigger the curiosity of field engineers to enable them to seek a solution for themselves as well as facilitate the petroleum engineering students to look beyond the academic syllabus and get a glimpse of what is there in store for aspiring mud engineers in the industry.

There are many areas/segments of mud engineering, which remain untouched in the book. Nano fluids, MMH, Exhaustive explanations on Rheology and so on and so forth, but as they say – Every good thing has to come to an end, which need not be complete but just a milestone in the long journey. The book precisely characterizes this concept and strives to include as many aspects as available, in the future.

DEFINITION:

Defining drilling fluid is probably as difficult as formulating, because it is so diversified in composition and performance that many in industry differ with each other as to how to define drilling fluid. In simplest terms without getting trapped in technical cobweb and broadly speaking
A drilling fluid can be defined as a complex chemical entity of continuous liquid/gas phase and suspended / dispersed / dissolved inorganic and organic chemicals tailor composed to perform a specific role/s during exploration and exploitation of hydrocarbons.

The definition need not be technically correct, but it helps everyone understand what drilling fluid is, especially for professionals who are from backgrounds other than drilling fluids.

CHAPTER 1

INTRODUCTION

The drilling fluid (popularly also called as mud) is related either directly or indirectly to almost every drilling operation. It is that complex and colloidal suspension, which travels in every corner of the hole to perform the functions, about which we shall speak in detail as we proceed further. Many aspire that mud formulation alone would not only solve all the problems, but could also make up for the lapses creeping into the system due to incorrect / sub-optimal operating practices. This simply cannot be the case. Drilling fluid engineering is an integral part of the drilling process and should be viewed as a supporting service to other facets of the operation. Having said that, this is also true that correct selection and execution of the drilling fluid goes a long way in the success of any drilling operation and incorrect selection / implementation of the fluid engineering program can jeopardize any drilling campaign despite following best industry practices!

Mud engineering is a judicious blend of science and art, to the extent that one not only MUST have practical experience of running the mud on a rig, but also have a clear understanding of the applicable principles of chemistry and physico-chemical features to qualify himself as a great mud engineer. One can find many good mud engineers who either run mud very efficiently on a rig, or have a very good basic knowledge of principles of mud engineering, but the great mud engineer is an amalgamated version of both the types, that helps efficient running of mud system as well as provide unconventional solutions to well-bore problems with his acquired practical knowledge and simultaneously, improving the effectiveness of the fluid by incorporating his theoretical knowledge.

PLANNING AND SELECTING A MUD SYSTEM

The first objective in planning a mud program is the selection of a mud system that will minimize the amount of lost time in the drilling operation while retaining the original potential of the rock to enable the flow of hydrocarbon to surface without any permeability impairment and if possible, help further smoothen the process of enabling the hydrocarbons to flow to the surface. Such mud will invariably be economical regardless of its overall cost or cost per barrel. Operationally, it is desirable to have a mud system that is flexible enough to allow changes to be made to meet the ever-changing requirements of drilling/ completion. The key to differentiating a good mud program and an ordinary mud program lies in the ability of the program to offer adequate provision for continuous and dynamic changes in response to the surprises offered by the well. Each possible change in the mud program should be thought of and included in advance. This will allow current treatment of the mud being in line with future requirements, which can be different than predicted.

The process of the mud system selection and finalizing the mud program begins with the acquisition of all relevant geological and offset well information. This includes pore pressure and fracture gradient profiles, formation characteristics, intervals of possible borehole instability, temperature gradient, chemical / physico-chemical and mechanical reactivity of the matrix, location of soluble salt beds and the possibility of sour gas or saltwater flows. Accurate, complete and timely availability of information is necessary for effective planning and flawless execution of mud engineering plan. This necessitates information sharing / communication between persons connected with the different parts of the operation.

The mud and casing programs should be integrated. At times, the solution to some of the problems related to drilling is to isolate the open hole with the help of casing or to alter the rock environment such that the open hole can be managed safely with just one type of mud and one set of parameters / properties of mud. This is especially true of cases, where two such situations co-exist, which require two completely opposite mud properties to be achieved at the same time. For instance, a loss zone and a high-pressure zone open at the same time leads to a conflict that cannot be resolved by the mud alone. The solution would often lie in either isolating the intervals with the help of casing or to chemically / mechanically alter the rock environment to allow both the situations to co-exist without causing any NPT and enabling operational continuity. Therefore, a mud program is almost invariably planned for each casing interval, with a contingency plan in place for deviations from the casing Program, should there be surprises!

After selecting a suitable mud program, consideration should be given to its flawless execution. The performance of a drilling fluid is determined more by engineers / managers who run and manage the mud than by the products used. Exotic mud systems, which are not understood by those running them, often result in failure.

In selecting a mud company, due consideration should be given to the technical ability of the personnel who will be working on the project and the quality of backup support available from the service providing company. There are instances where the quality and performance of their products is also as important as the technical and logistical services rendered by them. The cost of products should be considered only based on the overall value being added to operations and not on the cost of the sack. The objective is to minimize overall project cost and this should be aimed by focusing on the final dollar equivalent of deliverables from the mud than anything else. This can happen only when the performance of the fluid is continuously monitored by experienced staff who can then take corrective measures, if required, to further improve upon the accountability of fluid in use. Almost any type of mud, if properly planned and stringently implemented, can serve the purpose of drilling the well, but the right type of fluid is the one which does it without permanently damaging the rock and providing a stable as well as gauged hole which sustains its integrity for prolonged periods to provide enough flexibility to stretch an interval to its otherwise un-drillable limit and/or to accommodate unscheduled program without bore-hole getting unstable.

Running a mud system primarily consists of controlling the type and amount of solids in the mud, their chemical, physical and the physico-chemical environment. All the mud properties are controlled by controlling these compositional factors. Consequently, an adequate solids control program should be an integral part of every mud program.

The rig mud facilities should be designed to provide proper arrangements for solids control and mud mixing equipment. Rigs with insufficient mud pit space can make mud treatment and solids control ineffective. Inadequate rig facilities are sometimes a cause for a considerable amount of lost time, poor mud properties and increased mud costs. The situation is further complicated by the requirement of huge volume of mud to be kept in active circulation mode in the deep-water environment.

Accurate mud tests are equally important and necessary for proper control of the mud properties. We cannot have a successful drilling operation unless all facets of it are well planned and executed, including proper application of the drilling fluid.

An experienced mud engineer with prior exposure in same operating area / environment is a big asset to any drilling team and there is no substitute for this.

Another special feature of the book which deserves special mention is – It purposely does not include mathematics oriented discussions to a great length. Such information is available in abundance in many contemporary books. The key point which distinguishes the book from other books is its attempt to knit technical facts and field operations in such a way that it attains utilization for petroleum engineering students and field professionals alike, while explaining the technical facts in a common-man's language to make understanding and learning a fun.

If I should sum-up the intention of the book in a sentence, I would say

Never be a slave to calculated numbers: Use them judiciously by linking them with your observations, intuitions and confidence to cook a solution!

CHAPTER 2

FUNCTIONS OF DRILLING FLUID

There are many functions of a drilling fluid. The most basic of these are listed below:

1. Balance formation pressure and near well-bore stresses
2. Carry and bring cuttings, cavings and debris, if any (falling in the hole) to the surface
3. Prevent/eliminate/minimize damage to the rock meant for exploitation of hydrocarbons.
4. Clean the area underneath the drill-bit to facilitate uninterrupted excavation.
5. Cool and lubricate bit and drill string
6. Seal permeable formations
7. Stabilize borehole and keep it stable for prolonged periods.
8. Prevent the corrosion of tubular.
9. Provide a propagating medium for a log signal to be carried to surface.
10. Keep the cuttings in suspension when there is no circulation.

11. Plaster the walls of open hole with the liquid which also acts as a semi-permeable membrane to regulate the osmosis phenomena.
12. Transmit hydraulic energy to the bit / tool.

In addition to these functions, there are some aspects, with which the drilling fluid should not interfere. They are:

1. Formation evaluation
2. Completion operations
3. Production operations
4. Cementing operations
5. Reserve estimate

The list could have more applications, but even these applications are enough to indicate the complex nature of the role of drilling fluids in the drilling operation. Some of the expectations / roles of the drilling fluid contradict each other. It is therefore obvious that compromises will always be necessary when designing a fluid to carry out all these functions, which in some cases require fluids of opposing properties. The most important functions in an operation at that moment should be given the highest weighting in the design of the drilling fluid. This sort of judgment comes from drilling more wells in the same area with the same team who can catch up on all the flaws and apply their knowledge to overcome them in the next well.

Many of these functions are controlled by more than one property of mud and all of them are being discussed in more detail in forthcoming discussion.

PRESSURE CONTROL

The density of drilling fluid must be such that the hydrostatic pressure exerted by the mud column will prevent the flow of formation fluids into the wellbore and overcome the mechanical stresses near the exposed area of well-bore to prevent its failure, whether instantaneous or time dependent. This is the first requirement of any drilling fluid and it must be often prioritized over any other mud property or function. The equation for calculating hydrostatic pressure is:

Hydrostatic Pressure, psi = (depth, ft.) (mud weight, lb. /Gal) (0.052)

Pressure control would be rather simple, if it consisted only of balancing the hydrostatic and formation pressures in the static condition. Pressure is required to cause a fluid to flow and this pressure dissipates in the form of frictional losses along the entire flow path.

Consequently, the total pressure at any point in a circulating system is the sum of the hydrostatic pressure at that point and in the circulating pressure drop from that point to the exit point. Under normal circulating conditions, the pressure at any given point in the hole is the sum of the hydrostatic pressure at that point and the circulating pressure drop from that point to the flow line.

The total pressure at any point in the annulus is the hydrostatic minus the flowing pressure drop from the surface to that point in the annulus.

If the formation pressure is greater than the wellbore pressure under swab conditions, the formation fluid will flow into the well when the pipe is pulled. If the fracture pressure of a formation is less than the pressure at that depth under surge conditions, the formation will be fractured while running the pipe and lost circulation will occur. These factors must be duly considered when establishing the required density of a mud.

Normally the mud density will be run slightly higher than required to not only balance the formation pressure under static conditions, but also beyond breakout pressure. Breakout pressure, is that value of the pressure, below which the well-bore can collapse due to release of its own stresses near exposed wall of well-bore. This also allows for a safety margin under static conditions and offsets the same amount of negative swab pressure.

By the same analogy, if the surge or the circulating pressure drop causes the total pressure to exceed the fracture pressure of a formation, the pipe running speed or the circulating rate must be decreased enough to prevent fracturing from occurring. When it becomes impossible to meet minimum and maximum pressure requirements at realistic pipe moving speeds or circulating rates, it is perhaps the time to case the hole, or look for other means to strengthen the rock artificially at the weakest point of open hole.

After breaking circulation, the annular pressure drop will decrease for some time. This is due to "shearing down" the gel structure of the mud. The shear rate in the annulus is not high enough to break all flocculation bonds and the "bottoms-up" mud will remain abnormally high in viscosity. As this mud becomes cooler, while traversing up in the annulus, the viscosity will begin to increase.

HOLE-CLEANING

The ability to lift particles of various sizes out of the hole is one of the most important functions of drilling fluid. This is the only way by which the drilled or sloughed rock is carried out of the hole. To give a realistic view of the situation, following statement is very relevant

"In a 12¼-inch hole, about 130 pounds of earth material is generated for every foot of hole drilled. Ideally speaking the entire volume of this rock must be out of the hole in the first bottoms-up itself."

The mud circulation rate and rheology needs to be optimized to prevent any excessive build-up of drilled-solids getting re-circulated causing hole problems and deterioration in mud properties.

If the drilled-solids are fine and further dispersed into the mud, a substantial increase in viscosity will result. In extreme conditions, enormous build-up of solids and / or a very narrow margin between Equivalent circulating density and fracture gradient (popularly, termed as ECD) can cause the equivalent circulating density of the mud in the narrow annulus to exceed the fracture gradient and induce loss of circulation. Often in such cases the circulation rate can be increased to minimize the build-up of solids and thereby decrease ECD. However, higher circulating rate increases ECD to some extent and to find a suitable balance, either the drilling rate needs to be slowed down or the hole swept with viscous sweeps intermittently (often a combination of both the practices work better than each of them in isolation) without commencing drilling. Viscous sweeps do not flow upward without carrying the cuttings in most cases. This way the hole gets cleaned faster not only because of temporary halt in the generation of additional cuttings, but also by a marginal increase in hole-cleaning efficiency of the system. Therefore, a holistic approach of balancing the entire operations is often the answer to successful, safe and speedy drilling rather than aiming for highest achievable drilling rate, which could provide a temporary illusion of fast progress, but eventually tend to lead to complications and deterioration in over-all quality of bore-hole. The guiding factor which helps to optimize the process is to maintain a constant value of ECD and select that value of ECD which does not cause fracturing of the rock. This is one of the most important aspects of drilling through most of the deep-water wells, particularly the large diameter hole at shallow depth, where the gap between ECD and Fracture gradient of the rock could be as low as < 0.3 ppg!

A safe strategy to drill through such narrow margin holes is to maximize hole-cleaning by pumping viscous sweeps at regular intervals and optimize drilling rate to maintain a pre-decided the value of ECD which must be ~ 0.2 ppg less than the fracture gradient of the rock.

Although for shorter periods of time it is possible to obtain such high drilling rates in soft shale that cuttings cannot get wet and dispersed fast enough and thus prevent them from sticking together and forming "balls" or "slabs", it is necessary to watch the instantaneous rates also, as instantaneous higher drilling rates can be sustained only for a very short time, after which the cuttings will build-up and adversely affect the whole quality favoring conditions that possibly lead to rig NPT. Often the hole speaks for itself. Sometimes we don't understand the language or at times we get greedy for progress. Our greed has a potential to be counterproductive and yet, it might be forgiving also occasionally! Nevertheless, even if it is forgiving, it cannot be a part of our plan and one should carefully monitor the condition of cuttings at shakers to take a calculated risk and be prepared to accept the results which can be rewarding as well as punishing. This fine line between failure and success is often read correctly by the experienced engineers spending time on shakers, closely watching the cuttings reaching the surface. More important than the quality of cuttings is the ability of the engineer to soon identify changes in texture, size and shape of cuttings which are indicative signs that a course correction is required. If cuttings get too mushy, it clearly indicates that shale inhibition is insufficient. Under such circumstances one should focus more on adjusting overcome the situation rather than assuming that all my parameters are fine and I need not worry. Likewise, if suddenly the amount of cuttings increases or decrease reaching on shakers, the situation calls for close examination and finding correct remedial measure. Any deviation from regular trend MUST be carefully monitored and acted upon if required before continuing to excavate without any problems as well as ensuring that the occurrence of the problem is less likely. This is precisely the difference between a great mud engineer and a good mud engineer. A good mud engineer will always run the mud system in a way being told and a great mud engineer will respond to the signs shown by hole. The shale shaker area is the best place, to soon spot any upcoming anomaly and take preventive measure, rather than solving the problem after it has surfaced.

Another more common type of carrying capacity problem is the part-inability of the fluid to lift the cuttings or caving and carry them out of the hole. This problem is often difficult to detect because some of the smaller cuttings come out while the larger ones remain in the hole. If the hole is beginning to cave-in (exposed surface of open hole falling, leading to enlargement in size) the amount of shale coming across the shaker will appear to be normal, but large amounts may be collected in the hole. Sometimes the appearance of the cuttings will indicate poor hole-cleaning. If the cuttings are rounded, it may indicate that they have spent more than the required amount of time in the hole.

The condition of the hole during tripping is usually an indicator of hole-cleaning ability of the system. A fill at the bottom of the hole after a trip is an indicator of inadequate cleaning. The absence of fill does not mean that there is no hole-cleaning problem. Large amount of cuttings may be collected in the washed-out area in the open hole created due to falling of exposed area. Excessive drag while pulling up to make a connection may also indicate inadequate hole-cleaning. When the drill-pipe is moved upward, the swab effect may be sufficient to dislodge cuttings packed into a washed-out section of the hole. The sudden dumping of even a small amount of material is often enough to cause severe drag or sticking.

Hole-cleaning is a more severe problem with high-angle holes than in vertical holes. It is not only more difficult to carry the cuttings out of the hole, but they tend to settle only to the low side of the hole before causing problems. Consequently, more attention should be paid to hole-cleaning requirements in directional holes.

The ability of a fluid to lift a piece of rock is affected by the difference in density of the rock and the fluid. If there is no difference in densities, the rock will be suspended in the fluid and will move in a flow stream at the same velocity as the fluid. As the density of the fluid is decreased, the weight of the rock in the fluid is increased and it will tend to settle. The velocity of the rock will be somewhat less than the velocity of the fluid. The difference in velocities is usually referred to as a slip velocity. The shear stress that is supplying the drag force is a function of shear rate of the fluid at the surface of the rock and the viscosity of the mud at this shear rate. Many other factors - such as wall effects, inter-particle interference, and turbulent flow around the particles make exact calculations of the slip velocity almost impossible, although there are formulae available to estimate the number.

In general, hole-cleaning ability is enhanced by the following:

1. Increased fluid density
2. Increased annular velocity
3. Increased YP or mud viscosity at annular shear rates.

It should be noted that with shear thinning fluids it is sometimes possible to decrease annular velocity, increase the yield point, and increase the hole-cleaning. This is done to minimize hole-erosion. Where ever viscosity has been sufficient to clean the hole, the annular velocity should be maintained just below that for turbulent flow to minimize annular pressure drop and hole-erosion.

In deep-water locations, especially while drilling deviated holes of inclination beyond 30 degrees, it is often advisable to pump tandem sweeps of low and high viscosity of same mud weight as the circulating mud.

CLEANING THE DRILLED AREA UNDERNEATH THE DRILL BIT

Cleaning underneath the bit appears to require mud properties almost opposite from those required to lift cuttings from the hole. In this case we want the mud to have as low viscosity as possible. Since the fluid shear rates beneath the bit are greater than in the annulus, it is possible to have low viscosities at the bit and sufficiently high viscosity in the annulus to clean the hole. The shear-thinning property of the mud will allow both functions to be fulfilled. Flocculated mud and some polymer muds have this characteristic. Shear thinning can be described in simple terms as a property of mud by which it acquires Gel type structure and as force is applied to move it in a direction, it starts getting thinner and thinner with the increased application of force in the same direction.

COOLING AND LUBRICATING

The process of circulating the mud down the drill-pipe cools the bottom of the hole. The heated mud coming up the annulus is hotter than the earth temperature near the surface and the mud begins to heat the top part of the hole.

The maximum mud temperature when circulating, is cooler than the geothermal bottom-hole temperature. The point of maximum circulating temperature is not at the bottom, but about a third of the way up the hole. These facts are important to remember when attempting to predict mud behavior down-hole. A mud additive which is not completely stable at the geothermal bottom-hole temperature may perform adequately at the circulating temperatures. If flocculation due to temperature begins to occur during circulation, as evidenced by increases in yield point and gel strength at the flow line, then we can be sure that severe gelation will occur as the mud heats up after circulation is stopped.

In addition to cooling the well bore, the circulating mud also removes frictional heat and supplies a degree of lubrication. Cooling is especially important at the bit where a large amount of heat is generated. Sufficient circulation to keep the temperature below a critical point is essential in using a diamond bit. Cooling the bit and drill string is done automatically by the mud and not because of any special design characteristic. Mud has sufficient heat capacity and thermal conductivity to allow heat to be picked up, down the hole, transported to the surface, and dissipated to the atmosphere.

In deep-water environment, the column of seabed acts as a natural coolant and often the problem is not how to cool the mud, but how to warm it up to overcome the effect of low temperature on mud, and thus negate excess build-up of gel strength!

Lubrication is a very complex subject, especially when applied to the drilling operation. If a mud (except for silicate mud) does not contain a great deal of sand, it will supply enough lubrication to the drill string without any lubricant simply because it is a fluid that contains solids that are softer than the pipe and casing. Attempts to improve this basic lubricating quality of a mud are usually ineffective and expensive. Probably far greater benefits can be realized by keeping the abrasive content of a mud as low as possible.

Hole-symptoms such as excessive torque and drag, which are associated with the need for a lubricant in the mud, are often caused by other problems such as bit or stabilizer balling, key seats, and poor hole-cleaning.

SEAL PERMEABLE FORMATIONS

While drilling through permeable zones with large inter granular pore spaces which can result in seepage of fluid, it is desirable to include bridging and sealing material in mud to quickly seal off these micro-fractures by building a filter cake and plastering the same with the liquid layer in a way that a semi-permeable membrane type of effect is created to isolate the permeable and porous rock from excessive invasion of fluid.

When solids are added to a clear fluid, they will deposit on the wall of permeable formations as the fluid flows into the formation. This deposition of fluid will gradually retard the flow of the fluid. As the solids-laden cake grows thicker, the rate of flow decreases. This way, some sealing occurs.

Functionally, a mud must keep the loss of fluid to a formation small enough to prevent excessive flushing of the resident fluid away from the well bore. It must also prevent deposition of excessive thick filter cakes. This is a difficult feature to impart, but is also important from a drilling operations standpoint.

STABILIZE BOREHOLE

The borehole walls are normally competent immediately after drilling. By and large, no collapse or sloughing will occur if the formation pore pressure is balanced. In many cases, a considerable under-balance can be tolerated. Although there are few exceptions, but in general, the instability of the borehole is caused by the interaction between rock and drilling fluid. More specifically, chemical instability is caused by fluid filtrate being absorbed by exposed surface of the rock and the difference in the chemical environment of the rock fluid and mud, which triggers the movement of solvent either in or out of rock, depending upon the chemical environment of mud and formation fluid.

Absorption of fluid filtrate has two effects that combine to cause failure of the rock. First, absorption induces a stress in the rock. As more and more fluid filtrate is absorbed, the stress keeps growing until it overcomes the strength of the rock. At this stage, the stress is then relieved by failure of the rock. The second effect of the absorbed water is to weaken the shale by soaking. Again, as more water is absorbed, the rock becomes weaker and the stress becomes greater. The weakening accelerates the occurrence of rock failure.

The wellbore stress is partly balanced by the hydrostatic pressure on the wellbore. If the absorption of fluid filtrate has increased the stress and weakened the rock to the point of failure, increased mud density will counteract part of this stress and prevent rock failure. More water absorption will continue to increase the stress imbalance and weaken the rock until failure occurs again! For this reason, increasing mud density is only a temporary solution to most shale sloughing, resulting into disintegration and bore-hole collapse.

The only complete answer to shale sloughing is to prevent movement of solvent. This can be achieved in theory by creating as much similar chemical environment of mud and rock, as possible. Balanced-activity oil muds are presently considered the best means of completely preventing time-related bore-hole disintegration.

Special water base mud, in which the electrolyte content has been adjusted to minimize the reactivity with a specific type of shale, can slow the rate of absorption and thus delay the onset of sloughing. These types of mud are not universal in their ability to reduce sloughing in all types of shale. Since the absorption of fluid filtrate is not controlled only by the filtration characteristics of a mud, reduced filtration cannot not be the criteria for controlling shale swelling feature of the mud, although controlled filtration does provide some temporary relief by minimizing the available volume of solvent. Yet, reduction of fluid filtrate absorption is one of the most important means of control of shale sloughing. This includes both, the rate of absorption and time of exposure. The rate can be controlled by practical application of the right type of mud and the time of exposure of the shale to the mud must be reduced by minimizing the NPT. This means, to drill the shale and get it behind casing as quickly as possible. Increasing mud density can buy some extra time, ranging from a few hours to a few days. This is an only a time buying practice and not a technically preferred solution, but often it helps to case the open hole to an acceptable standard. In some cases, increased mud density must be used. These are cases, such as abnormal pressure zones, where the initial stress exceeds rock strength unless that stress is partially counteracted by increased hydrostatic pressure. Such instantaneous release of stress can, at times, destabilize the bore-hole no sooner than it is drilled. The fallen rock due to release of stress, often results in to stuck pipe situation, where the annular path is either completely/partly blocked, leading to a situation of no circulation in extreme case.

Most mud additives which supposedly aid in controlling hole stability, need not and do not reduce shale sloughing. They do aid in removal of the sloughed shale, reduce erosion effects, or slow the rate of disintegration and dispersion of the shale into the mud. In other words, these additives are remedial but do not work on the basic problem. There are times when this type of control is adequate to handle our drilling problems.

There are several purely mechanical aspects of bore-hole instability. Erosion by the drilling fluid is a factor in hole-enlargement. Erosional effects can be minimized by reducing the annular velocity.

Swabbing is another means of causing an already unstable zone to slough. Swabbing causes a reduction in hydrostatic pressure, which increases the stress imbalance at the wall. This may be sufficient to cause sloughing. While swabbing, the fluid motion is down the wellbore, which is a change from normal. This may have the effect of dislodging pieces of rock into the wellbore. These pieces could be either a part of the wall or have previously sloughed and being packed into a washed-out interval.

It is worth mentioning here that the short-term stability of wellbore is normally not an issue as any fluid would impart a reasonable stability to the rock. It is the time dependent hole-instability caused by various factors which is often a cause of concern and the system which offers a longer time frame for retaining the integrity of the hole is often the preferred mud system for practical application as far as drilling a hole is concerned.

CORROSION CONTROL

Corrosion is a reaction of metal with its environment in the presence of oxygen, moisture and Carbon di-oxide. Metals convert into their respective oxides by chemical reactions during the process of corrosion and can cause various types of corrosion about which we shall discuss in following description. We are concerned here with corrosion of tubing and other metallic components with which, the mud comes in contact.

The drilling fluid may not be basically corrosive, but it can become corrosive when common contaminants such as oxygen, carbon dioxide, and hydrogen sulfide get incorporated in it. Oxygen, and to a limited degree carbon dioxide, are incorporated into the mud as it circulates through the surface mud system. Hydrogen sulfide, carbon dioxide, and occasionally organic acids get incorporated in the mud from formations drilled, bacterial action and degradation of certain mud additives. The drilling fluid is a carrier of agents that promote corrosion; Likewise, the fluid can also become a carrier of materials that will counteract these agents and retard corrosion. Thus, corrosive agents introduced into the wellbore from the formation can be neutralized by the mud before they cause serious problems.

Oxygen is one of the main culprit, as far as corrosion of the drill assembly is concerned. The requirement for corrosion are an anode, a cathode, and contact with an electrolytic fluid to enable the current flow. The anode and cathode can occur on separate pieces of metal or on the same piece at different points. The anode is the point where corrosion occurs. Water muds are a part of this corrosion cell in that they provide the electrolyte contact.

Solids and some organic additives in muds appear to limit the amount of oxygen available to accelerate corrosion. The absence of these materials from brine and other clear water drilling fluids increases the severity of corrosion in these fluids. Special oxygen scavengers such as, Sodium Sulphite and hydrazine are sometimes used to minimize/eliminate corrosion. Efforts should be made to minimize the aeration of mud in the surface system.

Certain Cationic organic film formers can also be used to extend the life of drill-pipe. These inhibitors should not be added to a mud. They will react with the solids, becoming ineffective and causing possible mud problems. Various methods are used to get the corrosion inhibitor on the pipe rather than in the mud:

1. The drill-pipe is sometimes dipped or sprayed with a corrosion inhibitor solution before it is run in the hole.
2. Batches of inhibitor diluted to 5-10 parts of diesel oil are pumped down the drill-pipe periodically.
3. For better protection on the outside of the drill-pipe, the pipe is pulled through a wiper and sprayed with an inhibitor solution during trips out of the hole.

One of the most effective control of corrosion is to maintain a high alkalinity in the drilling fluids. This has a dual purpose. It slows the rate of corrosion and it reduces the rate of degradation of some organic mud additives (such as Lignosulfonate) that produce corrosive by-products.

Significant decreases in corrosion rate are apparent at a pH above 10.0. The rate of degradation of Lignosulfonate at high temperatures has also been shown to decline significantly at a pH above 10. For this reason, the use of caustic soda or lime is the most economical treatment of a water-base mud. If used properly, the treatment can greatly reduce major corrosion attack.

Hydrogen embrittlement is a type of corrosion that has become more prevalent in recent years. In deep drilling, the use of higher-strength steel and drilling of formations containing hydrogen sulfide, have combined to make this a major problem.

Hydrogen embrittlement is a process wherein, an atomic Hydrogen penetrates and is absorbed by the steel. Once inside the steel, Hydrogen atoms diffuse and seem to move to points of high tensile stress. When their passage through the steel is impeded, the atoms of Hydrogen combine to form Hydrogen gas molecule. In the process of combination and expansion, the Hydrogen molecules will crack the hard, but fine-grained steel. Any acid-former can possibly provide the atomic hydrogen necessary for embrittlement. Acid gases such as hydrogen sulfide and carbon dioxide are frequent culprits in drill-pipe embrittlement.

The tendency for hydrogen stress cracking in high-strength tubular goods is greatly reduced at elevated temperatures. It is believed that the increased temperature increases the mobility and diffusion rate of hydrogen and prevents its accumulation in the steel at points of stress.

An essential requirement for protecting high-strength pipe from hydrogen embrittlement is maintenance of the mud at high pH. This eliminates the development of an acid environment necessary for hydrogen embrittlement.

Even though an influx of hydrogen sulfide may be neutralized by a high pH mud, the sulfide will remain soluble in the mud. If the mud later becomes acidic; Hydrogen Sulfide will again be formed. In addition to being corrosive, hydrogen sulfide is highly toxic and must be eliminated from mud for reasons of safety to personnel. To remove any possibility of Hydrogen Sulfide reforming in a mud, the soluble sulfides should be precipitated in an insoluble form. Addition of certain compounds of copper, zinc, and iron is suitable for this purpose. Copper presents a secondary problem of causing metal loss corrosion when used in excessive concentrations. For this reason, Zinc and Iron compounds are preferred.

Thus far, this discussion of corrosion has been limited to water-base muds. The use of oil muds eliminates or greatly simplifies corrosion control problems. Since oil is the continuous phase, there is no electrolyte solution to carry current in a corrosion cell, if the water in the oil mud is sufficiently emulsified. Oil wetting agents in the mud keep the pipe wet with oil and provide corrosion protection even when the pipe is exposed to air. In an oil mud, it is possible to have high concentrations of lime to effectively neutralize acid gases. This greatly reduces the possibility of Hydrogen embrittlement. If corrosion control were the only consideration, oil muds would be superior to water muds.

FORMATION EVALUATION

A drilling fluid must carry out its basic functions in the drilling operation, and at the same time, allow retrieval of all necessary geological and formation evaluation information. Normally, this does not present any problems, that cannot be avoided by careful planning.

Both, the formation evaluation and mud programs must be consistent with one another. Mud type, whether oil or water continuous, will alter the type of logs which can be run. The salinity of water-base mud and the degree of flushing of the zone around the wellbore are important evaluation considerations. The mud type and condition are important from the standpoint of mud logging and oil/gas show detection. These points should be discussed with well evaluation personnel before final selection of mud and evaluation programs.

Filtrate, which invades the formation, displaces the formation fluid from around the wellbore. If the depth of filtrate invasion is greater than the depth of investigation of an electric logging tool, the log will give misleading information. A similar problem exists with a wire-line test, where, only filtrate is recovered or with a drill stem test, where unlimited filtration can distort interpretation of recovered fluids.

The depth of invasion is a function of the amount of filtrate lost, the porosity, and fingering of the filtrate due to heterogeneity of some formations. As far as mud formulation is concerned, very little can be done about the last two factors. A low porosity and filtrate fingering may allow deep invasion, which will seriously affect formation evaluation despite best efforts.

Due to the radial geometry of the hole, broadly speaking, the depth of invasion increases only equal to the square root of filtrate volume. In other words, a fourfold increase in filtrate volume will only double the depth of invasion. This shows that small changes in filtration rate will not appreciably affect the depth of invasion. Most of the problems of deep invasion are the results of little or no filtration control. When steps are taken to reduce the filter cake thickness to solve drilling problems, filtrate invasion is no longer a major problem.

The volume of filtrate lost to a formation is primarily a function of the dynamic filtrate rate and the time of filtration under dynamic conditions. Therefore, any attempt to limit the fluid loss to a formation should be aimed at reducing dynamic filtration.

Loss of whole mud in a pressure-induced fracture can lead to the same evaluation problems as a deep filtrate invasion. Lost returns with an oil mud may destroy the reliability of a log in the loss zone. This possibility should be avoided by careful pressure control practices.

Due to the non-conducting nature, oil base mud poses a limitation on the use of some logging tools, such as the SP and contact resistivity tools. They also disable the detection of oil show from cuttings, limiting the sidewall core and wire-line test interpretation and limit show detection from mud logging. The loss of these evaluation tools makes the use of oil muds in wildcat wells, undesirable. When drilling conditions demand the use of an oil mud, more testing will often be required due to the indefinite nature of other evaluation tools.

Use of high-salinity water-base mud also makes logging of all kinds more difficult. It is particularly severe on SP and induction curves, which are generally our primary log. Minor electrical leakage that causes no problem in a freshwater mud can become serious in a saline mud. Abrupt changes in salinity should be avoided while penetrating untagged objective sections.

A clean, stable borehole is essential to logging and testing operations. This often requires adjustment of mud properties prior to these operations. If this is pre-planned only a limited amount of time will be required to condition the hole. Stable mud properties will eliminate the need for conditioning between logging runs.

Use of sized bridging/sealing agents in mud can sometimes be a limiting factor in terms of extracting sample from reservoir with the help of MDT tools due to the coarser size of particles of the filter cake deposited on the external side and the solids intruded in the internal side of the exposed rock. Under those circumstances, it is often a compromise which needs to be made between two mutually opposing requirements of the hole. In deep waater environment the more important requirement of including sized particles to control losses takes priority over MDT tool's ability to extract sample most of the times, where-as reverse could hold true in areas where lost circulation is not a serious issue.

Coring is another method of formation evaluation that often requires special mud characteristics. The purpose for coring dictates the type of mud which should be used. It is almost impossible to obtain a core without an appreciable amount of flushing with the filtrate. Therefore, the measurements that are going to be made on the core dictate the type of the filtrate that can be tolerated.

Water-base muds are best suited for conventional core analysis or show detection because an oil filtrate from an oil mud will eliminate show detection, location of the oil-water contact, and oil saturation determinations. An oil mud is the best coring fluid to obtain irreducible water saturation from a core. In this case a water filtrate would introduce an error in the determination. When coring is done to obtain only lithology and geologic environmental data, it doesn't matter which type of mud is used.

Sometimes cores are obtained to study wettability and water flooding characteristics. In these cases, the filtrate should contain nothing that is surface active or that will alter the wettability of the rock. Special water-base fluids that contain no chemical treating agents are used. The Bentonite powder suspended in water is probably best for this application.

FORMATION DAMAGE

Broadly speaking, formation damage can be defined as a process which impedes the ability of the hydrocarbon bearing reservoir to flow to its natural potential.

Subsequent discussion shall focus on good drilling practices, mud design as well as mud maintenance during drilling the hydrocarbon bearing rock to minimize/eliminate damage to the rock.

Whether the rock is damaged, can be ascertained by reservoir engineers by skin factor of the drilled rock, which if > 0, is an indication of damage. Higher the value of skin factor, more is the extent of damage. Conversely if the skin value is negative, it means that the process of drilling has triggered the rock to flow more than its originally estimated recoverable reserve by primary technique. Zero value of the skin factor is an indication that the rock is neither damaged nor triggered for higher flow by drilling process.

Dos and DONTs for preventing formation damage

1. As far as possible, drill with just enough mud weight to keep reservoir content in its native place during drilling and the

well-bore mechanically stable. Geo-mechanical modelling studies help achieve this.

2. Do not drill at a rate faster than the overall ability of the drilling process to keep the hole free of cuttings. ECD monitoring is a viable technique to monitor this.
3. Focus on mud design to help build a compact and impermeable filter cake around open hole quickly to disallow filtrate invasion.
4. If information is available on inter-granular pore space, then use the same wisely to design a mud with particles in such a way that they do not invade the pore-spaces and cause fluid migration along with filtrate.
5. Mud weight more than optimum, will exert unnecessary extra pressure against the wall of open hole, resulting into filtrate invasion which will displace the reservoir content too far to activate the well during flowing back.
6. Retain the native wettability of the rock by selecting the correct type of mud system to disallow damage of rock, because of incompatibility between reservoir fluid and mud filtrate.
7. If the reservoir bearing open hole is to be cemented, then the slurry design should ensure that there is no mud / cement slurry lost during cement job. Software models help in achieving this.
8. The rheological profile of the mud should be conducive to hole flushing without exerting unnecessary additional ECD. Ideally the LSRY (3RPM*2/6 RPM) should be equal to the size of the open hole from which the fluid is expected to flow back.
9. While designing well cleanup programmed avoid use of too many fancy stuff (surfactants/solvents/corrosion inhibitors, oxygen scavengers) unless extremely essential. Question the logic of inclusion by assuming that default pill programmed does not require any of these chemicals. Plain brine of correct weight should clean the fluffy part of filter cake and create pin holes in compact filter cake to facilitate flow of oil/gas through them. This will facilitate the outflow of reservoir fluid during activation process.
10. Most important part of preventing formation damage is to select a mud system after conducting core flood tests on various formulations in a reputed core lab. Selection of completion design also is an important aspect of preventing formation

damage and completion engineer should be an integral team member to finalize the plan.

11. Monitor dynamic Fluid loss rather than static API F/L and use a PPT apparatus with ceramic disc of size matching with average pore throat size of the reservoir rock. The PPT must be kept below 2.5 cc and tests carried out at BHCT and not on predefined API temp., which are good for static measurements.
12. Formulate mud with fragile gels to negate the adverse impact of excess pressure on rock while initiating circulation after considerable gap.
13. While selecting brine weight always remember to apply temp. correction factor to arrive at a realistic weight at bottom and know the corresponding weight at surface temp.
14. The well bore cleanup plan should be drawn with great caution to prevent excessive pressure getting exerted on well bore while various density fluids entering the annulus.
15. Whole/partial mud loss into the producing rock is one of the biggest threats to formation damage and a fool-proof plan MUST be drawn to prevent loss of circulation as far as possible
16. While drilling through Limestone or fractured rock it is very difficult to prevent losses and ideally such formations should be drilled with floating mud cap technique or plain water (or brine of correct weight) to minimize damage. Such rocks are invariably acidized subsequently and use of water / brine of correct weight helps in ensuring that there is no incompatibility issue.
17. Do not tempt to drill at a rate faster than the ability of system to flush out cuttings no sooner they are generated. System efficiency can be improved by higher LSRV readings with fragile gels and optimized YP.
18. If situation warrants, flush the hole with low-hi vis pills without creating ay pressure fluctuation in circulatory system. Monitor ECD very carefully and never allow it to suddenly creep up. If required drilling can stop for a while and hole circulated to flush the cuttings and bring ECD in realistic range.
19. Do remember that with increase in depth there will be a nominal increase in ECD and this will not adversely affect the hole, but any abrupt rise in ECD must be addressed immediately and brought back in same range as prevailing before.

20. Last, but not the least – ALWAYS keep experienced and knowledgeable personnel on rig to protect the interest of the well.

RESERVOIR DRILL-IN FLUID

For the last three to four decades there is an increased awareness level as well as willingness to address formation damage elimination and minimization by thoughtfully designing the mud system with focus on means of eliminating mud related damage during drilling as well as during well flowing. With scientific approach, a system has been devised which addresses

1. Filtrate invasion minimization
2. Quick formation of impermeable and thin, yet compact filter cake
3. Thoughtful selection of mud products.
4. Use of technical data to select suitable mud weight and not over emphasize only on well safety even if it comes at a cost of damaging the well/rock.
5. As far as possible, elimination of additives which have known potential to damage the rock by chemical/physic-chemical mechanism.
6. Knowing the type of rock and using the same as an input in selecting the composition and the type of mud system.
7. Selection of mud parameters which avoid/minimize formation damage.

All these factors led to evolution of a type of mud which is intended to be used specifically only for reservoir section of the well and such system has been distinguished by the other types with the name of "Reservoir Drill-in fluid" system. The primary function of this system is to not only facilitate smooth drilling, but simultaneously attempt to protect the rock as much as realistically possible. One such essential feature of this system is that it is designed with focus on selecting the particles in mud of that size which does not intrude into the rock and make conscious attempt to maintain the size and concentration in as much the desired range as possible by continuous replenishment of sized particles in mud externally and compensate for loss of particles due to crushing action of bit while continuous circulation. At the same time, one should select correct configuration and size of shaker screens to dispel finer particles and keep inert solids in a range which does not adverse the rheology of the system and yet build the much-desired firm filter cake (internal as well as external).

A lot of emphasize should be laid in correct use of solids control equipment tom retain essential sized particles and only selectively discard harmful colloidal sized particles. This is being achieved by planning of re-introducing the correct sized particles into main stream from shaker discard.

Selection of various additives in RDF mud

The focus again, is on eliminating the additives which are generously used in drilling upper sections, like Asphalt, Bentonite to name few. Instead of them, bio-degradable starch as well as bio-degradable viscosifiers are used to build the skeleton of mud which is then converted to a RDF by administering correct sized particles of Calcium Carbonate and avoiding use of barite and instead achieve desired weight either by high density brine or by adding higher dose of Calcium Carbonate or a combination of both. The native wettability of the rock is ascertained and as far as possible, simpler version of water based mud is preferred over other types, unless severe shale instability issues warrant otherwise.

It is imperative to accept that the process of drilling will invariably damage the rock and efforts are made to minimize the impact as much as possible. One can aim for zero damage but that is not always achievable. The wisdom is to live with minimum damage and yet drill the rock safely and timely.

When to use RDF:

The decision to use RDF must be taken with all factors being given due weightage and selecting the system as well as components and properties which will not only minimize damage to the rock, but shall also facilitate safe and timely drilling. Wherever feasible, core flood tests should be conducted in lab with samples of probable mud type and system to find out which of them provide highest permeability regained figure. Out of the top few such formulations the ease of drilling and completion design chosen for the well should be reviewed. It may not always be possible to select a system which is zero damaging but the one which facilitates hassle free drilling and completion as well as damages the rock to the least, is often the chosen one.

With advancement in drilling fluids technology now it is possible on many / all occasions to use solids free design of RDF and yet get the benefit of drilling and completion with minimum complication. Solids-free RDF in reservoir means that whatever solids are there in system, they are purposely retained to facilitate damage minimization and other non-essential additives (Barite or other heavy density additive which has potential to damage the rock) are not included.

The idea is to negotiate with the well and not fight against its requirements. With the advancement of solids free Water base system with minimum number of additives it has been observed that obtaining negative skin factor is very much possible if the system is properly designed and executed.

CHANGING OVER TO RDF AND MAINTAINING RDF

Once a consensus has been arrived by team to drill the reservoir with a particular type of mud, it is imperative that the mud of previous system is completely replaced with RDF correctly formulated, including but not restricted to hole volume, surface volume. Before resuming reservoir drilling it is a good practice to build at least twice the open hole volume of the section planned for drilling to account for additional volume requirement, failing which often there is a temptation to blend the previous section mud. This can defeat the purpose and should be refrained from. Monitor particle size of the active mud with the help of particle size analyzer to be made available on rig with qualified person to operate the same periodically and help mud engineer to decide his tour dose of essentially solids of correct size. This aspect is often overlooked and is a cause of not being able to eliminate rock damage completely. Every mud company has a RDF of their own, but the generic composition is the same. The book stays away from saying anything about superiority of one over the other and leaves this to the team who confronts the situation.

A list of basic mud test equipment is being included in the book to give an idea about the minimum mud check which is required to understand the properties of mud and decide upon the remedial measures/ course-correction to be done as and when required. The list could go on and on, but the point which is being high-lighted is the necessity, upon which the lab can then grow as requirements arise from time to time.

SUMMARY

It is a bit unrealistic to conclude that use of RDF will eliminate formation damage but certainly, with right planning and correct execution the extent of damage can be brought with-in manageable limits. The concept still needs enormous support from the entire team for optimized results and success/failure ca not be entirely executed to any single individual.

CHAPTER 3

RELATIONSHIP OF MUD PROPERTIES TO FUNCTIONS

To design and maintain the mud to perform a given function / set of functions, it is necessary to accurately measure the mud properties which control its ability to perform those functions. For this reason, measurements are normally made on a variety of mud properties. Considerable research has been done to design tests and to correlate these tests with the functions of a mud.

Normally, the functions of mud carried out under down-hole conditions cannot be easily duplicated in a mud test. Over the years, attempts have been made to either simulate more closely the down-hole conditions or to predict the down-hole mud properties from surface condition measurements.

Measurements of both physical and compositional properties are made in a complete mud check. Some functions are controlled directly by the mud composition, but even those which we correlate with physical properties are controlled by adjusting the mud composition. Since we normally have a multifunctional requirement for a drilling fluid, we need an extensive series of both physical and compositional tests to properly monitor a mud system. Properties should be measured on the mud going in at the suction and on the same mud coming out at the flow line.

Consider it this way, the suction pit and the shakers are two parts of a volleyball ground. The Suction pit area belongs to mud engineer and the shaker pit area belongs to the other player, which is the borehole here! Mud engineer sends in his best service and the borehole tries to overpower him by altering the properties/condition of mud, which is detrimental to operations. Return mud at shaker shows how successful borehole had been. The mud engineer should defend his case now, before sending the mud back in the hole. This decides the treatment plan of the mud engineer. It is a very interesting match. You need not win this match, but just prolong it to the point where the open hole needs to be pushed to a depth, beyond which, casing needs to be lowered. Never fight with borehole, because you will lose even if you win. Always bargain for what it asks and what you think he needs, to remain healthy and capable of producing the hydrocarbons contained in it.

DENSITY

The starting point of pressure control is the control of mud density. The weight of a column of mud in the hole, necessary to balance formation pressure is the reference point from which all pressure control calculations are based. The required weight of the mud column establishes the density of the mud for any specific case.

Fortunately, density is one of our most accurate measurements. With a simple mud-balance, we can weigh the mud to the nearest 0.1 lbs. / Gal, which is equivalent to 5.2 psi per 1000 ft. Of mud column. Mistakes in measuring density account for most of the inaccuracies. The most frequent mistakes are:

1. Improperly calibrated balance
2. Entrained air or gas in the mud
3. Failure in filling the balance to exact volume
4. Dirty mud balance
5. Not keeping the balance on a perfectly flat surface

The calibration of a mud balance should be checked at two densities that span the range of densities to be measured. One convenient calibration check point is to measure the density of fresh water. At temperatures between 75° and 95°F, the density of water should range between 8.34 and 8.33 lb. /Gal. A completely suitable means of checking the calibration at the high end of the scale have not been devised. High weight muds can be used if their density is accurately known and if a representative sample can be transferred to the mud balance. Settling of solids in the sample makes it questionable whether a representative sample is used in the calibration check. Some high-density solutions have been used for this purpose, but they are highly corrosive. High density brines can be used to calibrate the mud balance on higher range. For the last two decades, Formate brines have come at rescuing this fallacy and can help calibrate mud-balance accurately up to mud weight as high as 2.0 SG

If a mud balance is found to be in error by more than 0.1 lb./gal at either end of the scale, it is advisable to obtain a new, correctly calibrated balance rather than re-calibrate or correct the readings of the old balance. The caps on mud balances should never be interchanged. A cap may weigh differently from the original and cause an error in density measurements.

When air or gas is trapped in a mud sample, it will decrease the measured density. This is especially troublesome in muds that have high yield points or gel strengths, but it can occur in any mud. There are devices which can be used to de-gas a mud sample. In less severe cases the sample can be stirred in a Rotational viscometer at medium speed to help "break out" the entrained gas. A pressurized mud balance can also be used to obtain more correct mud densities.

It is important to fill the balance 100 percent with mud to obtain an accurate density. An air bubble can be trapped under the lid and cause low density measurements or granular material may hold the lid up, resulting in overfilling and a high-density measurement.

The mud balance should be level and out of the wind. Air currents are often sufficient to tip the balance and affect the reading.

Increased temperature causes the density to decrease, but increased pressure causes an increase in density. Down-hole, these effects oppose one another and tend to equalize sometimes. Problems can result from large changes in measurement temperatures. If a new mud is carried to location to displace a circulating system, there may be a large difference in the temperatures at which the densities of the old and new muds were measured. The cooler mud may increase in density from

0.1 to 0.3 lb./gal when it returns to the flow line after the first circulation. To avoid this problem, the densities of both muds should be checked at the same temperature. In Deep-water locations with very low sea bed temperature, the effect of temperature is so prominent that a density difference can exist as high as +/-0.2 ppg between the densities measured at the surface and the actual density at seabed level. Software predictions often help in knowing the correct density in a mud pit to get the required density at bottom-hole conditions.

The density of mud is increased by adding Barite, a commercial grade of barium sulfate. Barite is used as a standard weighting agent because of its low cost, high specific gravity, inertness, and low abrasiveness. Commercial Barite is a mined product that undergoes very little processing other than grinding. Its specific gravity averages about 4.20. Pure barium sulfate has a specific gravity of 4.5, indicating that some impurities are present in the commercial grade. The impurities vary depending upon the source of the Barite. The color is not indicative of the purity or quality of the Barite. Soluble alkaline earth metals such as Calcium and Magnesium are among the more detrimental types of impurities and have been limited by API specification to a maximum of 250mg/l expressed as Calcium. The amount of Calcium introduced into the mud system from addition of API grade Barite should not be detrimental, but a Calcium check on the Barite should be run to assure that an abnormal amount of Calcium is not present. This can occur through poor quality control at the grinding plant or from cement contamination during transporting.

Muds maintained at densities higher than required to control formation pressures can cause a variety of problems. They decrease penetration rates, increase differential pressure sticking tendency, increase possibility of lost circulation, increase mud costs, cause possible formation damage and thus increase overall well cost. Consequently, the density should be controlled even in un-weighted muds. Of late with the awareness of formation damage control getting increased attention, often Calcium Carbonate is also being used as a preferred weighing agent due to its acid soluble nature, which cannot be achieved in Barite. The fact to remember here is – although we may use Calcium Carbonate preferentially over Barite, but in real cases how often do we acidize the rock to improve upon its damage caused by drilling!! Nevertheless, it is still a good practice to use Calcium Carbonate to the extent possible as it can be easily sized to the specific requirement and the mud so formulated, can help to arrest minor seepage losses while drilling through minor fractures/ permeable rock.

FLOW PROPERTIES

The rheological (study of deformation of matter is Rheology) properties of a mud describe the flow characteristics of a mud under various flow conditions. In a circulating system, mud flows at a variety of rates in conduits of different sizes and shapes. To know or predict the effects of this flow, we need to know the flow behavior of the mud at the various points of interest in the circulating system. To simplify the measurement procedure, we make only a limited number of measurements.

When a fluid flows, it exerts a frictional drag – called the shear stress – on the surface of the conduit. The magnitude of the shear stress depends on the frictional drag between adjacent "layers" of fluid traveling at different velocities, and the difference in velocities of adjacent layers next to the wall of the conduit. The difference in velocities between adjacent layers is called the shear rate. We are interested in the effect of the flow at the wall where both shear rate and shear stress are a maximum.

If the shear stress is known at all shear rates, the complete flow behavior of the fluid can be described. For simple fluids such as water, the ratio of shear stress to shear rate is a constant. In these fluids, called Newtonian fluids, measurement of shear stress at one shear rate is sufficient to predict the flow behavior at all shear rates. The ratio of shear stress to shear rate is the viscosity.

The viscosity is a single number that characterizes the flow behavior of a fluid. When solids are added to a Newtonian fluid, the irregular shapes of these particles resist flow in a manner not directly proportional to the rate of shear. The inter-particle interference to flow is more pronounced at low shear rates than at higher shear rates. Therefore, the ratio of shear stress to shear rate, or the effective viscosity, is high at low shear rates and decreases with increasing shear rate. This is known as "shear thinning." In other words, the increase in effective viscosity over that of water decreases with increasing shear rate.

The shear stress is directly proportional to the pressure required to produce the flow. In a Newtonian fluid, the smallest possible amount of pressure will cause the fluid to flow. In fluids that contain solids which link together to form a structure, flow will stop when the pressure or shear stress is reduced to a point which is less than the shear strength of the structure.

This point is called the yield stress of the fluid. These non-Newtonian fluids, when allowed to remain still for a period, continue to develop this semi-rigid structure and the shear stress required to initiate flow increases. This shear stress is called gel strength. The structure becomes more rigid with time, causing the gel strength to increase with time.

Basically, in the mud circulating system, we are interested in the flow behavior of the mud at the bit where shear rates are extremely high, in the annulus where shear rates are relatively low, and in the pits where shear rates are almost zero. We are also interested in the flow behavior inside the drill-pipe and drill collars, since this controls the amount of hydraulic horsepower that a pump can deliver to the bit.

Our basic design criteria for flow property control are - to:

1. Minimize viscosity at high shear rates to maximize penetration rates.
2. Maintain adequate viscosity at annular shear rates to lift cuttings and sloughing and clean the hole.
3. In the deep-water environment thumb rule says that the 6 RPM reading of the mud should be equal to the size of the hole in inches or at least 8 to 9, whichever is more.
4. Minimize viscosity at annular shear rates consistent with good hole-cleaning when lost circulation may occur due to excessive annular pressure drop.
5. Provide adequate gel strength at all points in the system to suspend Barite.
6. Prevent excessively high gel strengths from developing down-hole during long periods of quiescence to prevent high pump pressures required to break circulation, and high swab and surge pressures.

Obviously, this list of requirements has points of conflict. A degree of compromise will always be necessary. Several instruments are used to secure information about the flow characteristics of mud. A brief description of the various tests and interpretation of the test data follows.

Marsh funnel

The measurement of viscosity with marsh funnel is a crude method for measuring the consistency of a fluid. Although listed on the mud check sheet as viscosity, the Funnel viscosity is not in the true sense a viscosity.

The test consists of filling the funnel to the bottom of the screen with mud (~1500 ml) and timing how long it takes for one quart to flow out of the funnel. The time in seconds is reported as the funnel viscosity. Fresh water at 70°F will have a funnel viscosity of 26 seconds.

This test has the advantage of being quick, simple, and requiring very little equipment. It is useful in showing gross changes in the overall "viscosity" of a fluid, but it does not measure specific flow parameters. It can be changed by changes in plastic viscosity, yield point, gel strength, or density. For this reason, it should be used only to monitor a mud and not to diagnose problems or to prescribe treatment.

Rotational viscometer

This is a viscometer in which the fluid is contained between coaxial cylinders. The outer cylinder rotates at a constant speed and the viscous drag of the fluid on the inner cylinder or bob exerts a torque that is indicated on a calibrated dial. The torque is proportional to shear stress and the rotational speed is proportional to shear rate. The indicated dial reading times 1.067 is equivalent to shear stress in lb. /100 sq. ft. and the rotational speed in RPM times 1.703 is equivalent to shear rate in reciprocal seconds.

Two types of such vicometers in common use are, namely a six-speed model (600, 300, 200, 100, 6, and 3 RPM) and an eight-speed model (600, 300, 200, 100, 60, 30, 6, and 3 RPM). These instruments provide measurements of the actual flow parameters of shear rate and shear stress and provide a means of making gel strength measurements. With this information, we are better equipped to diagnose flow behavior and prescribe mud treatment than with the funnel viscosity.

After the shear stress/shear rate data are collected, they can be handled and reported in many ways. Traditionally, these data have been used to calculate the plastic viscosity and the yield point in the Bingham plastic rheological model, and these parameters have been reported on the mud check sheet. The difference in the rotational viscometer's dial readings at 600 and 300 RPM is the plastic viscosity, and the plastic viscosity subtracted from the 300 RPM reading is the yield point

Since the Bingham plastic model does not truly represent the shear rate/shear stress behavior of most muds, the calculated yield point is not equivalent to the true yield stress and the plastic viscosity is not a true viscosity. The wealth of experience the mud engineers have acquired in the use of these parameters to make them quite useful in predicting mud performance and diagnosing mud problems.

Plastic Viscosity

Although calculated from measurements at relatively low shear rates, the plastic viscosity is an indicator of high shear rate viscosities. Consequently, it tells us something about the expected behavior of the mud at the bit. One of our design criteria is to minimize the high shear rate viscosity. To accomplish this, we should minimize the plastic viscosity. A decrease in plastic viscosity should signal a corresponding decrease in the viscosity at the bit, resulting in a higher penetration rate.

The plastic viscosity is primarily a function of the viscosity of the liquid phase and the volume of solids contained in a mud. The viscosity of the liquid phase is increased by the addition of any soluble material. Many of the water-soluble polymers used for fluid-loss control are quite effective in increasing the plastic viscosity. Saturated salt water has twice the viscosity of fresh water. Diesel oil, which is commonly used as the liquid phase of oil-base muds, has three times the viscosity of fresh water. Both salt water muds and oil muds tend to have high plastic viscosities.

The volume of solids in the mud is the dry volume of solids plus the increase in volume due to hydration. The water of hydration becomes a part of the solid so far as its effect on viscosity is concerned. In other words, the plastic viscosity is increased by the addition of any type of solid; but solids such as clays, which hydrate, will further increase the plastic viscosity as their volume is increased by hydration. This makes the shale hydration and dispersion particularly detrimental. If these particles are large and relatively un-hydrated, their effect on viscosity is small. Time, temperature, and agitation tend to disperse and allow hydration of the individual clay platelets, which results in increased viscosities.

To combat the tendency of shale to disperse and hydrate, the "inhibitive" types of mud were designed. Materials such as lime, gypsum, Lignosulfonate, and polymers are added to inhibit the rate of dispersion and hydration. These materials do cause inhibition, but if the inhibited particles are not removed from the system, the solids content will continue to build. With time, the plastic viscosity will be as high, or higher than before and other mud properties such as filter cake thickness will suffer.

Minimum plastic viscosity can be achieved only to the degree that the mud is kept free of drilling solids. Plastic viscosity decreases with increasing temperature, due to thinning of the water. If the mud is checked at 130°F, the PV will be lower than at 120°F; if it is checked at 110°F, it will be higher. For this reason, all mud tests should be made at the same temperature, 120°F. With oil mud, the rotational viscometer recognizes the finely dispersed water droplets act is inert solids, thereby the PV values as shown by the rotational meter can be also termed as Pseudo number. This must be kept in mind while treating the oil base mud to reduce the PV, as no amount of chemical treatment can reduce PV which is being read by rotational viscometer by finely dispersed water droplets. By and large oil base mud will have a higher value of PV when compared to WBM, everything else remaining same. However, the net impact will be same in terms of performance for both the types of mud.

Yield Point

The yield point is a good indicator of flow behavior in the annulus.

When a fluid contains large molecules (such as a polymer solution) or colloidal particles, these molecules or particles tend to collide with each other, thereby increasing the resistance to flow. If these particles are quite long compared to their thickness, the inter-particle interference will be quite large when they are randomly oriented in the flow stream at low shear rates. As the shear rate is increased, the particles will "line up" in the flow stream and the effect of particle interaction decreases. If particles are electrically attracted to one another, the effect is quite similar. At low shear rates the particles link together, increasing the resistance to flow; at high shear-rates the linking bonds are broken and the fluid becomes more like water.

These two effects combine to determine the yield point of a mud. The electrical interaction of solids is controlled by chemical treatment, and the mechanical interaction is controlled by adjusting the type and amount of solids or polymer in a mud.

Clay platelets are both electrically charged and long compared to their thickness. This makes them quite good for increasing the yield point. Their ability to link together (flocculate) can be neutralized by the addition of certain chemicals such as tannins, phosphates, and Lignosulfonates. When this is accomplished, the yield point will be reduced to that caused by the mechanical interactions of the solids. The further chemical addition will not continue to reduce the yield point past this point. Complete de-flocculation is best identified by the occurrence of near zero gel strength. Removal of colloidal solids is another way to reduce the yield point. This reduces both the tendency to link and the mechanical interference of the particles.

When a higher yield point is desired, the addition of Bentonite is a cheap and effective means. It is most effective when well hydrated and dispersed in fresh water. This ensures a maximum amount of charged surface area exposed per volume of Bentonite added. The presence of deflocculating chemicals, defeats the purpose by neutralizing the ability of the platelets to link together. In some cases, flocculating agents or clay "extenders" are added to promote linking and cause even higher yield points to result. Such materials as soda ash, polyacrylates, and several different Calcium compounds are used for this purpose.

In summary, high yield points are caused by flocculation of clay solids or high concentrations of colloidal solids. Flocculation may be due to lack of sufficient de-flocculent, high temperature, or contaminants such as salt, Calcium, Carbonates, and bicarbonate. A high solids concentration will aggravate flocculation tendencies from any cause.

The yield point is primarily associated with two mud functions: the hole-cleaning capability and the pressure control characteristic of a mud. A higher yield point increases the carrying capacity of a mud and increases the circulating pressure drop in the annulus.

Associated with increased circulating pressure drop, is increased pressure surge and swab from pipe movement. Since increased yield point aids in hole-cleaning, but is detrimental to the problems of lost circulation and swabbing, a compromise must be reached. For this reason, no absolute guideline for yield point values can be given.

High mud weights improve hole-cleaning capabilities of a mud. Consequently, high yield points are normally not necessary in high-weight muds to ensure good cutting carrying capacity. Also, pressure control is often critical where high-weight muds are required. As a result, the need to minimize yield point, usually outweighs any advantages of maintaining a high yield point in high density muds.

Effect of temperature on rheology with special reference to deep-water

As expected the lower temperature will increase the viscosity by bringing closer the particles which will require additional force to overcome the resistance to flow by partly using up the available force to separate/dis-aggregate/distance out the particles enough to enable initiation of flow process. Under normal circumstances the sea bed temperature and ambient temperature are either closer or not too much different to cause any conspicuous difference in the final quantum of force required to overcome the effect of temperature, but with sea bed temperature as low as 2 deg C and an overhead of sea water as high as 4500 psi, the increase in viscosity is enough to necessitate suitable remedial measures in mud treatment / design policy to overcome this effect. This is often achieved by beginning the drilling just below sea bed with a mud of slightly lower viscosity in the pits and a higher value of LSRV to not to compromise on hole-cleaning in the annulus. Use of software model does provide an increased comfort level in deciding the range, but more important is a slightly conservative and cautious approach in real cases. Till the LOT is recorded just below the shoe, it is desirable to monitor hole very closely at shakers and purposely keep YP at a slightly lower value. Once the mud has equilibrated to formation temperature to a reasonable number to partly negate the adverse effect of low temperature the rheology can be gradually picked up. Till that time viscous sweeps can be administered in the system to not only ensure that hole-cleaning is not compromised, but also help system pick up additional desired rheology in a gradual fashion instead of any abrupt change. These changes must be closely monitored and executed in response to hole requirement and not based on a pre-decided number.

Gel Strength

The gel strength is a measurement of the shear stress necessary to initiate the flow of a fluid that has been quiescent for some time. It is caused by electrically charged particles that link together to form a rigid structure in the fluid. The strength of the structure formed is a function of the amount and type of solids in suspension, time, temperature and chemical environment. In other words. Anything which promotes or prevents the linking of the particles will increase or decrease the gelation tendency of a mud.

Gel strength is measured by beginning to turn the V-G meter at a very low speed and reading the peak dial deflection. This represents the shear stress necessary to break the gel structure and is recorded as gel strength in lbs/100 sq. ft. The gel strengths recorded on the mud check sheet are measured after setting times of 10 seconds and 10 minutes. These set times are arbitrary and have been selected simply to provide a standard basis of comparison of gel building characteristics of muds. The difference in the two measurements is an indication of the rate of gelation.

It is necessary for a weighted mud to have the gel strength of about 5 to 7 lbs/100 sq. ft. to suspend Barite. The Barite will settle in a mud that has no gel strength regardless of its viscosity. A high viscosity will simply slow the rate of settling. In a water-base mud that contains clay solids, the development of gel strength is a natural thing. Barite settling should never be a problem unless these clay solids have been made inert by excessive chemical treatment. In such cases, the addition of pre-hydrated Bentonite should solve the problem. Barite suspension in oil muds is somewhat more difficult since clay solids are essentially inert in oil. Even the oil-wet clays used in some oil muds to build viscosity are not particularly good at building Barite-suspending ability. It should be emphasized again that a high viscosity is not sufficient to suspend Barite. A minimum gel strength of 5 to 7 Ib/100 sq. ft. is necessary.

In water-base muds, flocculation increases gel strength and de-flocculation decreases gel strength. An increase in gel strength is normally the first indication of the beginning of flocculation. In fact, the 10-minute gel strength will usually begin to increase before the initial gel strength. If a pilot test with additional Lignosulfonate fails to reduce the gel strength, it signals either excessively high colloidal solids content or a chemical contamination problem. Dilution with water will temporarily solve either problem, but in a most expensive manner. The water chemistry test should be run on the mud filtrate to determine what contaminants are present. A check of the alkalinities may indicate that there is insufficient hydroxyl ion content to activate the Lignosulfonate or that an excessive concentration of Carbonate or bicarbonate is causing flocculation. When problems involving improper water chemistry are present, the mud treating costs can increase drastically, if correctly not diagnosed and timely acted upon.

In oil based mud, to design a mud of reasonable rheology and adequate gel strength is a challenge. The situation is sometimes further compounded in Deep-water environment, where the compressibility of base oil further thickens the mud without providing any genuine gel strength. Apart from it the tiny particles of emulsified internal brine phase behave like solid particles as far as rheological measurements are concerned, where as they do precious little to provide support to suspend the solids. The situation often calls for the addition of specialty additives which are low end rheology boosters to provide additional suspending ability to Oil base mud. Apart from it, organo-phillic clays start yielding only after getting adequately sheared in hole through nozzles of bit. With ever increasing water depth leading to requirement of larger volume of mud to drill the hole, this shearing and "seasoning" of the mud can sometimes be a tough situation to negotiate. One cannot over-treat the Oil base mud initially to obtain the desired level of rheology as the same overly treated mud will eventually yield so high that rheology shall shoot beyond desired level. If the oil base mud is treated to right concentration it does not yield soon and till the time taken by the clay to yield the oil base mud needs supplementary treatment of rheology boosters which will lose their property after a couple of circulations. Thus, a typical deep-water oil base mud should have two viscosity building components in it. One which quickly yields, but shall deteriorate after a couple of circulations and the other which will start yielding almost at the time when the quick rheology booster starts fading out. The quick fix rheology boosters are mainly low molecular weight polymers which provide a temporary "fillip" to low-end rheology but are susceptible to mechanical shear degradation. Thus, a combination of organo-phillic clay which yields after" seasoning" and a low-end rheology booster which yields quickly, but fades after sometimes, works better in the deep-water environment with oil base mud. The selection of base oil also helps in overcoming this situation. A base oil of low kinematic viscosity keeps rheology under reasonable control and arrests the tendency of oil to thicken with decrease in temperature.

SOLIDS ANALYSIS

The primary source of control of rheological and filtration properties of the mud is the amount and type of solids in the fluid. To properly manage solids, it is necessary that we know the amount of each of the various types of solids in a mud. A good estimate of the amounts of the various types of solids can be made from tests that are normally performed on a mud.

If we assume that all the suspended solids in a mud have a specific gravity of either 4.20 (Barite) or 2.6 (drilled solids and Bentonite), we can make measurements that will allow calculation of the relative amounts of these two types of solid. We need to know the volume percent water and its specific gravity, the volume percent oil and its specific gravity, the volume percent solids, and the mud density.

With the mud-retort, we can measure the volume percent oil and the volume percent distilled water in the mud. When there is a large amount of dissolved salt in the mud, the volume percent distilled water from the retort is less than the actual volume percent salt water in the mud. This dissolved salt increases the volume of water as well as its specific gravity. If the salt in the mud is predominantly chloride type, we can measure the chloride content of the filtrate and calculate the specific gravity and corrected water volume. Knowing the actual water volume and oil volume, the suspended solids can be calculated by difference. Using the mud density and the volume percent suspended solid, the relative amounts of Barite and low-gravity solids can be calculated.

Knowing the volume percent of low-gravity solids is important, but it is also desirable to know the relative amounts of high-yield Bentonite and low-quality drilled-solids that comprise the low gravity fraction. Since different types of solids have different cation exchange capacities, it is possible to differentiate between Bentonite and the low-quality drilled solids, as drilled-solids have lower exchange capacities. If we know the volume percent low-gravity solids and the cation exchange capacity (methylene blue test), we can estimate the percent Bentonite and the percent drilled solids.

Retort

A retort is a device used for measuring the liquid content of a mud. A measured sample of mud is heated until all the liquid components (both water and oil) vaporize. The vapors are then cooled through a condenser and the liquid / s is/are collected in a graduated cylinder. The most common retorts use either 20 or 50 cc of mud. The 100 percent line on the graduated cylinder should be the same volume as the mud sample. If the mud cup and the graduated cylinder are not exactly of the same volume, an error in measurement will result in incorrect measured values. Corrosion can cause metal loss, inside the mud cup, resulting in an increased volume, or the graduated cylinder may be incorrectly calibrated. If the mud cup volume is greater than the graduated cylinder, the calculated solids volume will be less than actual and conversely, the solids will read too high if the cylinder is larger than the cup. When preparing the sample for testing, the mud should be free of trapped air.

The volume percent of oil and water as read from the graduated cylinder are recorded on the mud check sheet. The sum of these two numbers subtracted from 100 is recorded as volume percent solids. If the mud contains as much as 20,000 mg/l chlorides, the total solids number should be corrected. In muds which contain no Barite, the total solids can be calculated from mud density. This is relatively more accurate than a retort analysis.

Correlation of mud solids composition with flow properties is important for proper interpretation of the flow properties. For instance, a high plastic viscosity is an indicator of high solids content, but it should be confirmed by solids analysis. The high plastic viscosity could also be caused by high viscosity of the base liquid. A high yield point often indicates a high concentration Bentonite, but it may be also caused by a chemical imbalance in the mud. A solids analysis and a filtrate analysis may both be necessary to diagnose a yield point problem.

The solids composition should also be correlated with filtration properties for similar reasons. Adequate Bentonite content is critical for good filtration control, and the drilled-solids content should be minimized to minimize cake thickness and insure good cake quality.

Recording the volume of oil and water in graduated cylinder must be done with following fundamental principle in mud. The water level should be recorded as a reading at the bottom end of reading and oil reading at the top end of the reading.

Methylene Blue Test

The methylene blue test is a method for measuring the Cation-exchange capacity of clay solids. The Cation-exchange capacity of clay is a measure of the ability of the clay to carry exchangeable Cations. Since it is through hydration of the exchangeable Cations that clay derives its swelling ability, the Cation-exchange capacity is a rough indicator of the rheological and filtration characteristics of a clay suspension.

When the methylene blue test (MBT) is run on a mud, the total Cation-exchange capacity of all the clays minerals present in the mud is measured. Since clay minerals other than Bentonite also have exchange capacity, the test is not a direct measurement of the Bentonite content. It is normal procedure, to report the MBT as the amount of Wyoming Bentonite that it would take to obtain the same methylene blue capacity.

An estimate of the actual amount of Bentonite in a mud can be obtained if we know the relative exchange capacities of Bentonite and the drilled solids. It has been found that on an average, the drilled-solids will have about one ninth the exchange capacity of Bentonite. Using this ratio of nine to one on exchange capacities, the clay solids fraction calculated from the retort analysis can be subdivided into Bentonite and drilled solids. This field procedure is quite helpful in determining the "quality" of the colloidal fraction of a mud.

It should be noted that the Cation exchange capacity is a measurement of a clay property. How that clay will perform in a suspension depends on the chemical content of the water. For this reason, we cannot assign ideal MBT values for various mud weights without specifying the water chemistry.

As salinity of the water is increased, it takes an increased amount of Bentonite to build the same viscosity. For good mud properties, it takes roughly twice as much Bentonite in sea water as it does in fresh water. In other words, the MBT values of seawater muds are going to run about twice as high as for equivalent fresh water muds.

Since the MBT is a measurement of the "reactive" solids in a mud, it should correlate with the yield point and gel strength of the mud. High yield point and gel strength values can be caused by either a high concentration of reactive solids or by flocculation due to temperature or contaminants. If the MBT values do not increase during a period of time when the yield point has increased, the problem must be due to chemical causes. Conversely, if the MBT values have been on an increasing trend with the yield point, excessively high clay content is indicated. Correlation of physical properties with composition is often necessary to properly prescribe treatment.

Methylene blue test result, when multiplied by 5 gives the approximate dose of Bentonite in ppb (pounds per barrel)

FILTRATION

Filtration occurs when a permeable formation is exposed to the mud at a pressure higher than the formation pressure. The pressure causes the filtrate to flow into the rock and deposit mud solids on the walls of the borehole. Thus, filtration accounts for two distinctly different types of phenomena- filtrate invasion and filter cake deposition.

Most of the problems caused by filtrate invasion are not drilling related, but are formation evaluation and completion problems. Excessive fluid loss may cause flushing of the zone around a wellbore to the extent that logging and formation test information is incorrect. This is normally not a problem with weighted muds where filtration control is necessary for control of filter cake deposition. In clear water or low solids muds, excessive flushing may pose problems.

Another problem is the invasion of filtrate in the formation can greatly reduce the formation permeability. The resistance to flow occurs close to the wellbore due to the radial geometry. Consequently, the volume of spurt filtrate lost is not so important as the type of filtrate. When a mud composed with non-damaging filtrate is used to drill a well, equal care should be taken on subsequent operations such as cementing or acidizing to prevent damage from these liquids.

From the standpoint of the drilling operation, the filter cake is of more concern than the volume of filtrate. The filter cake has a direct bearing on such problems as differential pressure sticking, torque and drag and poor primary cement job. Our aim should be to minimize the thickness and permeability of the deposited cake.

There are two types of filtration: dynamic and static. Dynamic filtration occurs when the fluid is being circulated and static filtration occurs when the fluid is at rest. Dynamic filtration differs from static filtration in that the flow of mud tends to erode the cake as it is deposited by the filtration process. The filter cake builds until the rate of deposition equals the rate of erosion. When it reaches an equilibrium thickness, the rate of filtration becomes constant. This contrasts with static filtration where the cake continues to grow thicker with time, causing the rate of filtration to continually decrease. After equivalent filtration times, a static cake will be thicker than a dynamic cake and the static rate of filtration will be less than the dynamic rate. Consequently, we should control static filtration to minimize filter cake thickness and dynamic filtration should be controlled to control filtrate invasion.

Dynamic Filtration

When filtration begins with a freshly exposed rock surface, the flow of filtrate into the rock is relatively large. Solids carried by the filtrate are deposited as a filter cake. As the cake grows thicker, the rate of filtrate flow decreases and the rate of solids deposition also decrease. If the mud is circulated at a steady rate, a point will be reached at which the flowing mud will carry away the solids as fast as they are being deposited to the cake surface by the flow of filtrate into the cake. Now, the cake thickness and the filtration rate become constant.

The erosion rate is a function of annular velocity, mud viscosity, and flow regime. An increase in annular velocity or mud viscosity will cause an increase in erosion rate, resulting in an increased filtration rate. Changing from laminar to turbulent flow will also increase the erosion rate. In turbulent flow, doubling the velocity will cause the filtration rate to be increased by a factor higher than in laminar flow. Therefore, ensuring that the flow regime is laminar in the annulus and that the velocity is as low as practical will ensure greatly reduced dynamic filtration rates

The viscosity of the mud is important. An increase in viscosity will cause a proportionate increase in erosion rate. Fluid-loss control agents that cause large increases in mud viscosity may cause the dynamic filtration rate to be increased.

The deposition rate is controlled by the rate at which solids are carried to the cake by the flow of filtrate. When the solids content of a mud is increased, the same volume of filtrate will carry more solids to the cake. This will cause an increased deposition rate and will result in a thicker equilibrium cake and reduced filtration rate. If these solids are the type that form low permeability filter cakes, a small increase in equilibrium cake thickness will cause a significant decrease in filtration rate. Bentonite appears to be the most effective additive in reducing dynamic filtration rates. Sizable decreases in filtration rate can be obtained with increasing Bentonite concentrations up to about 25 lbs. /bbl. At higher concentration, Bentonite causes very rapid increases in mud viscosity, resulting in increased erosion rates and minimizing further reductions in filtrate invasion rate.

Materials such as Lignosulfonate, lignite, oil, starch, and CMC, which are used to reduce static filtration, have little effect on dynamic filtration. These additives reduce the cake permeability, causing a decrease in the equilibrium cake thickness rather than a reduction in filtration rate. As a result, reducing the static API fluid loss with these materials need not necessarily cause a corresponding reduction in the dynamic filtration rate.

Static Filtration

Static filtration control is necessary to control the characteristics of the filter cake deposited down-hole. It is the cake, which is the source of filtration-related drilling problems. We are interested in the thickness of the cake, its permeability, slickness, and texture. Filtrate volume is only one of the indicators that can be used to evaluate filtration characteristics of a mud. Therefore, we should concern ourselves with all the cake characteristics rather than only with the filtrate volume.

It is always desirable that the filter cake be thin. This is the most important requirement placed on the filtration characteristics of a mud. The next characteristic in importance is that the cake has a low permeability. This too, is always desirable and usually the case with a thin cake. Although these characteristics are related to low fluid loss, other factors can also cause a low fluid loss. High solids content can cause the fluid loss to be low, but result in a thick cake. Depending on the quality of these solids, the cake permeability may also be high. This is obviously not a desirable situation but one that can occur, if we place too much emphasis on fluid loss alone!

It is difficult to measure the cake thickness with a high degree of accuracy. The basic problem is in determining where the gelled mud ends and the filter cake begins. This is especially true in un-weighted or low-density muds. High-density muds form filter cakes of higher solids content, so the boundary between mud and cake is more easily distinguished. A standard method for measuring the cake thickness should be adopted and this measurement made and recorded on all filtration tests with the greatest accuracy that can be accomplished. This measurement is more important than the filtrate volume as far as mitigation of drilling problems is concerned.

When we have attained our initial goal of minimum fluid loss and minimum cake thickness, we have basically solved the filtration problems for most situations.

The slickness of a cake is related to the frictional drag that can develop during differential pressure sticking. A relative measurement of this property can be obtained with the "stickometer". The stickometer test is particularly useful in evaluating the effects of special additives used to prevent sticking.

The texture of a filter cake is a characteristic for which no direct test is run. It is so closely related to the other cake characteristics that it will probably never be a problem if the other cake properties are in good shape. For instance, a fluffy filter cake will probably also be excessively thick. A dry, brittle cake has a high solids content and shall have a high sticking coefficient.

When all the cake characteristics are considered, a much better picture of the filtration characteristics of a mud will be obtained. Specific mud problems can be spotted that otherwise could go undetected while treating the wrongly diagnosed problem.

Two static filtration tests are normally run-a low-temperature test run at surface temperature and 100 psi filtration pressure, and a high-temperature test run at 300°F and 500 psi filtration pressure. These two tests span the range of temperatures from surface to bottom for most wells drilled. They also serve as reference tests so that changes in the mud can be detected and the filtration properties of a mud can be correlated with other jobs where the same tests were run.

The low-temperature test is quicker and more easily performed than the high-temperature test. For this reason, it is normally run more frequently to detect any sudden changes in the filtration behavior of a mud. It should not be used exclusively when the bottom-hole temperature exceeds 180°F or when there is any reason to believe that the temperature down-hole may detrimentally affect the mud.

The 300°F test is used to evaluate filtration behavior when the down-hole temperatures get high enough to significantly affect the filtration behavior. This can occur from temperature induced flocculation, the degradation of a fluid-loss control additive, or any unpredictable temperature reaction. Even though the 300°F test usually does not duplicate the down-hole temperature, it is an indicator of whether a mud is stable up to a temperature of 300°F. If the bottom-hole temperature exceeds 300°F, a filtration test should also be run at this temperature. When the expected maximum bottom-hole temperature is considerably less than 300°F, a fluid-loss test should be run at this temperature. Treating a mud to have good fluid-loss control at temperatures higher than necessary is expensive.

When the temperature of the mud is increased, the fluid loss will increase due to a reduction in viscosity of the filtrate. If a mud is otherwise thermally stable, the fluid loss will increase in the same proportion as the decrease of the square root of a filtrate viscosity. When a mud is heated from 75°F to 250°F, the viscosity of the filtrate is decreased by a factor of four, resulting in the fluid loss increasing by a factor of two. Fluid-loss tests can be run on a mud at two different temperature values, but at the same filtration pressure, and the ratio of the two values compared with the ratio of the relative values of this curve. If the test value indicates a greater change in fluid loss with temperature than predicted from the curve, the mud is probably thermally unstable. This is one of the best tests for thermal stability.

Unfortunately, the low-temperature and high-temperature API tests cannot be compared in this manner since they are run at different pressures. Pressure also affects fluid loss, depending on the compressibility of the filter cake. Normally, the 300°F-500 psi fluid loss will be 3 to 4 times greater than the low temperature-100 psi fluid loss. If the factor is greater than four, temperature-induced flocculation or breakdown of one of the mud components is probable. If the factor is less than two, a mistake in measurement is indicated. It could mean that the filtrate volume in the high-temperature test was not multiplied by two before reporting. This is necessary, since the filtration area of the HTHP cell is half that of the low-temperature cell.

The temperature, at which the low-temperature test is performed, also makes a difference. For water-base mud in the temperature range of 70°F to 140°F, the fluid loss will increase about 10 percent for each 15° increase in temperature. An effort should be made to run all tests at the same temperature.

The figure quoted here is only for illustration purpose.The focus is to display the trend.

Additional information, which is useful in diagnosing filtration problems, can be derived from filtration tests at different pressures and the same temperature. The effect of pressure on fluid loss is largely controlled by the colloidal solids content of the mud. If the solids in the filter cake are deformable under pressure, such as hydrated clay particles, increased pressure will cause a decrease in cake permeability, which in turn minimizes any increase in fluid loss. If the solids are rigid, such as sand and silt particles, the permeability will not reduce under pressure and the entire effect of pressure will be reflected in the fluid loss. The theoretical maximum for this effect is for the fluid loss to increase as the square root of filtration pressure. For a freshwater Bentonite suspension, the 500-psi fluid loss will be marginally higher than the 100psi fluid loss, due to the compressible nature of the filter cake. These two cases represent the extremes that will normally be obtained with a drilling mud. From this the quality of the colloidal solid content of a mud can be judged.

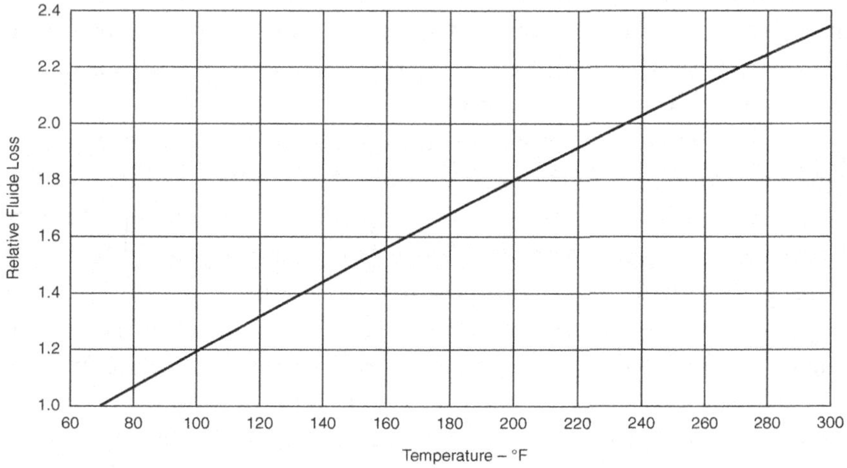

WATER CHEMISTRY

Knowledge of the water chemistry of mud is essential in diagnosing and curing the mud problems. A mud is basically solids dispersed in liquid with organic and inorganic chemical compounds added to impart specific properties to the suspension. Clay solids are dominant in their influence on the physical properties of a mud and the chemical environment, to which, they are exposed and these variables are critical to the performance of mud.

Other than the indirect influence on the physical properties of a mud, water chemistry is also important to corrosion control, effects on formation evaluation and formation damage, and detection of salt water influx.

The water chemistry tests normally made are:

1. pH and alkalinity
2. Total hardness, or Calcium and Magnesium
3. Chlorides

Several other tests are run in special situations. A potassium ion determination may be run when using a potassium-based drilling fluid. A check for soluble sulfides is used when hydrogen sulfide is present, and still other tests are run to determine concentrations of specific mud additives.

Most of these tests are run on the mud filtrate. This filtrate should come from the low-temperature filtration test.

pH and Alkalinity

The pH of a solution is a measure of its hydrogen ion concentration. For each hydrogen ion (H^+) concentration, there is an equilibrium concentration of hydroxyl (OH^-) ions. By measuring the hydrogen ion concentration, we are, in effect, also measuring the hydroxyl ion concentration. In pure water, the H^+ and OH^- concentrations are the same. This is the neutral point, or a pH of 7. When acid is added to water, the pH decreases on a scale from 7 to 0. When a base (caustic) is added to water, the pH increases on a scale from 7 to 14. The pH scale is logarithmic. This means the hydrogen or hydroxyl ion concentration changes by a factor of ten for each unit change in pH. In other words, a solution at pH 10 has 10 times the hydroxyl ion concentration of a solution at pH 9.

Maintaining adequate pH in a mud is important for following reasons.

1. Corrosion rates are suppressed at a pH above 10.
2. Hydroxyl ions neutralize the H^+ ions from H_2S and prevent hydrogen embrittlement. A pH above 10 should be maintained if H_2S is anticipated.
3. Organic additives such as lignite and Lignosulfonate require a pH > 9.5 to function effectively.
4. The rate of thermal degradation of Lignosulfonate is reduced at a pH above 10.
5. Magnesium is precipitated from sea water at pH 10.
6. Calcium solubility is reduced as pH is increased.

When all these factors are taken into consideration, the mud pH should be kept in the range of +/- 9.5 / 9.0. A pH value of > 10.5 should be avoided unless lime mud is being used. Very high pH is detrimental to most organic additives and can cause flocculation of the clay solids. This is basically the problem encountered due to cement contamination.

The high yield point, gel strength, and fluid loss that occur are largely due to the sudden increase in pH to very high levels, rather than to the small amount of Calcium introduced into the system.

There are two basic ways of measuring pH: either a pH meter or pH paper test strips. A pH meter is preferred. The pH paper does not have the inherent accuracy of a meter and becomes unreliable in the presence of high salt concentration or dark filtrate. A Sodium-compensated electrode should be used on the meter when measuring pH of high salt content solutions.

The alkalinity of a solution is related to pH in much the same manner as heat capacity is related to temperature. Alkalinity is a measure of the amount of acid required to reduce the pH of a solution to a specified value. Different solutions require different amounts of acid to reduce the pH just as different substances require different amounts of heat removal to reduce the temperature. This is due to the presence of buffering ions and substances that react with the acid in addition to the hydroxyl ions in solution.

Two filtrate alkalinities are normally measured as the Pf and Mf. The Pf alkalinity is the volume cc of 0.02 N sulfuric acid necessary to reduce the pH of 1 cc of filtrate to 8.3, which is the end-point color change of phenolphthalein indicator solution. The Mf, or total alkalinity, is a measure of the amount of the same acid necessary to reduce the pH of the filtrate to 4.3, which is the color change end-point of methyl orange indicator solution.

The significance of titrating to these specific values of pH is due to the presence of Carbonate (CO_3^{--}) and bicarbonate (HCO_3^{-}) ions in many mud filtrates. The Carbonate ion reacts with acid to form bicarbonate. This reaction is completed when the pH is reduced to 8.3. Also at this pH, essentially all the hydroxyl ions have been neutralized. As a result, Pf alkalinity is a measure of the combined contributions of hydroxyl and Carbonate alkalinities. As the pH is further reduced to 4.3, the acid reacts with bicarbonate ions to convert them to $C0_2$ and water. From these two alkalinity determinations, it is possible to calculate the amounts of hydroxyl, Carbonate, and bicarbonate ion in solution if no other interfering ions are present. Unfortunately, in many mud filtrates organic acid and buffering ion such as chromate are present and contribute to the measured Mf alkalinity. There are other, more detailed titration procedures that can be used to circumvent this problem. It is worth adding here that the pH scale of 0 to 14 is only an arbitrary range. In real situations pH can be higher than 14 as well as less than zero depending upon the normality of the acid/base.

High concentrations of Carbonate or bicarbonate ions in a mud cause high gel strength development, which cannot be reduced by the addition of thinners. This leads to dilution of the mud system with water, to control the flow properties. The alkalinities should be watched carefully to avoid this problem. The beginning of the problem is indicated when the Mf begins an increasing trend and the pH decreases or requires a larger than normal amount of caustic to hold constant. If this happens in conjunction with an increasing trend in the 10-minute gel strength, which cannot be controlled by normal treatment with Lignosulfonate. This is the right time to begin slow additions of a soluble form of Calcium, such as gypsum or lime. The Calcium will combine with Carbonate to precipitate it as limestone and thus remove the source of the problem.

When the pH of a mud is run in the proper range and the Carbonate and bicarbonate concentrations are minimized, the mud properties should remain stable and responsive to chemical treatment.

In addition to the filtrate alkalinities, an alkalinity determination is made in the mud also (Pm). This test is done in the same way as the Pf test, except that the whole mud is used rather than filtrate. The purpose of this test is to determine the mud's reserve alkalinity. There are sources of alkalinity, which due to the limited solubility of materials such as lime or cement or absorption of hydroxyl ions in solids, are carried as solids in a mud. As the hydroxyl ions in solutions are used up, more hydroxyl goes into solution from these sources and stabilizes the pH

The water fraction of a mud sample will have the same concentration of hydroxyl ions as the filtrate. If the volume fraction of water in a mud is multiplied by Pf, the result should be the contribution of the water phase to the Pm alkalinity. This number subtracted from Pm is the contribution to alkalinity from the solids. If this number is multiplied by 0.26, the solids' contribution to Pm alkalinity is calculated as equivalent free lime concentration in lb./bbl.

$$\text{Free lim e, lb./bbl} = .26\left[\text{Pm} - \frac{\%\,\text{water}}{100}(\text{Pf})\right]$$

Calcium and Magnesium Determinations

Calcium or Magnesium ions in a mud have a pronounced effect on the behavior of the clays. These ions replace Sodium ions on the clays, reduce the degree of hydration, and promote flocculation and aggregation of the clay particles. This leads to increases in yield point, gel strength, and fluid loss of the mud. Many of the organic treating agents are also sensitive to Calcium and Magnesium concentrations. For these reasons, it is necessary to monitor the concentration of these ions in the mud filtrate.

To determine the combined concentrations of Calcium and Magnesium in the filtrate, a total hardness test is made. The result of this test is reported as Calcium in ppm. If the actual Calcium and Magnesium concentrations are desired, it is necessary to run a second titration using another indicator, specific to Calcium. This test gives the Calcium concentration. The Magnesium concentration can then be calculated. When Magnesium is known to be present, both determinations should be made. This is necessary since treatments to remove these two ions from solution are different.

The solubility of a Calcium salt such as gypsum or lime is increased by increased salinity and decreased by increased pH or temperature. As a result, the addition of these materials to a mud either as contaminants or intentional additions will result in various Calcium ion concentrations in the filtrate.

Caustic (Na OH) is sometimes used to suppress the level of soluble Calcium in a mud. This technique is often used in lime muds to control the Calcium concentration.

The usual method for precipitation of Calcium is the addition of soda ash (Na_2CO3). The Calcium is precipitated as Ca $C0_3$ (limestone).

One lb/bbl of soda ash will precipitate approximately 1000 ppm of Calcium. If an overtreatment of soda ash is used to remove Calcium from solution, a Carbonate problem will occur. Overtreatment will result in high yield point, high gel strength, and fluid loss. In fact, it is possible to go from one problem to the other without realizing what happened, since both the Calcium and Carbonate ions produce nearly the same changes in mud properties. To avoid this problem, a small amount of Calcium (approximately 100 mg/l) should be left in solution

Calcium can also be removed from solution with lignite. However, at very high lignite concentrations, an influx of Calcium will form a precipitate that can cause an increase in plastic viscosity and yield point. Prolonged use of lignite in such a situation should therefore be avoided. Its use should also be suspended when adding Calcium to remove Carbonates, since it will interfere with this reaction.

Magnesium can be removed from solution by the addition of hydroxyl ions. At a pH above 10, essentially all the Magnesium is precipitated as the insoluble Magnesium hydroxide. This precipitate is very fine and has an extremely large surface area. If precipitated in large quantities, it will form a rigid gel. Consequently, concentrations above a few thousand mg/l cannot be precipitated without causing major increases in mud viscosity. Due to its large surface-area, it is also capable of physically extracting organic treating agents such as Lignosulfonate and CMC (Carboxy Methyl Cellulose) from solution. These materials should be added to a mud after addition of caustic.

Chloride Determination

Salinity is another important part of the water chemistry of a mud. It is measured by titration of the filtrate for chloride ion concentration and is reported as chloride concentration in mg/l or as Sodium chloride (Na Cl) concentration in mg/l. The Sodium chloride concentration is equal to 1.65 times the chloride concentration.

The salt content of a mud has a great effect on the behavior of the clay solids. The degree of hydration of Bentonite is reduced by increased salinity. This results in decreased plastic viscosity and increased fluid loss. When clays are hydrated in fresh water, addition of salt will cause flocculation and increased yield point and gel strength.

Monitoring the chloride content of a mud is a means of detecting a salt water flow or the drilling of salt. It is sometimes possible to detect an increasing trend in chlorides as the formation pore pressure increases. This is one of the several other indicators of abnormal pressure. Makeup water should also be titrated for chlorides to determine whether an increase in filtrate-chlorides has come from the formation or from the makeup water.

When the chloride concentration is 20,000 mg/l or higher, Pre-hydrated Bentonite should be added for fluid-loss control. Dry Bentonite added to such muds will not hydrate sufficiently to give good fluid-loss control.

PILOT TESTING

Pilot testing essentially means that a possible combination of additives and/or their doses is first mixed in the lab and tested under bottom-hole conditions to predict the behavior / response of the mud to ensure that the proposed combination will work. Although such tests are often put prior to pumping specialty LCM, it is usually wise to run pilot tests before attempting to make any major change in mud properties. By pilot testing the proposed additions to the mud, costly mistakes in treating the total system can be avoided. Pilot testing can also be used to evaluate in advance the performance of a mud under anticipated conditions, such as long-term exposure to high temperature or influx of various contaminants.

Any time a mud problem occurs; pilot testing should be used to confirm the diagnosis. There is often more than one possible cause for a problem. A series of pilot tests designed to cure each possible cause will indicate the best solution. Pilot testing can also be done in advance of increasing the mud density. Mud problems or Barite contamination problems can be discovered before any great amount of harm has been done.

A pilot test should mimic the conditions down-hole as closely as possible. Temperature, pressure and shear are all important conditions that affect mud performance. Even the cycling of the conditions that the mud undergoes during circulation may affect its performance. The rate and order of addition of the mud additives also influence mud properties.

A pilot test accurately indicates the performance of mud under the conditions that are used in the test. Unfortunately, our pilot test information will not be completely accurate in all cases because the test conditions are not duplications of identical down-hole conditions. The information should serve as a quite useful guide.

Normally, temperature is the most important condition that affects mud performance. An oven equipped to roll or rotate mud samples will allow heat aging of mud samples under conditions of very mild agitation. Static ageing can also be studied with the same oven if we are concerned with Barite settling or packer mud performance.

The rolling or rotating test is more representative of the mud being circulated. It is also more rigorous than a static test. It supplies mixing, which is necessary for some reactions in the mud to go to completion.

Properties should be checked both before and after heat ageing, to evaluate initial and aged effects of mud treatment. A sample of the base mud should be carried through the pilot testing conditions to provide reference data.

We should routinely conduct pilot tests in critical situations, looking ahead to higher temperatures and mud densities as we drill deeper. With this data, we should be able to keep ahead of impending mud problems.

There are some specific situations where a pilot test is not something which is desirable but is a MUST!!

One such occasion is when we chose to pump cross-linking polymer LCM pill through the tubulars in hole to control massive losses. It is essential that the pumped components of the pill start gelling up after the pill is out of string and is just at the mouth of the loss / thief zone. Any early setting of the pill can be detrimental and likewise, any delayed setting also could defeat the purpose of the pill.

To test all the properties and parameters of the mud, a host of equipment and apparatus are required. The list of such equipment as such can be too long to be included here and yet be incomplete. However, staying with the concept of the book for starters, a minimum list is described here which is a must for any place to qualify for a mud test lab.

S. No	Equipment name	Used for
1	8 speed rotational viscometer	To measure rheology
2	Heating cup with thermostat	Maintain content temperature while measuring rheology
3	Digital weighing scale	To accurately weigh samples for mixing mud
4	pH meter	To measure pH
5	Retort kit	To check solids content
6	API F/L Apparatus	Check API F/L under normal conditions
7	HTHP F/L Apparatus	Check HTHP F/L under HTHP conditions
8	API Mud balance	To measure mud weight
9	Pressurized mud balance	To measure mud weight after pressurizing the mud for more accurate measurements
10	Marsh funnel with cup	To check funnel viscosity of mud

Continues

S. No	Equipment name	Used for
11	Sand content kit and MBT kit	To check sand content and CEC of mud
12	ES meter	To measure electrical stability of oil base mud
13	Hamilton beach mixer	To mix pre-hydrated bentonite
14	Thermostatic hot plate with magnetic stirrer	To heat mud and simultaneously stir
15	Static oven	For drying samples and sometimes for static ageing
16	Roller oven for dynamic hot rolling	To check mud after subjecting to required bottom hole pressure and temperature
17	Ageing cells	To fill the mud under pressure and keep in roller oven for dynamic ageing
18	Nitrogen cylinder	To apply pressure in fluid loss measurement and pressurizing the ageing cells
19	Glasswares, reagents, mud additives	For managing daily routine tests
20	Silverson mixer	To mix Oil base mud
21	Lab blender	To mix WBM
22	PPT Dynamic fluid loss measurement apparatus	TO check dynamic HTHP fluid under bottom hole conditions.

CHAPTER 4

RELATIONSHIP OF MUD COMPOSTION TO PROPERTIES

For effective control of the properties of mud, we need to know how specific changes in mud composition will affect the properties. There are many ways in which the mud property can be adjusted to a given value. We are usually faced with maintaining several properties within acceptable limits. This means that there is one best way to adjust a mud property and still maintain control of the other properties. We must understand the interrelationship between mud composition and mud properties to be able to do so. The liquid and solids content of mud forms the foundation which ultimately determines the performance of the mud. Chemical treatment should be thought of as a means of making small or selective adjustments in mud properties and not as the primary controlling component of a mud.

SOLIDS

The primary source of control of theological and filtration properties is the control of the amount and type of solids in the fluid. The importance of controlling the solids in a mud cannot be overly stressed. This is the basis of all mud treating. Undesirable solids increase drilling costs. They cause a reduction in the penetration rate, promote lost circulation and differential pressure sticking and cause high mud costs due to treatment and dilution to maintain necessary mud properties. The normal mud cost of drilling a well is linked very closely to the solids control practice used.

In general, our aim is to minimize plastic viscosity for increased penetration rate, provide sufficient yield point or effective annular viscosity to clean the hole without causing unnecessary high circulating pressures and provide sufficient gel strength to suspend weighing material and cuttings. To fulfill these requirements, we must adjust the flow properties of mud to handle each situation that is encountered. It is obvious that ideal mud properties for one problem may increase the severity of another. For this reason, there is no standard recipe for an ideal mud composition.

For all except a very few special situations, Barite and Bentonite are the only solids needed to provide the density, viscosity, and fluid-loss control requirements of a mud. Formation solids that enter the system from the drilling process provide no function that could not be better satisfied by the addition of Barite and Bentonite.

Inert Solids

The solids in a mud can be roughly divided into two types, inert and active. The inert solids are those that have very little tendency to react with each other or to a change in environment. These are non-swelling solids which are essentially electrically uncharged and easily dispersed. They consist of such materials as sand, silt, limestone, felspar, and Barite. Except Barite, these solids are undesirable in mud. They increase the plastic viscosity and crowd the system so that more desirable solids cannot be added. They do little to reduce fluid loss and cause the filter cake to be thicker. Since they are not hydrated, they cause the filter cake to be dry and brittle with a high coefficient of friction. These are poor qualities and they promote differential pressure sticking, an undesirable feature.

As the total volume of solids in suspension is increased, the mechanical resistance to flow is increased. This causes an increase in the plastic viscosity. As the solids become crowded, the plastic viscosity increases quite rapidly.

Active Solids

The active solids are those that have electrically charged surfaces and are reactive to their environment. Due to their charge distribution, they can interact with one another to form a gel structure within a fluid. Most of the active solids swell in water.

Clays are the primary source of active solids in a drilling fluid. Due to their characteristics, they exhibit a dominant role in determining the rheological and filtration properties of a mud. For this reason, it is necessary to understand basic clay chemistry to properly control water-base muds.

In most muds, there are two primary sources of clay solids: Bentonite, which is added intentionally, and native clays which enter the mud through dispersion of drilled solids. The native clays are a combination of several clay minerals that exhibit varying degrees of hydration and dispersion. Different shales contain different amounts of the various clay minerals. Consequently, some shales exhibit a greater "mud-making" ability than others.

The amount of water adsorbed on clay platelets (and hence swelling) is controlled by both the type of clay and the exchangeable Cation on the clay. Bentonite has a smaller particle size, higher charge density, and greater ability to swell in water than the other types of clays. This makes it a better viscosifier and fluid loss control agent.

When a Bentonite particle is added to water, it absorbs water on the surfaces of the individual platelets that make up the particle. This increases the distance between platelets and causes the particle to grow and expand. In this expanded state, either thermal or mechanical agitation will cause the individual platelets to disperse in the water.

The water adsorbed on the platelet becomes a part of the platelet and greatly increases its size.

We started with one particle and through the process of hydration and dispersion created many particles whose total volume and surface area are many times greater than the original dry particle. Since the apparent volume of solids has increased, the plastic viscosity will be increased. Due to the increased surface area of the electrically charged platelets, the linking of the particles is enhanced. This will increase the yield point, gel strength, and low shear rate viscosity. Since these platelets are small, thin, and highly hydrated, they are exceptionally good for fluid loss control and prevention of differential pressure sticking.

These characteristics make Bentonite, hydrated with fresh water, the best single mud additive that we have. By controlling the amount, we use and treating to control its natural tendency to flocculate, we can construct a mud which will basically fit many requirements.

We should minimize the amount of native clay solids in a mud and rely on Bentonite for control of the flow properties. This allows us to carry a minimum amount of solids in a mud and simplifies treatment of the mud. The amount of Bentonite required should be based on the required yield point. Gel strength and filtration control. Additions of Bentonite should be made slowly and both the "in" and "out" properties checked. Bentonite increases in yield with both time and temperature. If the mud going in is adjusted to the desired properties, it will come out at higher properties due to further dispersion from the down-hole temperature. An increase in both, plastic viscosity and yield point will be observed from temperature dispersion and hydration of clay solids. This contrasts with the symptoms of temperature flocculation where only the yield point and gel strength are increased.

Large amounts of solids in a mud will decrease the fluid loss, but will also cause thicker filter cakes to be deposited. This is contrary to our basic objective of maintaining thin filter cakes. To avoid this problem, we should always attempt to maintain a minimum amount of total solids in a mud. We should minimize solids that form high permeability cakes and add solids that form low permeability cakes.

Basically, this means we should minimize the amount of formation solids in a mud and add Bentonite. The formation solids are normally poorly hydrated and dispersed, and control fluid loss only by building excessively thick filter cakes of poor quality. For reasons of economy, muds in the range from 10 to 11 lb./gal are often built with native solids. Many times, this is a false economy, since filtration related problems often occur.

Bentonite is one of the best filtration control agents we have. It is quite temperature stable and considering its low cost; it seems only natural that it should be the basis of filtration control in water-base muds. When used properly, it has the capability of solving most filtration problems. The ability of Bentonite to control filtration is based on its naturally small, plate-like particles, and high degree of hydration. Since these characteristics depend on the chemical environment in which Bentonite is placed, most of our efforts in treating a mud are based on control of this environment.

Effect of Salinity on Clays

The ionic content of water affects the hydration, dispersion and flocculation behavior of clay particles. As the salinity increases, the degree of hydration decreases. This is caused by the Cations in solution pushing the exchange Cations closer to the surface of the clay platelet. This causes the water layer bound by the clay surface and the exchange Cations to be thinner. As a result the plastic viscosity will be less and the fluid loss will be greater than when the clay is hydrated with fresh water. Since the water layer is thinner, the platelets can come closer to one another, and, tendency to flocculate will be greater.

Increased Bentonite additions are required to help offset the effects caused by increased salinity. Dry Bentonite added to muds with chloride concentrations above about 10,000 mg/l will not hydrate sufficiently to give good filtration control. In these muds, Pre-hydrated Bentonite should be used.

Hydration of Bentonite

Pre-hydrated Bentonite should be prepared in fresh water. Any salt, Calcium. Caustic, or chemical treating agent will decrease the amount of hydration and dispersion that can be obtained. The hydration of Bentonite increases with increased time. Temperature, and agitation. If possible, about 6 to 8 hours should be allowed to ensure sufficient hydration. High shear mixing aids in hydration and dispersion.

After the hydration period is complete the slurry should be treated with 2 to 8 lbs. /bbl of Lignosulfonate and the pH increased to 10 by adding caustic. This causes the Bentonite to maintain a high degree of hydration after it is subjected to a high chloride environment. Good fluid-loss control can be achieved with this technique, even in salinities up to saturation.

When Pre-hydrated Bentonite is used, the concentration required in the mud is approximately that which is required in a freshwater mud. It should be prepared with as high a concentration as possible and still remain pumpable. Otherwise. It may be a source of excessive dilution when added to a weighted mud. It is usually possible to mix 35 to 45 lbs. /bbl. After addition of Lignosulfonate, it becomes less viscous and more easily pumped. When mixing subsequent batches in the same tank, care should be taken to completely empty the tank each time. Otherwise, the Lignosulfonate remaining in the tank will inhibit the hydration and dispersion of the next batch.

Effects of Calcium and Magnesium on Clays

Calcium and Magnesium have a greater Cation exchange replacing power than Sodium. When added to a Bentonite suspension in relatively small concentrations, they have the ability to displace part of the Sodium from the exchange sites on the clay platelets. Due to their greater charge. Calcium and Magnesium are held closer to the surface of the clay platelets. This causes the adsorbed water layer to be thinner and decreases the repelling forces between the negatively charged platelets. For this reason, relatively small amounts of Calcium or Magnesium cause great changes in the performance of clays. When Bentonite is added to water containing more than a few hundred ppm of either of these ions, the degree of hydration is greatly reduced and the tendency to disperse is inhibited. As a result, very little viscosity is developed and the fluid loss is quite high.

At one time, gyp and lime muds were commonly used to take advantage of the inhibiting feature of Calcium. Lime muds contain from 100 to 300 mg/1 Calcium in the filtrate and gyp muds contain 400-1200 mg/1 Calcium. Due to their inhibitive nature, these muds will tolerate a high concentration of formation solids before becoming too viscous to pump. The high solids content and high fluid loss of these muds made them poor from the standpoint of differential pressure sticking. Also, higher solids content and flocculation tendency of Calcium causes severe gelation problems at high temperature. Because of these problems, lime and gyp muds declined in use. When the Calcium or Magnesium concentration in a mud is suddenly increased, the mud undergoes some drastic changes. The clays dehydrate and flocculate. This causes a decrease in plastic viscosity and an increase in yield point. Gel strength, and fluid loss. The severity of these problems is a function of the concentration of clay solids, the amount of Calcium introduced into the system, and the concentration of Lignosulfonate in the mud. In lightly treated muds 100 to 200 mg/l of Calcium may completely flocculate the mud. This often occurs from cement contamination in surface muds.

Flocculation from any cause is always aggravated by a high solids content. Maintaining mud with a minimum amount of clay solids is the first line of defense against Calcium contamination. If the mud has a high concentration of Lignosulfonate, it will behave like a Prehydrated Bentonite suspension. The Lignosulfonate will protect the Bentonite and minimize the amount of dehydration and flocculation that occurs. This protection is limited to, Calcium concentrations below about 2000 mg/l. Usually a saltwater flow will contain both Calcium and Magnesium. The Magnesium is precipitated by the hydroxyl ion. A rapid decrease in pH will occur. This necessitates a high starting pH to minimize the effects of Magnesium influx. The precipitate that forms will physically extract, organic treating agents. Fresh Lignosulfonate should be added to replace that which is removed.

Lignite, also offers some protection from an influx of Calcium. By and large one lb. /bbl of lignite will precipitate-out about 200 mg/l of Calcium. If the Calcium influx is not too great, this is a good way to handle it without upsetting the mud. However, The precipitate will build viscosity if more than about 6 to 10 lbs. /bbl of lignite is precipitated.

When the total hardness reaches a level where precipitation is no longer practical or when the influx is sustained for a prolonged period of time, (such as drilling gypsum or anhydrite), it is necessary to convert to a Calcium-based mud. In this state, the clay platelets have aggregated face-to-face like a peck of cards. With a smaller total number of particles, the yield point and gel strength are less and the fluid loss greater. The mud may require addition of materials that will build viscosity and decrease fluid loss in the presence of Calcium. The concentration of Calcium in the mud and the bottom-hole temperature will dictate what materials can be used. At temperatures below 225°F, polymers such as starch, CMC, and Xanthan polymer can be used to build viscosity. Starch and CMC will also reduce the fluid loss. At high temperatures, materials such as attapulgite and asbestos are required to build viscosity. These materials will not control fluid loss.

Cement Contamination

Cement affects mud properties in essentially the same manner as lime. As cement goes into solution, it raises the pH to values above

12. The Calcium concentration is normally in the range of 100 to 300 mg/L. These values are fixed by the limit of solubility of the cement. Consequently. Most of the cement remains as small particles in the mud. This is evidenced by a high Pm.

In cement contamination, both the Calcium and the ultra-high pH are responsible for the flocculation of clay solids. Therefore, both need to be controlled. The standard treating agents are Sodium bicarbonate and SAPP. Both of these materials reduce the pH and precipitate or sequester the Calcium ions. Overtreatment with either material is as harmful as cement to the mud properties. The pH alkalinities, and Calcium should be monitored to avoid overtreatment.

When the Pm is reduced to about > twice the Pf, the treatment of Bi-carb/SAPP should be suspended. Lignite and Lignosulfonate are both useful in treating a cement-contaminated mud. Both are acid and will reduce the pH, and both will deflocculate and thin the mud. In addition Lignite will also precipitate Calcium. Pretreatment of the mud with these additives is also quite beneficial and recommended. They will not upset the mud properties as will SAPP and bicarbonate. Do not use alkaline materials such as soda ash or Sodium tetraphosphate.

Since most of the cement is carried in the mud as particles, treating will only remove that which is in solution and allow more of the cement particles to go into solution. Consequently, the problem will reoccur repeatedly until all of the cement has gone into solution and been treated out chemically. This is a very slow process and often leads to eventual overtreatment. If the cement particles are removed from the mud mechanically, the mud can be restored to original condition much quicker. Fine-screen shakers should be used to remove as much cement as possible. Then the centrifuge can be used to remove the small particles as well as the contaminated water and clay solids. Often, it is desirable to dump as much of the highly contaminated mud as possible.

Emulsified Oil

Emulsified oil acts as small, inert solids in a mud and increases the plastic viscosity. When a mud becomes crowded with solids. In the case of a higher weight mud, the addition of oil will cause significant increases in plastic viscosity. This effect largely determines the amount of oil that can be added to a mud. By removing as much of the drilled-solidsas possible, some room can be made for oil. When mud weight exceeds

17 lbs. /Gal, only 2 to 4 percent oil can be added without producing excessively high plastic viscosities and yield points.

Significant decrease in fluid loss is achieved with increasing oil concentration up to about 5 volume percent; at higher concentrations fluid loss declines more gradually. Emulsified oil droplets act in much the same manner as colloidal particles in plugging fine pores in the filter cake and reducing cake permeability.

Oil reduces the friction between the filter cake and pipe more than other additives and fluid-loss control agents. The oil droplets apparently occupy space on the surface of the filter cake that would otherwise be occupied by solids that have a much higher coefficient of friction on steel. This unique quality makes oil one of the best additives that we can have in mud to prevent differential pressure sticking.

In order to function effectively as a differential pressure sticking additive, the oil must be well emulsified. This requires a great deal of shear to reduce the oil to fine droplets, and a stabilizing film around the droplets to keep them from coalescing. Stable emulsions can usually be formed without the aid of a chemical emulsifier when dispersed clays, lignite, and Lignosulfonate are present. Operationally, it is best to add the oil slowly and directly into the pump suction. In this manner, a high degree of emulsification is accomplished in the first circulation cycle with uniform distribution and no loss of oil.

When oil is added slowly and continuously, it often appears to increase penetration rate and prevent bit balling. These effects seem to occur only when the oil makes its first pass through the bit and before it becomes completely stabilized in an emulsion.

TREATING AGENTS / MUD ADDITIVES

After the solids content of a mud has been properly adjusted in type and amount, selective adjustments in mud properties can be made by special treating agents. These additives fit roughly into three categories: deflocculants, fluid-loss control agents, and viscosifiers.

Deflocculants (Thinners)

Any material that will deflocculate a clay suspension is normally referred to as a thinner. Flocculation is caused by a positive charge site at the edge of a clay platelet linking to the negative surface of another platelet. The primary effect of a thinner is believed to be a neutralization of the positive charge, which destroys the ability of the platelets to link together in this manner. This renders the platelets completely negatively charged and causes them to repel one another, or to avoid contact. Thus, the resistance to flow of the suspension is reduced.

Thinners are used to reduce the yield point, gel strength, and low shear rate viscosity. They are not used to reduce the plastic viscosity since it is a function of the total volume of solids in suspension and not concerned with the linking of these particles.

The tendency of a mud to flocculate is increased by:

1. A high solids content, especially a high active solids content.
2. A high electrolyte content such as salt or soluble Calcium compounds.
3. High temperature.

To be effective, a thinner must be capable of performing under these conditions. Two of the first thinners used in muds, phosphate (Sodium tetraphosphate) and quebracho (Sodium tannate), have deficiencies in all of these areas. Phosphate is a very efficient thinner at low temperature in freshwater muds. The addition of 0.1 to 0.2 lb. /bbl will completely deflocculate a mud. It will not perform at high salt concentrations or in the presence of Calcium. It is actually a sequestering agent for Calcium and is thus depleted in a Calcium environment. At temperatures above 180°F, it reverts to orthophosphate, which is a flocculating agent.

Quebracho is also an effective thinner, requiring only 1 to 2 lbs. /bbl to deflocculate a mud. For best results, it should be used at a pH above 11. It is effective in lime muds or cement-contaminated muds, but is not effective at higher Calcium levels or in high salt content muds. It is not capable of deflocculating at temperatures above 225°F.

The suitable all-purpose thinner is chrome Lignosulfonate. It has a much greater tolerance for salt and Calcium than the other thinners and will deflocculate at temperatures in excess of 350°F. It is altered by high temperatures, and has a rather short life expectancy at 350°F. It is thought that polymerization takes place at high temperature, making the material less soluble and less effective as a deflocculant. This makes it difficult to assign a value to its temperature stability. At 350°F, degradation takes place and carbon dioxide and hydrogen sulfide are liberated. These by-products actually appear to be more detrimental to the mud than the loss of active Lignosulfonate. A high pH, in the range of 10 to 11, helps to cancel the effects of CO2 and H2S and increases the time that Lignosulfonate will function effectively at the elevated temperatures.

The concentration of Lignosulfonate necessary to deflocculate is a function of the severity of the factors that promote flocculation. In an uncontaminated freshwater mud at temperatures below 200°F, about 1 to 2 lbs. /bbl is sufficient. At 300°F, approximately 4 lb./bbl is required. In seawater muds. From 6 to 10 lbs./bbl is necessary to effectively control yield point and gel strength.

Sufficient Lignosulfonate concentration is indicated when the gel strengths are low and do not change appreciably between measurements at the suction and flowline. If the gel strength measured at the flowline is significantly higher than the gel strength measured at the suction, flocculation is occurring down-hole. This could be caused by either a Lignosulfonate deficiency or a contamination problem. If the gel strength does not respond to the small Lignosulfonate addition, a problem in water chemistry is indicated. The filtrate analysis should be carefully studied to help diagnose the problem. At times the high clay concentration coupled with high solids concentration leads to fragile gel, which quickly develops and also diminishes equally quickly. Thus, there is no real adverse effect on mud performance. On the contrary the gels which disappear as well as appear soon with then circulation resumption and suspension are desirable to hold solids as well as cuttings in suspension, a basic feature of mud, popularly called as thixotropy. On the contrary, a gel which is small to begin with, but keeps increasing with time at a rapid rate, is detrimental as the gel-breaking pressure during resuming circulation after long hours of stagnation can result in loss of fluid in rock due to high gel-breaking pressure. They are often termed as progressive gels and all efforts are to be focussed on minimizing progressive gel. The best defense is to keep inert solids ALAP.

Lignite is another product that is sometimes used as a thinner. It is more resistant to thermal degradation than Lignosulfonate, but it is less effective as a deflocculant at high temperatures. This fact is sometimes not recognized. When the flow properties are measured after allowing the mud to cool, the yield point and gel strength will be considerably less than if they were measured at a down - hole temperature. This can lead to the mistaken belief that the mud is also deflocculated down-hole. Lignite is not an effective thinner at temperatures above 250°F.

Fluid-Loss Control Agents

Several types of materials are used to control fluid loss in water-base muds. Basically, these materials reduce the fluid loss either indirectly by acting on the clay solids to improve their filtration control characteristics, or directly by acting as a colloid or water thickener. In most cases, they affect the flow properties as well as the fluid loss.

Basically, Lignosulfonate and lignite are used to improve the filtration control performance of Bentonite. When Bentonite flocculates, it becomes less effective as a filtration control agent. Lignosulfonate and lignite deflocculate the Bentonite, thereby reducing the fluid loss. Contaminating ions, such as Calcium and Magnesium, not only cause flocculation, but also dehydration and aggregation of Bentonite. Lignosulfonate appears to protect the Bentonite from these harmful effects and lignite precipitates the ions, thereby removing the source of the problem. Consequently, these additives are quite effective at high temperature or in the presence of Calcium and Magnesium. Since their ability to control fluid loss is tied directly to the colloidal solids content of a mud, care must be taken to provide sufficient Bentonite that is well hydrated and dispersed.

Lignite is different from Lignosulfonate in that it also acts as a colloidal material. In high weight muds where additional Bentonite cannot be used because of viscosity considerations, the addition of lignite will reduce fluid loss without causing excessive increases in viscosity.

Causticized, or pre-treated lignite is more expensive than ordinary lignite but is not substantially different in performance. The solubility of lignite is increased as the pH is increased. This makes the causticized lignite appear more soluble in water than ordinary lignite; caustic added with the lignite will attain the same results.

Some lignite products are neutralized with potassium hydroxide and blended with a chromium or zinc compound. This supposedly causes them to inhibit shale dispersion and increase the temperature stability of mud. Both potassium and the heavy metals are detrimental to fluid-loss control.

Natural lignite has a pH of about 4.5, which makes it a far superior material for treating a cement-contaminated mud than the causticized lignites. In general. It is cheaper on a cost-performance basis than the specially prepared lignite products. A number of water-soluble polymers are used to control fluid loss in situations where it is impractical or undesirable to use Bentonite. All these materials have a limited range of application and should be considered special-purpose products.

Starch is primarily used in brine muds at relatively low temperatures. Starch is one of a few products unaffected by water salinity or hardness. It thickens water, thereby deriving most of its fluid-loss control. As a consequence, it increases the plastic vicscosity and tends to reduce penetration rate. A starch solution thins with temperature at the same rate as water. This leads to reduced viscosity down-hole and a loss of hole-cleaning ability.

Thermal degradation begins at about 180°F and increases to a prohibitive rate at a temperature of 250°F. Starch is used in concentrations of 2 to 6 lbs. /bbl. Due to its viscosity-building ability, it is often used as the source of both viscosity and fluid-loss control. It is subject to bacterial degradation and must be used with a preservative except in saturated salt water or at a pH above 11.5.

CMC (Carboxy methyl Cellulose) is another polymer type material used for fluid-loss control. It comes in three grades: technical, regular and high viscosity. The technical grade is a low viscosity material that contains salt and is about 75 percent pure. The regular and high-viscosity materials are nearly 100 percent pure and differ only in the chain length of the polymer. High viscosity CMC has a longer chain length than regular CMC and will build about twice as much viscosity. The ability of the three grades to control fluid loss is essentially proportional to their ability to build viscosity.

At low concentrations, CMC deflocculates clay suspensions in much the same manner as Lignosulfonate. This results in a reduction in fluid loss as well as yield point. As the CMC concentration is increased, the viscosity of the water is increased and further reduction in fluid loss is achieved. If a high yield point and lower plastic viscosity are desired, addition of CMC is detrimental, since it has the tendency to deflocculate and to build plastic viscosity.

As water salinity is increased it becomes more difficult to hydrate and expand the CMC molecule. Extremely high shear mixing is required. At salinities above 50,000 mg/l chlorides, CMC rapidly becomes less effective as a fluid loss control agent. It also becomes ineffective at Calcium concentrations above 100 mg/l.

CMC, like other organic additives, has a time rate of thermal degradation which makes it difficult to identify a breakdown temperature. As temperature is increased above 250°F, the rate of degradation accelerates very rapidly until almost complete degradation is achieved in 16 hours at 300°F. Fluid-loss control at temperatures above 250°F should be considered only as short-term control.

When CMC is thermally degraded, CO2 is formed. In time, a Carbonate and bicarbonate concentration can build in the mud and lead to a high yield point and gel strength. Small additions of lime may be necessary to control this problem.

In summary, the various grades of CMC share a common shortcoming with starch. They build viscosity in order to control fluid loss. This factor severely limits their use in high weight muds. The thermal instability and sensitivity to salinity and water hardness also limits the use of this material.

Another type of water-soluble polymer used for fluid loss control is polyacrylate. It is one of many different polymers that can be grouped under the basic classification of vinyl polymer. These materials are somewhat more thermally more stable than CMC or starch, but they are extremely sensitive to Calcium. Calcium concentrations above 200 mg/l will cause excessively high gel strengths that cannot be controlled. These polymers are similar to other vinyl polymers used as clay extenders. At low concentrations (up to about 0.1 lb./bbl) the, are flocculants. At higher concentrations they deflocculate and increase the viscosity of the water, thus reducing the fluid loss. To avoid flocculation, they are added at higher concentrations (0.5 lb./bbl) on initial treatment. When used in this application, they actually break the mud over to a deflocculated system. Both clay solids and Calcium must be kept to a minimum when using these polymers.

Poly-anionic Cellulose is another very popular class of fluid loss control agent which finds wide application in several conditions of moderate salinity and temperature. Although HTHP compliant PAC polymers do find a place in the form of effective F/L control agents, but they often need a temperature stabilizing agent like Mono-ethanol amine to perform beyond 160 deg C for longer duration.

Viscosifiers

Viscosifiers are necessary in the muds for two primary reasons: to suspend Barite and to increase cuttings carrying capacity. These two functions are carried out by increasing the yield point and providing gel strengths of about 2 to 4 lbs. /100 sq. ft. Normally, Bentonite performs these functions without the aid of supplemental additives.

Special-purpose viscosifiers are usually limited to situations where Bentonite is not well suited (due to high electrolyte content) of where a low-solids mud is desired. Some of these materials such as starch, CMC, and Xanthan polymer build viscosity directly. Others, such as the vinyl polymers, act indirectly by increasing the yield of Bentonite.

The use of starch and CMC as fluid-loss control agents was discussed in the previous section. Their limitations as viscosifiers are essentially the same as for fluid-loss control. Temperature appears to have an even greater adverse effect on their viscosifying ability. A very rapid loss of viscosity occurs at temperatures above 200 °F.

Xanthan polymer is another material that increases viscosity when added to water. It is a long-chain polymer that increases the yield point much more than the plastic viscosity. This is a very desirable characteristic for a viscosifier. Good hole-cleaning can be attained without high viscosities at the bit and reduced penetration rates.

Xanthan polymer is not greatly affected by electrolyte content. It will build viscosity nearly as well in salt water as in fresh water. One difference is that it cannot be effectively cross-linked in salt water. It thermally degrades quite rapidly at temperatures above 250°F. Like other polymer solutions, it decreases in viscosity with increased temperature. The viscosity at 250°F is only one-fourth of the viscosity at 75°F. This can account for greater loss in hole-cleaning ability even before degradation occurs. It is sometimes used with small amounts of Bentonite (up to 10 lb./bbl) to help offset this temperature-thinning effect. Bentonite will flocculate with increased temperature, making up for the loss in viscosity of the polymer.

Unlike starch and CMC, Xanthan polymer has poor fluid-loss control ability. If an API fluid loss below 30 cc is desired, a fluid-loss control additive is needed. Small amounts of Pre-hydrated Bentonite help in this respect as well as in improving the temperature stability of the mud.

The vinyl polymers derive their viscosifying ability by cross-linking clay particles, causing an increase in yield point and gel strength. If one molecule is attached to two or more particles, a state of artificial flocculation exists and increased yield point and gel strength will occur.

The degree of flocculation is quite sensitive to polymer concentration and to clay content. As the concentration of polymer is increased in a dispersed Bentonite suspension, the yield point and gel strength will increase until maximum cross-linking occurs.

In a Bentonite suspension of 10 to 15 lbs. /bbl, approximately 0.05 lb. /bbl of Maleic Anhydride-vinyl acetate copolymer is required to give maximum cross-linking. Approximately 0.1 lb./bbl will deflocculate the system and give minimum yield point and gel strength. As formation clay solids or Barite are added, the maximum and minimum cross-linking concentrations of polymer are increased. At the maximum cross-linking point, the yield point and gel strength will be greatly increased as the solids content is increased. This necessitates careful control of solid content, either by mechanical removal or dilution.

When high molecular weight Sodium polyacrylate is added to a system for fluid loss control, its effect is additive to that of the Maleic Anhydride-vinyl acetate copolymer. If the system is at the point of maximum cross-linking, the high molecular weight Sodium polyacrylate will cause deflocculation as though more Maleic Anhydride-vinyl acetate copolymer had been added. As a result, relatively low yield points and gel strengths will result. This is not totally the way a weighted polymer mud is run. It is an extremely sensitive system and consequently very difficult to run with stable mud properties. It is quite sensitive to temperatures above 225°F. At elevated temperatures, severe flocculation may occur down-hole without significant differences in properties at the flow line temperature being observed.

The use of these polymers to extend the yield of Bentonite is limited to fresh water. They will not work in salt water or in the presence of Calcium at concentrations above 200 mg/l. If Calcium is introduced into the system, extreme gelation will occur.

Shale inhibtors:

No book on mud engineering can be complete without mentioning about shale inhiition control. While we talk about shale inhibiting chemicals it makes sense first to understand what shale inhitbiion is and why do we need to inhibition in shale? Shale, or broadly speaking, any charged molecule or a group of such charged molecules with edges of surface carrying charges on them, has a tendency to react with water for two main reasons

1 Shals are devoid of water because their original water molecules got expelled from them at the time they were getting subjected to pressure upon them by overlying rock.
2 Being charged in nature they behave like open wires of electricity, ready to conduct electricity in presence of a conducting medium available to them for doing so.

Not all the times and all the shale contains broken edges of charge. Therefore despite the popular fear of shale being chemically reactive,instances are common when shale does not swell in water at all!

Thus, there is only a certain degree of the entire content of the rock, which is chemically active.

Without drifting from the main point of shale inhibitors, if we now talk about shale inhibition, we can say that any additive which disallows / inhibits chemical interaction of shale and water can in principle serve as a shale inhibitor. Another equally important aspect of shale composition aforementioned is the fact that they are devoid of water which they contained once, a apart of their native composition. So, when a shale, which is thirsty for few million years, sees / comes in contact with water contained in WBM as its continuous liquid phase, just tries to usurp as much water as it could. After few initial gulps, it slows down because its thirst has been quenched. Depending upon how much water did it drink in those gulps, it will expand itself to create room for the additional volume. The packed layers of buried shale then start seperating and distancing from each other. This precisely is shale swelling in its simplest terms. In reality the mechanism is more complex as physico-chemical forces of osmosis also come into play, in addition to re-aligning of open chargeed sites of shale. This realignment can bring similarly charged molecules closer initially, only to further facilitate the swelling of shale because of electrostatic repulsion between charged sites. As nature always does, after an initial havoc, it settles. Same is the case of sweeling also. Once the clay's thirst of water is quenched, it will not further react with the rock as fast as it had in first place. However, subsequent interactions do not stop completely. There are transfer of Ions from and towards the shale and mud, depending upon the chemical environment. This gives rise to delayed bore-hole destabilization due to chemical interactions. Having got an overview of the chemical / physico-chemical changes, let us now touch upon forces purely of mechanical nature, which also destabilize shale and water cannot be held responsible for this destabilization. People travelling by local and domestic trains during peak hours get through this ordeal daily! Yes, whenever a train stops at any platform the stop is always very short and dozens of people have to get out of board while equal number of people trying to get inside the train. Both rush to their capacity, but the fight lasts for only as long as sthe train is stationary on the platform. Now draw an analogy with drilling – as soon as the bits takes away a big chunk of rock from otherwise stable rock, there is a void created, rsulting into realignment of stresses around well-bore. This release of stress by rock in the immeddiate vicinity of drilled piece of rock can trigger falling of rock, which was not cut by the bit, till the stresses ralign with-in themselves and buy peace. This relignment is synonimous to a state of peace inside the locaal train and at platform, as soon as the train leaves the platform

We are now better prepared to understand what a shale inhibitor should be to stop the destabilization of well-bore, number-one NPT causing factor in oil-well drilling.

Any additive which stops the gushing of water towards rock, no sooner than it is drilled can act as a shale inhibitor. We can either increase the viscosity of water in mud to slow down its mobility.

Any polymer, which has a higher degree of affinity to absorb water can compete with the shale to keep the shale intaact. In addition to it, if the ionic environment of the drilling mud and rock (if known previously) can be made similar to each other then despite water coming in contact with shale- shall not be gushing due to ionic balance. Sodium chloride,potassium chloride and calcium chloride are the commn examples of ionic compounds shwing shale inhibition effeciency. It would not be improper or out of place to mention here that Potassium chloride still enjoys the status of number one shale inhibitor. In first case we introduced a competetor to shale for its water affinity and in second case, we changed the chemical environent in which water exists, making it useless for the shale to absorb. In yet another game, probably the biggest (if chemical swelling inhibition is the only criteria to assess the suitability) innovation emerged from a very radical step. Shales swell in water. Let us not use water as continuous phase for mud preparation! Oil base mud is born!

When we stick to basics and understand the problem completely before rushing to act and are prepared to negotiate a win-win situation, we have already solved half of the problem. Rest is only a matter of execution. If chemical instability of the rock was ONLY single problem in oil industry, then I can say, we have conquered nature,but that is far from true! For more clarity on advantages and disadvantages of oil base mud, refer to the chater of oil base mud. Coming back to shale inhibitors, we can conclude that inorganic halides, organic polymers are capable of doing the job to the extent that most of the wells can be drilled with acceptable loss of time,if there are other operational problems,mechanical break-downs etc. With the advancement in understanding of Chemistry, over a period of time many new class of additives have joined the wagon. They are

Glycols

Poly-amines

Asphalt and so on and so forth....

Briefly touching upon HTHP shale inhibitors, inorganic halide and oil are unaffected by temperature. One of the tough technical challenges facing the industry today is to develop HTHP shale inhibitors which can be used in regular dosages like other lab additives and yet, be as effective as at low temperature to disallow any continuous maintenance dose to counter the thermal degradation. Breakthroughs in the form of temperature limit getting pushed up beyond 200 deg C keep hitting the market to the whole scenario of shale management, rather than only focussing upon shale inhibition.. There is a simultaneous need which keeps arising to further push the limit because mankind keeps venturing into those unknown territories which were once considerd technically unfit to drill. The race is ON and for sure, someday sooner than expected – there will be a chemical which will surpass the requirement. We just need to think a bit differently

Fndamental point in all shale inhibitors is – delay the process of shale and water coming in contact, by whatsoever way one can. High performance water base mud part of the book deals with this issue in detail.

Emulsifiers, Surfactants, and Lubricants

A number of special-purpose additives are used in mud to emulsify oil, reduce torque and drag, and improve the lubricating qualities of the mud. In general, these additives have very limited utility.

Chemical emulsifying agents are normally not necessary in a water base mud because of the ability of colloidal solids, lignite, and

Lignosulfonate to stabilize an oil-in-water emulsion. Only when these materials are not present in a mud, the addition of an emulsifier should be considered. Due to the difficulty of emulsifying oil in a brine, an emulsifier may be necessary in some cases.

Detergents have been extensively used in drilling muds for a variety of reasons ranging from general mud conditioning to hole stabilization. They supposedly reduce torque and drag, eliminate balling of bits and stabilizers and a number of other things. Detergent is such a weak surface-active agent that it probably cannot perform any of these functions at the recommended concentrations. At quite high concentrations (3 to 5 lbs. /bbl), there is some evidence that it helps prevent shale cuttings from sticking to each other and to bits, stabilizers, and the drill string. Lower concentrations appear to be a complete waste of money. Even at higher concentrations, the cost may outweigh the benefits.

A 30-Mol ethylene oxide adduct of phenol, a more powerful surface-active agent, has also been used extensively in mud. It is a nonionic surfactant that is adsorbed on the surface of clay solids. In this condition, the clay solids are less active and appear to disperse less easily. This helps to prevent gelation and may also help in preventing balling. At temperatures above about 200°F it is insoluble. As it comes out of solution in a high solids mud, high viscosities may develop which cannot be controlled. In general, surfactant muds are quite sensitive to temperature and are somewhat unpredictable.

A number of special lubricants on the market are supposed to reduce torque, drag, and pipe sticking. Claims for these products usually are far greater than their performance. Only after careful field testing and analysis of performance, they should be placed in general use. Even then, the product should be checked periodically to see if it has undergone any structural change during storage, rendering it less effective. Simple physical changes in material are often an indication of lost utility. Uneven consistency, presence of thick lumps, change in color are indicators of the probable loss of effectiveness. Most torque and drag problems are mechanical in nature and are unlikely to disappear upon addition of any mud additive.

In summary, all these products are special-purpose additives. They should be used and evaluated only for a specific reason. They are high-cost materials and should never be used indiscriminately. They do not take the place of good mud and drilling practices.

CHAPTER 5

DRILLING CHALLENGES

A number of major drilling challenges have a solution coming from proper management of drilling fluids, yet it should not be considered the exclusive controlling factor. The solution to these challenges often requires adjustments of mud properties, and operational procedures such that all facets of the problem are relieved.

LOST CIRCULATION

Lost circulation is defined as the loss of whole mud to a formation. It results from two basic causes:

1. Seepage loss to vugular zones, coarse gravel beds, and shell deposits.
2. Hydraulic fracturing.

It is necessary to identify which of these two causes of mud loss we are trying to combat, because the means of their prevention and cure are quite different.

Loss of mud to vugular or cavernous zones and to coarse beds of gravel or shell is a natural consequence of the solids in a mud being unable to bridge the openings in the rock. Consequently, the mud is able to flow through networks of the voids and get lost from the circulating system. In shallow, unconsolidated sands or gravel beds, the addition of Bentonite to thicken the mud is often sufficient to stop or restrict the losses to acceptable levels. If this does not succeed, coarse, fibrous material is added. Cottonseed hulls, the least expensive coarse material available, is adequate for this application. When coarse material is required, it is best to mix batches of larger volume as a pill rather than to add the coarse material to the entire mud system. This reduces waste of the material and allows higher concentrations to be placed opposite the thief zone. High concentrations of lost circulation material (LCM) in the entire mud system may thicken the mud to the point of causing pressure-induced fracturing and thus compounding the problem. When inexpensive muds are in use, it is often cheaper to drill with partial returns or seepage losses than to try to stop the loss with LCM, if formation being drilled through is not the prospective candidate for production. When adequate water supply is available, it is sometimes possible to drill long intervals without returns. The voids in cavernous formations are often large enough to accept cuttings and thus allow the hole to be cleaned without returns to the surface. In order to drill without returns, the open-hole section must be competent enough to "stand up" with a reduced mud column. In this situation control of formation pressure is not possible and must be taken into account.

When other means of lost returns control have not worked, cement is sometimes used. It should not be used unless the depth of the thief zone is known. The bottom of the drill string should be 30 to 50 feet above the thief zone, so that the cement will move down the hole instead of up around the pipe. If partial returns are being obtained, the rams should be closed just prior to displacing cement from the pipe, so the cement will be forced down the hole and not allowed to "flash set" around the pipe.

It should be noted that some lost circulation zones cannot be cured within economic reason. In these cases, economics dictate that the well be plugged and abandoned. In many cases, the rig has been skidded and a second hole drilled without difficulty.

The most prevalent type of lost circulation is that caused by pressure-induced fracturing. When the combination of mud weight and pressure due to circulating the mud, or pressure surge from pipe movement, causes a total pressure opposite a formation which is greater than the strength of the rock, a fracture will develop in the rock. As the mud flows into the fracture, it propagates the fracture and allows a loss of mud.

A great deal of progress has been made in preventing hydraulic fracturing by increased knowledge of formation pore pressure and fracture gradient. Advanced information on these two parameters allows the use of minimum mud weights and better casing programs. This is the first and most important method for preventing fractures.

Lost circulation often occurs while circulating, while attempting to break circulation, or while running pipe in the hole. This is a result of the increase in total pressure exerted on the formation, due to the pressure required to move the mud up the annulus. The pressure due to the frictional drag of the moving mud stream is increased by the increased flow rate or increased viscosity of the mud as it exists in the annulus.

The highest mud viscosities in the annulus occur after a trip, due to gelation of mud under static conditions at elevated temperature. Running the pipe into the hole under these conditions may cause excessively high pressure surges unless the peak pipe running speeds are carefully controlled. The running speeds should decrease with depth in the hole. In extreme circumstances, it may be necessary to circulate out the gelled mud at one or more points in the hole before getting back to the bottom. The mud should be treated to minimize down-hole gelation before making a trip.

Returns are often lost while attempting to establish circulation after trips and on connections. If the hydrostatic pressure plus the pressure required to break the gel of the mud and initiate flow is greater than the fracture pressure, fracturing will occur. When the fracture pressure is known, the maximum allowable pressure to initiate circulation can be determined. With the use of a standpipe choke and pressure gauge, the pressure required to break circulation can be monitored and controlled. The pipe should be rotated a few minutes to break the gel structure of the mud as much as possible. Then, with the choke open, start the pumps at idle speed while continuing to rotate the pipe. Slowly close the choke until the predetermined pressure is reached. If circulation is established, continue to close the choke in steps until fully closed. The standpipe pressure should never exceed a predetermined value at each circulation rate. The standpipe pressure should be monitored until after "bottoms up". The pressure imposed on the hole while circulating can also be monitored and controlled by this method.

The usual remedy for an induced fracture is to wait 2 to 4 hours for the formation to heal. However, this age old concept has now almost disappeared, justifiably so – considering the high operating cost of latest generation rigs. Reduce the mud weight if possible. If the formation does not heal, the next step is to spot a LCM pill (usually 50 barrels or more, depending on hole size) containing 10 to 50 lbs./bbl of various sizes and types of lost circulation material. Bit nozzle sizes should be taken into consideration when selecting the concentration and type of lost circulation material to be used. Relatively coarse material should be used if possible, since its purpose is to bridge the fracture.

If the LCM pill is not successful, some type of special high fluid-loss slurry with 30 to 50 lbs. /bbl of LCM is used to try to prop the fracture open and form a bridge to seal the fracture. A squeeze pressure equal to the final mud column weight is usually used. These slurries have met with varying degrees of success, but they usually do not have a lasting effect. These types of slurries will often seal the hole, for long enough to record log/s, or run and cement a string of casing.

The types of lost circulation discussed thus far are those that occur in the open hole. Losses can also occur at the casing seat or through holes in the casing. If mud is able to channel up behind the casing to a weaker zone, fracturing of this zone and loss of mud can occur. The usual procedure is to drill out 10 to 15 feet below surface or protective casing and test to a pressure approximately equal to the anticipated final equivalent circulating density. If the casing seal fails to hold, squeeze cementing is usually successful.

Subsequent testing in the open hole is also useful in determining the integrity of the hole. A pressure integrity test after opening the first sand below casing will indicate the highest mud weight that can be used without fracturing. When subsequent pressure integrity tests are run, it is not unusual for the "leak-off" pressures to increase. A good filter cake will strengthen the borehole. This may account for some lost circulation problems healing with time.

If a hole develops in the casing, mud can be lost to a lower pressure zone behind the casing. This type of lost circulation must be identified before taking remedial action. If the loss is assumed to be at the casing seat or in the open hole, and if a high water loss slurry or cement is spotted, it will move up the hole and stick the pipe. This possibility must be eliminated before attempting a cure for any lost returns problem.

Lost circulation problems vary from area to area, and no universal solution has been found. The following general rules should be kept in mind:

1. Drill with lightest weight mud practical.
2. Control mud properties to achieve lowest practical plastic viscosity, yield point, and gel strengths. Special emphasis should be placed on minimizing flocculation at down-hole temperatures.
3. Control pipe running speeds to minimize surge pressures.
4. Use standpipe choke method to initiate circulation. Reduce circulating rates when necessary.
5. Avoid use of precautionary lost circulation material.
6. Do not use fine lost circulation materials. They do no more good than drilled-solidsand simply add to the viscosity of the mud.
7. Screen out coarse material as soon as possible after a batch treatment.
8. Set the protective pipe as deep as necessary to minimize lost circulation problems.
9. Test casing seats to maximum mud weight anticipated.
10. If returns are lost with mud weights 4 or 5 lb. /Gal higher than the formation pore pressure in which the last casing was set, a liner or casing string will generally be required.
11. Locate the point of loss before spotting any type of pill that could set up around the pipe.

For more detailed description Pl. Refer to the exclusive coverage in a separate chapter.

PIPE STICKING

Pipe sticking is one of the more costly drilling problems that we encounter. It occurs from any of these several different causes:

1. Differential pressure sticking.
2. Accumulation of cutting or cavings in the hole, due to inadequate hole-cleaning.

3. Plastic flow into the hole, such as salt.
4. Key seats in crooked holes.
5. Mechanical obstructions.
6. Sudden settling of Barite, due to such things as Barite stripping from gas flow or inversion of oil-mud emulsion.
7. Collapse of weak or unconsolidated formation, due to sudden reduction in hydrostatic head from lost circulation.

Obviously, the prevention or solutions of these various causes of stuck pipe are different and require correct identification before taking remedial action. Unfortunately, it is not always easy to identify the specific cause of pipe sticking and an incorrect assumption may lead to treatment which is detrimental. For instance, differential pressure sticking may be assumed when inadequate hole-cleaning is the actual problem. If lignite and Lignosulfonate are added to decrease the fluid loss and relieve the differential sticking problem, they will thin down the mud and further worsen the hole-cleaning. Consequently, we must be careful not to jump to an incorrect conclusion and take action that will magnify the problem. Probably the best method of identification is by a process of elimination of all other possible causes.

Differential Pressure Sticking

The most prevalent form of pipe sticking is a differential pressure sticking. It is the most significant problem related to the mud filter cake.

When the pipe becomes partially embedded in the filter cake, it becomes a part of the cake as though, it is one of the solid ingredients of mud! It is held to the cake by the pressure difference between the wellbore pressure and the pore pressure in the cake under the pipe. The force required to pull the pipe free is a function of the area of pipe in contact with the cake, the pressure holding it to the cake, and the coefficient of friction between the pipe and cake.

When the pipe first contacts the cake, the pore pressure at the surface of the cake is the same as the wellbore pressure and the pipe is free. As filtrate is squeezed from the cake, the pore pressure in the cake, which is under the pipe, decreases until it finally reaches formation pressure. This causes the differential pressure holding the pipe on the cake to increase to such a point that the resulting frictional drag exceeds the available force to pull the pipe free. Consequently, the time that the pipe is motionless is a very important controlling factor as to whether the pipe sticks to such a point that it cannot be pulled free.

The total differential pressure is usually imposed on the pipe in less than two hours. Often, only a few minutes are required for the pipe to stick to the point it cannot be pulled free. This makes it imperative that the pipe is kept in motion. Rotation of the pipe is best in that it continually pulls new mud between the cake and the pipe and prevents a buildup of the differential force.

As a cake is being deposited, the filtration rate decreases as the cake grows thicker. This means that the rate of sticking is greatest during the early growth period of a cake. During a short trip, the cake may be scraped off and a new cake begins to deposit. The probability of sticking on this newly developing cake is greater than before. Extra care should be taken to keep the pipe moving.

The area of contact is a function of depth of burial of the pipe in the cake and, consequently, is a function of the cake thickness. For this reason, we want the cake to be as thin as possible. This means the cake should be of low permeability and deposited from a minimum-solids mud. For this purpose, drilled-solidsshould be maintained at a minimum. Lignite, Lignosulfonate, and other fluid-loss control agents reduce the cake permeability and the resulting cake thickness.

As the mud weight has increased the amount of Barite in the mud is naturally increased, resulting in filter cakes with increased coefficient of friction. The cake thickness also tends to be greater due to the higher solids content. These two factors make it imperative that high-density muds carry a minimum amount of drilled-solidsand as much Bentonite as possible, consistent with good flow properties. Also, the mud should be maintained in a deflocculated condition at bottom-hole temperature in order to minimize the high-temperature fluid loss.

Oil is often added to mud to prevent sticking. It not only reduces the fluid loss, but also decreases the coefficient of friction of the cake. Emulsified oil droplets are deformable under pressure and apparently spread, under the force of the pipe, against the cake and lubricate the interface. Diesel oil, although a poor lubricating oil, works quite well. Special additives can be added to diesel oil to improve its performance. Oil must be well emulsified in order to function properly. This is usually not a great problem. Lignite added at the same time will aid in emulsification. For best results, the oil should be added slowly to the pump suction. 3 to 5 percent oil is sufficient.

For environmental protection, several nontoxic substitutes for diesel oil have been placed on the market. These materials consist of alcohols, vegetable oils, specially refined mineral oil with low toxicity, surfactants, and other water-soluble lubricants. In general, they are very expensive and must be used in quite low concentrations. As a result, their performance does not compare to that of diesel.

The identification of differential pressure sticking is somewhat difficult. It occurs with little or no warning. The pipe is usually free of any undue torque or drag until it has been left motionless for a short period of time. If it is possible to pull the pipe free, it should be rotated as soon as possible in order to break the pressure seal with new mud being pulled across the face of the cake. It should be possible to circulate freely when the pipe is differentially pressure stuck. This is in contrast to some of the other forms of sticking.

The existence of sands in the hole at pore pressures less than the hydrostatic pressure always makes differential pressure sticking a possibility. As the differential pressure overbalance on a sand is increased, extra care should be taken to avoid sticking the pipe. Figure below shows a suggested range of HTHP fluid-loss values at various differential pressures. The location of these sands should be kept in mind. To the extent possible, the drill collars should not be allowed to stay idle for an appreciable period of time opposite such sand, since the majority of pipe sticking occurs at the drill collars.

Many cases of differential pressure sticking occur while taking a kick.

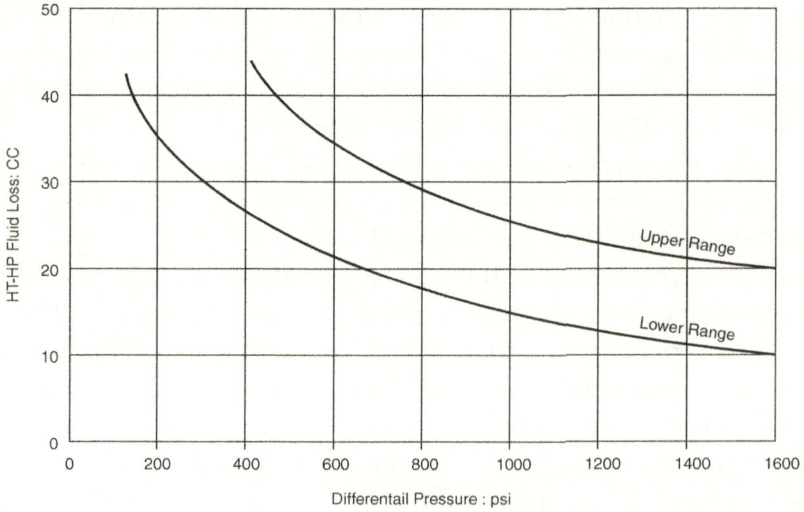

Obviously, this is a very critical time for sticking the pipe since increases in surface pressure on the annulus are directly translated to increased differential pressure. Often the pipe sticks due to being motionless after shutting in the well and before increasing the mud weight to kill the well. Moving the pipe as much as possible after shutting in the well would prevent some cases of sticking. The pipe should periodically be rotated slowly and reciprocated. Care should be taken to avoid excessive wear at one point on the pipe.

When the pipe has become stuck, one of the first steps to free the pipe is to spot oil or oil mud over the stuck interval. The oil apparently compresses the cake away from the pipe and penetrates between the pipe and cake. In this manner, the stuck area is gradually decreased until the pipe can be pulled free. The time required for freeing the pipe may range from 1 to 24 hours. There have been cases where the pipe came free after 72 hours. However, in majority cases the probability of getting loose after 36 hours is very small.

One of the most prevalent causes of failure to free the pipe is because the oil was not spotted over the entire stuck interval. Hole enlargement can cause considerably larger volumes of spotting fluid to be required than the calculated hole-volume. This should be considered when sizing the volume of spot to be used. If possible, the entire open hole should be filled with the spotting fluid. Sufficient spotting fluid should be left in the pipe to allow pumping 1 to 3 barrels every hour while the soaking procedure is in progress.

Wetting agents added to the oil sometimes appear to speed the soaking process. If the differential pressure can safely be reduced, this will also aid in freeing the pipe. When the spotting procedure fails to free the pipe, the next step is usually a complete displacement of the system with an oil mud. This accomplishes two purposes. First, a more complete removal of the water base mud from around the pipe may be accomplished by circulating an oil mud. If this occurs, the pipe may be freed. Secondly, an oil mud reduces the possibility of sticking the wash pipe on a wash over operation if this course of action is taken.

Pipe Sticking Due to Poor hole-cleaning

When cuttings and cavings accumulate in the hole, they cause excessive torque and drag and possibly eventual sticking of the drill string. This is essentially a result of inadequate hole-cleaning, and is usually associated with borehole instability.

The condition of cuttings and sloughing coming across the shaker should be checked. If the edges of the larger pieces are rounded, it indicates they have been tumbled into the hole and are not being lifted efficiently. The amount of shale coming across the shaker may tell how badly the hole is sloughing, but it is not a good indicator of how much is being left in the hole. Severe problems can occur even though there are no clear-cut indications from the amount or appearance of the shale being carried to the surface.

The occurrences of increased torque or drag are normally the first indications of hole-problems due to inadequate cleaning. This condition will normally persist, appearing to move down the hole as drilling progresses, and it will grow in severity. Sometimes, sudden increases in pump pressure will be observed as bridges form and obstruct the flow of mud up the annulus.

Sometimes a trip can be made without difficulty, but bridges and fill will be encountered when going back into the hole. This is a positive indication of inadequate hole-cleaning that may lead to stuck pipe on the next trip. The absence of fill does not eliminate the possibility of trouble, because the accumulation may be up the hole in a washed-out section.

When it is apparent that the hole is not being cleaned, the yield point of the mud should be increased. The amount of increase required is unknown because the sizes of the pieces of rock to be lifted are unknown. Some threshold level of lifting capability is required for each size, which increases as the size of the rock increases. The hole will not clean up until this threshold level is reached. In general, the yield point may need to be increased from 2 to 4 times the normal value.

The lifting ability of a mud increases as mud weight increases, since the buoyed weight of the particle decreases. The weight of a piece of rock in a 15 lb. /Gal mud is only about half of its weight in a 10 lb. / Gal mud. Consequently, the yield point required in a 15 lb. /Gal mud is only about half that of a 10 lb. /Gal mud.

An alternate method for increasing the lifting capacity is to increase the pump rate. This should be done with caution since it may lead to increased sloughing or lost returns. It probably has application only when the present pump rate is abnormally low.

There are many methods for increasing the yield point of a mud. It should be done in such a way that the plastic viscosity is not appreciably increased. Increasing the plastic viscosity is a very inefficient way to increase the lifting capacity and causes decreased penetration rates. We are interested in increasing the yield point at the temperatures that exist at down-hole. The yield point of polymer muds decreases with increased temperature. Polymers are also susceptible to thermal degradation at temperatures above 225°F. For these reasons, it is desirable to use HTHP compliant polymers. Hole-cleaning problems often occur when using polymer muds even when the yield point at the surface appears to be adequate, if the polymer in use is not stable at BHCT.

The addition of Bentonite is a quite efficient method for increasing the yield point. This is especially true when it flocculates. Thinners should not be used when attempting to increase the yield point. In light-weight muds, flocculation can be promoted by the addition of lime or soda ash. This causes high yield points to be developed with low plastic viscosity and allows for both good hole-cleaning and fast penetration rates.

Pipe sticking in SOBM

Although not very common, but the string does get stuck in SOBM also. Instantaneous instability of the rock leading to its collapse around BHA or/and excessive thick filter cake, are the principle reasons. Since lubrication is not an issue in SOBM, it is very tricky to decide the composition of the pill as well as its volume. Decreasing the mud weight is not a very good option whenever rock has revealed mechanical instability. Not many known cases are available in the literature, but on a specific occasion, the string was successfully freed by changing the wettability of the oil wet particles of filter cake, thereby dislodging them from the wellbore to create additional annular space for the fluid to flow through. This almost invariably will spoil the annulus SOBM as the emulsion of SOBM will also weaken, rendering it of no use before re-conditioning the entire circulatory system to a strongly emulsified SOBM of the desired parameters. Often mud engineers scare to suggest/implement this for the fear of uncertain results and additional hard work of re-conditioning almost > 2000 bbls of SOBM. In addition to it, there is no certainty of string freeing up, which further demotivates the drilling team as well. If the choice is between snapping the string and trying to free the string by surfactant pill, later option must be exercised. A strongly water-wetting surfactant (HLB number > 15) increases the possibility of dislodging all oil wet solids from filter cake. One can always learn this trick from the standard practice of changing the wettability of rock prior to pumping cement and applying to a situation, not heard commonly.

Pipe sticking due to mechanical wellbore-Instability

A rock can become unstable, no sooner than it is drilled, if it has to, due to mechanical reasons. Such instability is instantaneous and often catches us unaware. Misaligned near wellbore stresses results in dislodging a huge chunk of uncut rock which falls in the annulus near the BHA area. Being already constricted in hydraulic diameter, this area is vulnerable to getting stuck. Falling of big pieces of rock is like adding fuel to the fire. To free the such stuck pipe, mud weight needs to be raised to a level enough to overcome the tendency of the rock to fall apart. Ideally the mud weight should have been higher in the first place itself. Such mismatches are due to unpredictability of bottom hole conditions right always and at times, we pay for them in the form of NPT for freeing the pipe.

UNCONVENTIONAL AND INNOVATIVE PILL

It is worth-mentioning here about a real-life scenario where the hole would instantaneiously collapse, no sooner than it is drilled, and pipe getting stuck. The situation would improve apparently by raising the mud weight by as high as 0.5 to 1ppg.. While making a a short-trip to prove the hole, the string pulled free upwards.. However while reaching to previously drilled depth, the tottom portion was always found full of freshly fallen rock, threatening to hold the pipe again if a forced attempt is made to reach bottom without more remedial measures. To further add to the woes of the team, the boulders falling in the hole were too big to be brought back by mud of conventional or predictable properties! The team was left first to innovate ways of bringing the boilders to surface and then think of overcoming the mechanical instability. All known types of viscous, super-viscous pills did not prove useful. They were either too viscous to be pumped or not enough viscous to bring the boulders back to surface. After a lot of brain-storming an unconventional composition of pill was suggested which did not get approval from the team in first place. The concept was to have very high viscosity/Yp but simultaneously improve the lubricity of the mud to a level where it can be pumped down-hole with manageable pressure and avoid triggering down-hole losses even before reaching the troubled spot When situation reached to a point of getting worse as time elapsed and no other feasible solution in sight, eventually rig team agreed to pump that unconventional pill which comprised of **EP Lube and Lime** in active mud to flocculate the same at as high level as would be required and yet, substantially increase the lubricity by incorporating EP Lube in it.

CHEMISTRY BEHIND THE EP-LUBE LIME PILL:

After failing to pump the conventional pill with as much carrrying capacity as required to bring the boulders to surface, the focus wass how to make the pill pumpable as it was evident that the pill would bring boulders to surface if we can send it to the boulders. This has to be achieved and yet, the lifting capacity should not be compromised. Simple addition of EP Lube was imparting lubricity to the pill, but a compromise had to be made on flocculated state of pill. This was then aptly compensated by lime addition to a small amount. The point here to notice is – unconventional things first have to be visualized in mind to make them happen, then in reality.

In yet another instance a rare occurrence of severe mechanical instability was observed at a depth of only 300 m from surface on a land location. After grouting false conductor and completing drilling of top-hole, it was noticed that the entire drilled interval collapsed as evidenced by the fact that the surface casing failed to go below 50 mts. Thrice, the casing had to be pulled after every unsuccessful attempt to clear the falling boulders of rock. With no solution in sight the unorthodox decision of filling the entire hole with cement slurry soon after completing the drilling was taken and executed. The set cement was then drilled and this strengthened hole enabled lowering of casing through it. Thus, cement slurry was again pumped to cement the annulus behind the cemented rock and casing OD!

These examples open our eyes and compel us to re-consider: Is it right to always attribute every well-bore in stability as to caused by mud? Let us think.

Geo-models and various aspects of inter-relationship of diferent pressure gradients co-existing in a rock.

Broadly speaking, unstable bore hole is a result of mechanical, chemical as well as physico-chemical forces altering their magnitude either instantaneously or with the passage of time. The rock is stable with all forces balancing each other, till it is not penetrated. At the first action of a drill bit penetrating the stable rock, it removes a part of the matrix from its original place, where it was contributing to the stability of the entire rock. The void created in the rock provides additional available space for the remaining matrix to release a part of overburden pressure upon it till it regains the static equilibrium. This realignment of the rock releases some of the stress force acting on the entire rock. In the event of uneven realignment of the forces the rock loses a part of its shape and the stabilization of the stress can result into falling of nearby rock immediately, or with time, by invasion of solid and / or liquid (both) and interact with the content of the rock. This interaction varies in magnitude and governed by the composition of mud as well as the rock and its trapped fluid, if any. However, the situation is not so simple as it appears, because with continued drilling more and more forces realign, thereby impacting the final impact. The pressure exerted by the column of mud over the open rock balances the situation in terms of imparting instant mechanical stability. However, the time dependent gradual realignment of forces due to mechanical, chemical and physical-chemical interaction changes the scenario and hole can gradually become more and more unstable. With advancement in Geo-mechanical modeling and evolution of mud system which do not interact with rock and designing of mud to near zero invasion, these forces can be kept under control by selecting the right mud weight and adopting best operational practices to buy extra time till the bore hole reaches the planned depth where the casing is set. The fulcrum of well-bore stability always targets to delay the destabilization process, as long as feasible, or at least to that time before which the casing is set.

The pre-drill pore pressure predicts the curve shown below is a generic display of pore pressure, fracture pressure and overburden pressure changing with depth and helps professionals plan their activities and casing setting depth at various intervals to eventually have a stable bore-hole drilled with minimum or preferably no NPT.

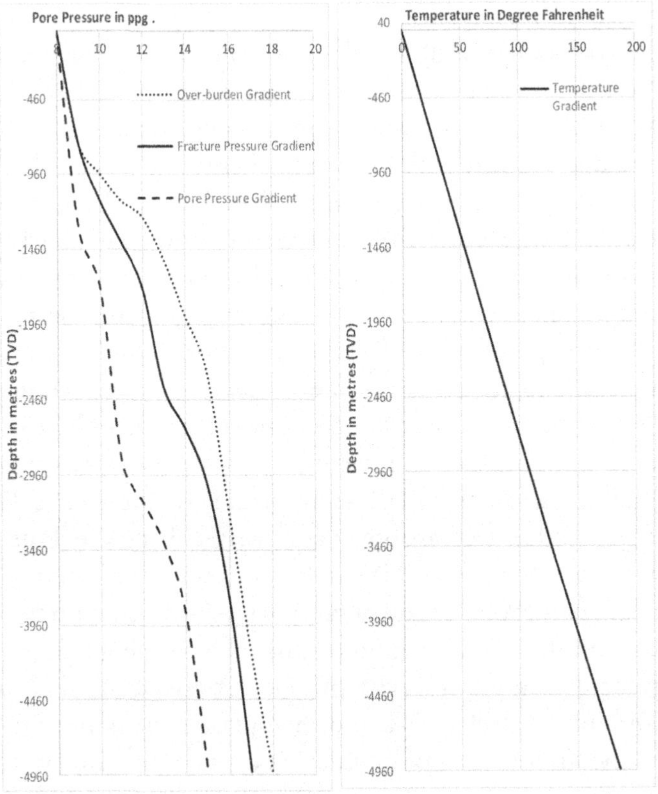

Predrill Geo-pressure & Temperature Plot of Typical Deep-water Well

Brif description of the curve:

Essentially the curve aims at providing advance intimation of possible change in values of pore-pressure, fracture pressure and over-burden pressure in line with increase in depth by geo-mechanical modelling based on the study conducted by companies specialized in such work and adding the value with the help of off-set data, if and where available. Based upon the known, (**predicted with the help of equations which, has many assumptions**) values of various types of forces which compensate each other to stay in harmony wih each other, the drilling engineer prepares the well plan in conjunction with various other known parameters (like kick tolerance, mud properties, rig capacity etc). The guiding force behind this plan is most of the times based on keeping the casing strings as less as realistically possible and yet, do not compromise with the safety standards of overall operations. By and large it is considerd safe to keep drilling without lowering casing till a depth is reached where the ECD is still less than atleast 0.5 ppg value of the LOT / FIT at the shoe depth of the previous shoe. There is no strict rule about the number of 0.5 ppg. Many factors other than drilling also play an important role here. One such example is, the E&P company may like to drill further without lowering casing, but the drilling contractor may just not oblige, stating safety criteria and uncertainty of the numbers, which are anyway predicted and not real! In their own ideologies, both could be stated right as, at one hand, no one would ever favour compromising with safety of the personnel onboard, but simultaneously it is a proven fact that with advancement in drilling the rock keeps attaining competence by solidification of void pores of rock. These are those solids which are a part pf mud as well as some very finely churned rock itself which does not reach to surface and instead choses the eassy route of entering the rock matrix through the open pores while traversing up the annulus. Strictly technically speaking mud enginners have a big role to play here by designing a mud which is comprised of solids of size that can fill-up all the pores of the drilled rock and then plaster the same with the help of an impervious fine layer of liquid, that oozes out of mud due to application of pressure on it during circulation. To further strengthen this fine liquid layer, the particles of mud do deposit over the layer causing a filtercake. Thus, the concept of internal and external filtercake gets recognition. This gives birth to Designer mud system

WHAT ARE DESIGNER MUD SYSTEMS?

These are the specialized designs of the mud, tailore-made for the hole / section under question, by incorporating available reservoir data to help a software model select the concentration and size of designer particles in mud. The most commonly used material as designer particles are $CaC0_3$ as bridging agent and resilient graphite as sealing agent. Comparatively Graphite being lighter, can go into mud in more volume even at a lower dose than Calcium Carbonate. The fluid is normally so designed that the volumetric ratio of CaC03 with resilient graphite is 2:1 and the total combined dose of designer particles is ranging between 32+/-6 ppb, depending upon real need as well as accepting the fat that too much dependence on designer particles can force us to compromise with other equally essential properties of mud. One such limitation can be the mud weight itself! There might not be enough pore pressure in the rock to support the increased mud wt. despite not adding any other weighing material. These are advanced stages of engineering designs where experience of drilling the rock is a key factor, that no book can teach – nor can any engineer mug up.

Modeling borehole stability is a complex problem. The instabilities are subdivided into two main types: chemical and mechanical. In the first case, the instability is caused by formation reaction to the drilling fluids and is controlled with the mud type and its properties. Where in the second case it is a function of in-situ stresses, pore pressure, rock strength and are controlled by the annulus pressure or mud weight. Most of the time the problem is due to a combination of both the types. In many cases, oil based mud handles only the chemical instability of the well.

To minimize the mechanical instability, mud of the right density should be determined in the pre-drill well bore stability analysis. The analysis contains: study of the mechanical and elastic rock properties, rock strength, identifying principal stress magnitudes and directions and building a pore pressure model. Image and caliper logs in conjunction with cavings morphology are used to identify the failure mechanism. Combined with the results of core analysis, along with geological and basin knowledge, a deformation model can be chosen. This model will best represent the rock behavior once the hole is drilled. Using the well inclination and azimuth, the far-field principal stresses are converted into the stresses around the wellbore (Hoop stresses). This is used to gain an understanding of the principal stresses at each point around the circumference of the borehole with depth. A preferred failure criterion is then applied. This calculation gives an indication of the minimum required mud weight to avoid certain modes of failure.

Various remedial/preventive measures can minimize the impact or eliminate the well bore instability when effectively implemented in drilling fluid management.

1 Keep fluid loss ALAP to delay the pore-pressure transmission phenomena to delay the instability of well-bore.
2 Keep the correct mud weight, which is slightly more than predicted break out pressure, which can be obtained from Geomechanical modeling of the area.

3 Do not allow un-flushed cuttings / inert solids to accumulate in the circulatory system to a level where it increases the ECD/ ESD to close to the fracture strength of the rock. The golden rule is to pre-decide an ECD based upon the highest required mud weight and pump discharge for the particular hole size and stringently maintain that number regardless of the fact whether we have any additional margin to let the ECD creep up.

4 Keep filter cake as sleek as possible to not to have a messed-up environment of cuttings instantly falling in the hole due to sudden release of stressed matrix, thereby endangering stuck-string.

5 Keep rheology of mud just enough to disallow any accumulation of cuttings.

6 Drilling at an ROP which does not allow ECD to creep up is the most professional technique to drill through unstable borehole area or bore hole with very narrow margin between fracture strength and break out the strength of the rock warranting minimum mud weight mandatory requirement.

7 As and when possible, measure the activity of drilled-cuttings and adjust the activity of the mud to match with this value and prevent destabilization of rock due to osmosis.

8 With every 1 ppg increase in mud weight, simultaneously also increase the activity of the mud by 0.01 after 11 ppg mud weight. For example, if the Activity of mud of 11 ppg mud is 0.77 and if the weight is raised to 12 ppg, simultaneously also decrease the activity from 0.77 to 0.76 by adding more electrolyte to mud**. This action offsets the impact of increased filtrate invasion due to the increased hydrostatic head of promoting reverse osmosis to some extent.**

9 Pay close attention to size/shape and texture of cuttings coming at shakers to judge whether to increase the mud weight or alter rheology to minimize hole erosion. At times, the near well-bore stresses get released with sudden and immediate falling of well-bore no sooner than it is excavated and the hole itself becomes stable after that, requiring no changes in the mud. Therefore, before raising the mud weight it is important to ensure that the decision is not based upon a small-time observation. Take some time to observe the behavior of the hole and then act as per the requirement of the hole. Often, we are tempted to raise the

mud weight at the first sign of caving, which can sometimes be more harmful.

10 As far as possible, select the trajectory which is likely to cause the least release of stress so that the well-bore instability is prevented at first place.

11 Selecting right BHA and build-up also helps to minimize the tendency of well-bore instability.

INFLUX FROM THE HOLE

An uncontrolled intrusion of hydrocarbon into well-bore for whatsoever reasons, can be classified as an influx / kick in the hole. Depending upon the type of mud and the type of influx coming out (oil/gas/water, all together) the mud engineer plays active and important role in ensuring that the contaminated mud is isolated and re-circulated only after reconditioning the same to the revised parameters as may be necessary to regain the control on well. This brings up another responsibility that an adequate volume of uncontaminated mud is available to continue to keep the hole full of mud all the times with revised mud weight and other properties. Other than this, Industry protocol does not leave room for mud engineers in the process of regaining the control. Therefore, the discussion here shall be focused on the adverse aftereffects. Broadly the discussion will focus on SOBM and WBM as these are the two types which cover most of the commonly used systems.

KICK CONTAMINATION MANAGEMENT IN SOBM

Since this is a two-phase system, the adverse effects are more severe. In extreme cases, the entire mud coming out can disintegrate to the extent that both the phases separate out, leaving behind a huge lumpy mass of mushy solid, creating blocked circulatory path. Under normal circumstances the gas / oil dissolves in the mud and gas then separates out as mud travels up in the annulus. The contamination of oil disrupts the OWR and can thus thin down the mud causing hole-cleaning issues. Water intrusion is the toughest one to handle as excess water breaks the emulsion rendering the mud completely useless, often requiring complete isolation and huge reconditioning. At times, if the water contamination is too high, it is technically as well as economically sensible to simply keep this mud in a quarantine area and add an equal volume of fresh mud of right parameters. Since dumping of SOBM is not possible due to environmental restrictions, the ideal way to consume this mud is to administer smaller volumes in the main stream of larger circulatory volume to bring down the impact to a level, where it can be handled by cautious and continuous treatment of the circulatory system. This is a frustrating process as the contaminated mud will sit in storage tanks, disallowing complete utilization of available total pit space. Simultaneously, an equal volume of fresh mud also needs to be introduced. Those who have handled such situation can gauge how difficult it is! Not to discourage the use of SOBM, but such situations often compel one to think, why did we use this type of mud! The treatment should to be planned in a very systematic way and prioritize that mud property, which needs immediate attention, which here is obtaining a stable emulsion first. Adequate mechanical shearing and inclusion of base oil along with wetting agents regain the skeleton. At this stage, Electrical Stability gives a clear idea for the next course of action, which is often weighing up the mud to desired mud weight for bringing the well under control. If one is lucky, the rheology can provide respite to do its role of keeping the Barite in suspension without any treatment. In case it is not, then use of Barite and low-end rheology modifier / quick rheology booster treatment go together. If the contaminated mud can be kept in continuous circulation mode in two pits equalized with each other, and hopper obtaining suction from one pit while return falling in another pit, then the process can be expedited due to additional impact of extra shearing as well as larger volume of mud being prepared in one batch. The larger volume requirement on Deep-water locations further adds to the woes. By and large a 500 bbl of contaminated SOBM, when blended with equal volume of fresh mud and treated, it can take 6 to 8 hours minimum for preparing 1000 bbls mud to qualify for re-administration in active stream. In order of priority following steps need to be taken to ensure that every minute spent on re-conditioning is well accounted

1 Isolate contaminated mud in a dedicated pit and as soon as the pit has minimum volume to provide suction to hopper, turn ON the hopper while continuing to receive additional mud. If excess fresh volume is already available after replenishing the hole requirement (Always keep at least 500 bbls batch ready

for instant transfer in active) then start receiving contaminated mud in the same pit and transfer base oil. These concurrent activities not only save, but the mechanical shearing and the additional base oil will soon strengthen the ES to a level, where one can treat the mud with Barite and wetting agent.

2 By the time the pit is full to its capacity (fill only 80 % of its capacity) stack F/L control additive and organo-phillic clay sacks near the hopper for a second course of treatment. Silos of Barite needs to be topped up to save time.

3 Once this treatment is done, a sample of mud needs to be checked for ES, HTHP F/L and rheology immediately. The time taken for this check is effectively utilized in shearing the mud with gun lines and high pressure shearing line along with a hopper. One can safely take back the mud into main stream if it shows the ES > 300 to start with, if operational constrain demand. Invariably, there is a pressure to feed fresh mud in the hole because the influx will contaminate every returning mud at the shaker.

4 The single most important property is the correct mud weight. Everything we do with mud is to facilitate the quickest possible mode of weighing up the mud and keeping the Barite nicely suspended in the mud.

5 Having achieved this stage, the focus can now be on F/L and desired YP as well as 6 RPM readings.

6 We cannot forget that we need to add bridging and sealing agents also and now is the time to do so.

This activity should be on concurrent mode because keeping the active mud in good shape to continue to take the kick out is the primary responsibility of mud engineer

KICK CONTAMINATION MANAGEMENT IN WBM:

Water base mud does not suffer as much adversity as SOBM due to the simple reason that the gas gets diffused out, as contaminated mud travels up in the annulus and gets subsequently eliminated from the system with the help of a D-gasser. Crude oil, although a kick contaminant, becomes a useful ingredient in WBM by its contribution in bringing down the fluid loss. Drop in weight also requires the simple addition of Barite. Having said that, weight management in both the cases, is equally crucial and important. The only relief to the mud engineer in this case is, the entire focus remains on weight management and there is no need to worry about any wetting characteristic change, emulsion breakage.

SIGNS OF KICK /INFLUX ENTERING THE WELLBORE

Kick detection at the earliest stage makes kick management very simple in comparison to delayed detection. Following indications are pre-cursor to kick enters the well-bore.

Sudden increase in ROP, technically called as "drilling break" is a very strong indicator of the possible entry of hydrocarbons/ well-bore fluids entering the open hole. The simple fact that rock matrix which holds hydrocarbon fluids become lighter due to decreased mass because of lighter fluid (gas/oil) occupying the pore space. There may be other factors also contributing to it, but in exploratory drilling, it is a standard practice to conduct flow-check, every time a drilling break is encountered.

1 There will be an increase in volume of active system due to intrusion of extraneous fluid
2 At times, the intensity of kick can be so strong that it can completely lift the entire annular mud like a piston and the resultant mud reaching at the surface can start over-flowing from shakers itself.
3 Brittle and needle shaped cuttings reaching at shakers is also an indication of geo-pressured matrix
4 A decrease in mud weight at shakers is also an indication of influx entering the well-bore.
5 The pump strokes can abruptly increase due to void created in the annulus.

There are several factors other than the quoted which also provide evidence of influx entering the well-bore. With the advancement in technology and availability of mud logging unit, kick detection has become very easy nowadays, resulting into remedial action being taken in time to avoid disaster.

If a Geo-pressured rock is anticipated during drilling, then it is always recommended to drill with a fixed and slow ROP to hit the rock top. This practice helps in detecting the kick at the slightest increase in ROP. Conducting flow check and monitoring Shut-in pressures give a confirmatory evidence about kick entry. Fingerprinting and flow check often requires patience while drilling with SOBM as the fluid takes longer time to complete stop due to compressibility, but the time spent is worth when compared to the repercussions of being negligent in such matters and jeopardizing safety of the crew as well as a rig.

CHAPTER 6

MANAGING SOLIDS IN MUD

Solids management is the most important function in mud treatment. Control of mud properties center around the treatment necessary to counter the adverse effect of the continuous influx of undesirable drilled-solids in the active mud system.

Undesirable solids increase drilling costs in many ways. In addition to increased mud costs, if not properly controlled, they lead directly to problems as lost circulation, differential pressure sticking, and reduced penetration rate.

There are basically three mechanisms for reducing the undesirable solids content of a drilling fluid. They are, dilution, screening, and settling or centrifuging. Dilution is accomplished by adding liquid to reduce the concentration of solids. Solids can also be reduced by passing the mud over a screen and discarding those particles that will not pass through the screen openings. Solids can be separated from a fluid by settling or by applying centrifugal force. This is accomplished in settling pits, hydro-cyclones, and centrifuges.

All these methods are used and have a place in mud treating. The method that will work best and most economically is a function of mud type, density, viscosity, and cost. The degree of solids control necessary, is a function of the severity of the drilling problems and the type of formation being drilled. This will also influence the method of solids control that should be employed.

Our basic aim in solids control is to reduce the formation solids content of a mud to a practical limit that will allow us to maintain the filtration and flow properties in a range, to best handle all the drilling problems. This means that we must pay special attention to the control of the particles less than 2 microns in size, since they are dominant in determining the mud properties. Unfortunately, the small particles are the most difficult to remove. To reduce the concentration of these particles, it is necessary to either dilute or build mud volume or to centrifuge the mud. When centrifuging, these small particles are discarded along with a portion of the liquid phase, which contains a large portion of the chemicals in the mud. Both dilution and centrifuging are expensive. Centrifuging is normally considered to be less costly in mud weights above 12 lbs. /Gal.

The problems and cost of removing the small particles make it imperative that we remove as much of the large particles as possible before they are reduced to the fine size range. This is accomplished primarily by screening weighted muds and by using De-sanders and de-silters on un-weighted muds.

The cost of reducing the solids content in a 10 lb/gal mud is less than that of a 12 lb. /Gal mud This means that we should begin weighting up a mud which contains a minimum amount of drilled solids. This will always ensure a minimum cost and better mud properties.

Dilution

The addition of water or the phase which constitutes the continuous phase of the mud to a mud system directly results in the cost of converting that water or continuous phase into the mud of the proper density and chemical content. The volume of new mud which is built will increase the total volume of the mud system, thereby reducing the concentration of drilled-solids in the mud.

The first rule for minimum dilution cost is to maintain the mud system at optimum volume. A volume of mud equal to the volume of new mud to be built should be removed from the mud system prior to beginning dilution. This results in a maximum percent dilution from the addition of a given volume of new mud. The mud discarded prior to dilution is also of less value than the mud discarded after dilution, because the former contains more of the undesirable solids.

This leads to rule number two. The mud should be diluted to the desired solids content in one step rather than a series of smaller steps. The mud discarded at each step contains some of the new mud from the previous dilution. Consequently, the total cost to reduce the solids content to a predetermined level will be considerably more if done in smaller stages than by a single dilution or displacement of part of the mud system.

If the mud system can increase in volume, the costs will be even higher. If the system can increase in volume by 25 percent and then discarded back to the original volume, the cost will be > twice higher than partial displacement cost. The most expensive dilution method is to allow the volume to continually increase.

Obviously, the expense of dilution can be minimized by proper application of these principles. The dilution program should be planned rather than left to chance. The solids content can be allowed to build during the intervals where careful control is not necessary and then reduced by massive dilution before proceeding into more critical situations.

The cost of controlling solids by dilution is a function of mud weight and the amount of drilling-solids incorporated into each barrel of mud.

In other words, as mud weight increases, the cost of a barrel of new mud increases; and as the allowable concentration of drilled-solids in the mud is decreased, the barrels of new mud required to maintain a given percent drilled-solids increases. This gives us two more methods for reducing dilution costs. First, dilute the mud and reduce the drilled-solids content at low mud weights rather than waiting until the mud weight has been increased and it becomes necessary to dilute to achieve the desired properties. This is simply a matter of planning ahead. If the mud contains more drilled-solids at the low mud weight than can be tolerated at the highest expected mud weight, the solids should be reduced now to a level that will allow increasing the mud weight without subsequent dilution.

The solids content outlined in a mud program should be realistic. If it is lower than needed, the mud maintenance costs will be forced to be higher than necessary. If it is not consistent with other designated mud properties it will lead to confusion.

Screening

Since screening is a relatively low-cost process that does not waste any appreciable amount of Barite or chemical, it is the most desirable means of solids control. The only cost of screening a mud is the cost of screens. Effort should be made to maximize the amount of drilled-solids removed by screening. This means that 100 percent of the mud stream should be screened through as fine a screen as possible without screening out an excessive amount of Barite. API specifications of Barite call for a maximum of 3 percent Barite retained on a 200-mesh screen. This means that screens approaching 200 mesh can be used without wasting an excessive amount of Barite.

A woven wire cloth made from stainless steel is normally used for mud screening on shakers. This wire cloth is usually specified by mesh. The mesh is the number of openings per linear inch. Unfortunately, the mesh designation is not sufficient to describe the screen. The wire size, which is used also influences the opening size and the percent open area of the screen. The finer the wire diameter the greater will be the screen capacity. Screen life will be less and the opening size greater -- both of which are undesirable. There must be a compromise between opening size, capacity, and screen life.

Screen designations have also been complicated by screens which have rectangular rather than square openings. Such a screen may have 80 openings per inch in one direction and 40 in the other direction. This screen would be specified as an 80 x 40 mesh screen. Obviously, it would have a greater capacity than a square 80 mesh, but would allow some particles to pass that would be retained on a square 80 mesh.

Drilled solids may exist in the mud system in any size ranging from the finest dispersed Bentonite, 0.05 microns, up to the largest cutting or sloughing. The purpose of screening is to remove as much as possible of the drilled-solids larger than the Barite particles. In this manner Bentonite, Barite, and the liquid phase of the mud containing the chemicals are salvaged; the coarse, drilled-solids are discarded. Since only a very small amount of mud that adheres to these larger particles is lost from the system, screening is a very economical way to remove drilled solids.

The drilled-solids normally reduce in size with time and continued circulation in the mud system. When they enter the Barite size range, it becomes impossible to separate them from the Barite until they are further reduced in size and become smaller than the Barite particles. At this stage, a centrifuge can be used to separate a portion of the liquid fraction of the mud containing these small particles along with the chemical and Bentonite. This operation is expensive and removes only a small fraction of the total drilled solids. This makes it imperative that we remove as much of the drilled-solids as possible the first time they reach the surface.

Recent improvements in shale shakers allow the use of much finer screens than what was possible with the conventional shakers. These new shakers are normally somewhat larger and operate at a higher vibrational frequency than conventional shakers. Some are designed so that the speed and amplitude can be easily changed to fit the needs of various muds.

Careful attention should be given to maintaining the shaker in good working condition and ensuring that the finest screens that will handle the entire mud stream are always used. Every pound of solids that is missed by the shaker will cost several times as much to be treated or removed from the mud system by other means. **It is important here to mention that there may be an occasion where it is more beneficial to retain solids in mud (deep-water shallow depth permeable rock) and deliberately the coarser screens are deployed. Therefore, the decision to select the right size of the screen should be based upon a holistic view of the situation and not just to expel the solids.**

Muds that retard dispersion or lift the cuttings out of the hole in a minimum length of time also aid screening efficiency. This ensures that the particles remain as large as possible, thus allowing the maximum amount to be removed by screening. Oil muds are particularly good at inhibiting dispersion of shale particles. Normally, a fine-screen shaker will control the buildup of drilling solids content in an oil mud with only a minimum amount of dilution being necessary.

Settling

Theoretically speaking, with the increased use of SOBM and simultaneous increase in Deep-water drilling some of these concepts used critical review and adjustments made in accordance with requirements of well. For example, plain dilution or discarding solids laden mud are not practically feasible steps in SOBM and often this compels us to let un-desirable solids' concentration be a little higher than ideal. Non-reactivity and lubricity of SOBM partly compensates for this fallacy and the system is still as efficient as desired.

Solids can also be separated by taking advantage of the fact that solids of different sizes and densities settle at different rates. The settling velocity of a particle increases as a function of increased gravitational force, increased particle diameter, increased difference in density of particle and fluid, and decreased fluid viscosity. The exact functions of these variables depend on particle size, rate of settling, the shear behavior of the fluid, and inter-particle interference. There is no single equation that can be used to calculate settling rates under all conditions.

Routing the mud through settling pits is the first attempt to separate drilled-solids from the mud. When the mud could sit undisturbed in the pit for a long period of time, the larger solids settled to the bottom.

In practice, settling pits are not very useful except when drilling with clear water. Even in this case, flocculants are sometimes needed to aggregate particles into a larger size that will settle faster.

To overcome these deficiencies, devices such as centrifuges and hydro cyclones were built. These devices increase the gravitational force acting on the particles, thus increasing the rate of settling.

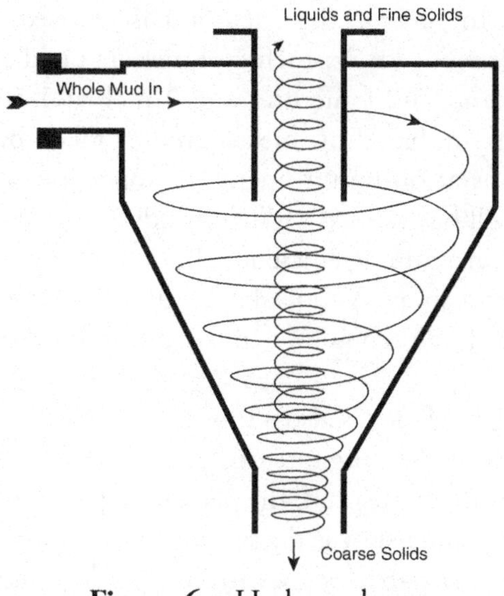

Figure 6 Hydrocyclone

Hydro cyclones

Hydro cyclones are a broad classification of settling devices that include the de-silter, De-sander, and centrifuge. These are all basically the same device in different sizes which separate solids at different cut points. A cross-sectional diagram of a hydro-cyclone is shown in Figure. The mud is pumped into the cyclone through a tangential opening at the large end of the cyclone. This results in a rotating motion of the fluid, much like a water spout. The resulting centrifugal force causes the particles to concentrate on the wall. The dense layer next to the wall travels downward and is ejected at the bottom of the cone. The remainder of the mud travels upward and exits through the overflow end.

At the overflow end, a short pipe extends down into the hydro-cyclone past the mud intake. It stops mud from short-circuiting directly to the overflow. In effect, it taps the light portion of mud in the cyclone, allowing only the dense sludge to spirally slide down to the underflow. If mud is fed to the cyclone at too high a pressure, some of the mud will exit the overflow without having time for the solids to segregate. If the feed pressure is too low, poor separation of solids occurs. This makes it imperative that the hydro-cyclone be operated at the manufacturer's recommended pressure. Four-inch de-silters are generally designed to operate at 40-45 psi, and six-inch de-sanders at 35-40 psi.

The cut point of a solids-separation device is usually defined as the particle size at which half of the particles go to the underflow and half to the overflow. In general, the cut point is reduced as the hydro-cyclone diameter is reduced. Six-inch cyclones are rated at 30-35 microns cut point, 4-inch at 15-20 microns, and 2-inch at 5-10 microns. These cut point ratings are based on measurements taken on minimum-viscosity suspensions at optimum operating conditions. As a result, they are lower than will be attained on field muds.

For maximum removal efficiency, the underflow discharge should be in the form of a spray. A rope-shaped discharge means that some of the solids which should come out in the underflow are forced to go to the overflow.

The removal efficiency is also a function of the viscosity and density of the mud. Since the settling rate of a particle is decreased by increased viscosity and density of the mud, the percent of particles of any given size which reaches the underflow is decreased. In other words, an increasing amount of the larger particles goes to the overflow and are returned to the mud system. This means the de-silting should be started before the drilled-solids build up in a mud and increase the viscosity and density.

Solids that are not removed from the system at the first opportunity tend to be reduced in size as they are re-circulated. As they reduce in size, they not only become too small to be removed by a de-silter, but they also quickly build viscosity which reduces the efficiency of the de-silter in removing the larger particles.

When water is added to the mud, it should be added upstream from the de-silter to reduce the mud viscosity and improve the efficiency of the de-silter. The addition of chemical, Bentonite or other mud additives should be added to the system downstream from the de-silter so that none is discarded by the de-silter.

De-sanders or fine-screen shakers should be used ahead of the de-silters to remove the larger particles that can clog or overload the de-silter. Clogged discharge on the underflow of a cyclone probably accounts for more inefficient de-silter performance than all other problems.

The de-silter unit should be sized to handle 125 to 150 percent of the mud circulating rate. Each hydro-cyclone will process from 50 to75 gal/min. The suction, feed, and overflow lines should be sized with generous diameters and minimum lengths in accordance with good centrifugal pump installation practices.

When operated in the conventional manner, a large fraction of the particles in the Barite size range is discarded by De-sanders and de-silters. It follows that large quantities of Barite will be discarded when these devices are used on weighted muds. Normally, this will cause prohibitive costs of mud maintenance.

De-silter screen Combination (Mud Cleaner)

Since the use of a de-silter on a weighted mud is prohibitively expensive (due to loss of Barite), and since ultra-fine screens are incapable of handling the entire mud stream, another type of solids removal device is necessary. A unit consisting of both devices eliminates their individual limitations.

If a mud is de-silted, the underflow will contain a fraction of the Barite in the mud as well as undesirable solids. An ultra-fine screen shaker has the capacity to handle the reduced volume of the underflow. If screens with openings slightly larger than the largest Barite particles are used, only the drilled-solids will be retained on the screen. The remainder of the underflow containing the liquid and Barite can be returned to the mud system along with the de-silter overflow. The net result is to be able to de-silt a mud without losing an excessive amount of Barite.

Since the efficiency of a de-silter is reduced by increased viscosity and density of the mud, this unit will lose efficiency as mud density increases. This is not to say that its efficiency will necessarily be reduced below that of alternate means of controlling the solids. It is best suited for the low-to-intermediate weight range of 10-15 lbs. /Gal. This is fortunate, since the centrifuge becomes less efficient when compared with dilution in the low mud weight range. The mud cleaner should be started at the beginning of weight-up and continued if the rate of drilled-solids removal will justify the rental cost.

A screen size of 140 to 160 mesh is recommended for this unit. This will remove solids larger than about 100 microns without removing much of the Barite. Screens of 200 to 250 mesh have been used. Barite consumption increased substantially.

The unit should be shut off for one or two circulation cycles when adding Barite. Otherwise, considerable amounts of the newly added Barite will be removed from the system. It appears to take this long for the Barite to become dispersed in the mud.

In screening, de-silting, or use of this combination unit, only the coarser particles are removed. Particles that enter the mud system in a finer size are not removed by these devices and continue to build in concentration. Despite the buildup of these solids is slower than if the larger solids were also left in the system and allowed to break up and disperse, some type of control will eventually be necessary. In weighted mud, the use of a centrifuge will usually be required to remove the colloid-size particles, which have the most pronounced effect on mud properties.

Centrifuge

The mud centrifuge is designed to remove small, drilled particles while saving Barite. It consists of a conveyor screw inside a cone-shaped bowl that rotates at approximately 1800 to 2000 RPM. The conveyor screw rotates in the same direction as the bowl, but at slower RPM.

Mud with dilution water is metered into the centrifuge in the conveyor screw. The mud is thrown to the wall of the bowl by the centrifugal force where it forms a pool. A circular row of ports at the large end of the bowl allows the liquid portion of the mud at the top surface of the pool to flow from the centrifuge. As the mud moves toward these overflow ports, solids settle toward the wall, where they are scraped by the conveyor screw toward the underflow ports. In this manner, the mud gets

Divided into two streams, the light overflow and the heavy concentrated underflow slurry.

The primary factors that determine the minimum size of the particles which will settle to the wall and be discharged at the underflow are:

1. Bowl speed (centrifugal force)
2. Mud viscosity
3. Mud feed rate (retention time in the bowl)

Increased bowl speed increases the centrifugal force acting on the particles and increases their settling velocity. The bowl speed should be maintained at 1800 RPM to effectively settle Barite particles in the 3-4-micron range. It is often seen that the centrifuge operators run the centrifuge at slow speeds to increase the life of the bearings. This reduces the efficiency of Barite recovery and increases mud costs.

The total feed rate to a centrifuge must be limited. If the rate is too high, the mud will flow to the overflow ports before sufficient time has elapsed to allow all the Barite particles to settle. The maximum feed rate for efficient operation is a function of mud weight. As mud weight is increased, the feed rate should be decreased. Feed rates may range from 20 gal/min at 12 lb. /Gal to 6 gal/min at 18 lb. /Gal mud weight. In the final analysis, the feed rate can be decreased if Barite recovery is not desired.

It is a common misconception that the underflow from a centrifuge consists only of Barite. The drilled-solids are also subjected to the same settling forces. Those approximately 1.5 times larger than the smallest Barite particles which settle to underflow will also settle at the same rate. Therefore, the underflow will contain the coarser drilled-solids as well as Barite.

The overflow consists of the diluted mud that was fed to the centrifuge, minus those solids which settled out of it. Its density will depend on the amount of solids that did not settle to the underflow. The density will increase as the amount of Barite left in the overflow increases.

A good, quick means of checking the performance of a centrifuge is to simply measure the density of the overflow. Generally, a density of 9.0 to 8.7 lbs. /Gal indicates good performance. However, if the brine phase itself weighs more than the suggested number, then brine phase density can be a criterion to evaluate the performance of the centrifuge. Density values too higher than brine weight or the numbers mentioned indicate that a substantial amount of Barite is being discarded. The bowl speed, dilution rate, and mud feed rate should be checked and appropriately changed.

When the centrifuge is operating correctly, the underflow will contain ~ > 90% percent of the Barite in the processed mud. The remainder shall get discarded in the overflow. The underflow will also contain about 15-30 percent of the original feed mud. As a result, it will contain a like amount of the Bentonite and chemical in the original mud.

When the underflow is returned to the active mud system, it must be rebuilt into the mud of the proper density containing the desired amounts of Bentonite and chemical. This means that additions of water, Bentonite, and chemical must be made. Since 70-85 percent of the Bentonite and chemical are discarded in the overflow, this amount should be replaced to keep from "stripping" the mud. If too much Bentonite is added, the flow properties will increase and dilution will be required. Consequently, Bentonite additions should be carefully monitored to control the flow properties

Since the centrifuge processes only a small fraction of the mud returns and removes half or less of the drilled-solids in the mud which is processed, it is not capable of reducing the total solids content of a mud while drilling. Consider the case where one-third of the mud system is centrifuged in one day. If the mud contains 6 percent drilled-solids and one-half of the drilled-solids are removed from one-third of the mud, a one percent reduction in drilled-solids would result. One percent additional solids could be added to the system in one day of drilling. At the end of one day the mud would still contain 6 percent drilled solids.

This analysis shows that the centrifuge may arrest the drilled-solids content from increasing, but will not effectively reduce the concentration. The drilled-solids content of a weighted mud must start at a low level if it is to be maintained at a low level.

Since the centrifuge selectively removes the smallest particles, it is quite effective in controlling the flow properties. Small hydrated clay particles are almost totally responsible for excessively high plastic viscosities, yield points, and gel strengths. By removing these small particles and adding an adequate amount of Bentonite, the filter cake properties are also improved. These are the primary reasons for centrifuging a mud.

Use of solids recovery mode shakers:

Of late while drilling through very soft, young and un-consolidated rock matrix in ultra-deep-water environment, there has been a paradigm shift in the outlook of engineers regarding solids in mud. Once considered un-desirable have suddenly caught the attention of many in industry for the benefit they accrue to the mud system by forming a plastered filter cake around the wall of wellbore to stop seepage losses in micro-fractures and eliminate ballooning, especially in an oil base mud environment.

It is now an established fact that inert solids, once termed as undesirable can now find a meaningful place in mud system, when rightly engineered. The concept of stress-caging is based upon sizing the solids and placing them in a fracture which is engineered to selective size. Under such circumstances, it is often desirable to deliberately put a coarser screen on shakers to retain these essential inert solids even at a cost of a slight increase in PV. However, with the passage of drilling and continuing to incorporate more drilled-solids it is only a matter of time when the mud properties start showing up the adverse effect of increased inert fine solids intruding the mud either directly from excavation process or by attrition of sized solids making them useless.

It is at this stage innovative concept took birth in the industry by which the three deck shakers were so designed that the middle deck preferentially picks up all the essential inert solids of a finite size and divert them back into the circulatory system to enable retaining their concentration, thereby extending their utility over a longer period.

In addition to this, the re-inclusion of essential inert solids also helps in maintaining the required dose of these solids as otherwise no amount of replenishment treatment can match the attrition level and the best designed mud too can fail due to inability to maintain the desired level of sizing inert solids. The word inert is purposely mentioned here to ensure that they are not un-desirable. At times, this can mean that the low gravity solids / inert solids/drilled solids could be as high as 7 % in the system!, while customarily speaking one would be tempted to keep this number < 3 %!!

CHAPTER 7

PRODUCT DESCRIPTION

WEIGHTING AGENTS-

Barite

Barite is a naturally occurring mineral which is mined and ground to API specifications. It is predominately barium sulfate, but contains some impurities. The minimum specific gravity allowed by the API is 4.2 compared to a 4.5 specific gravity for pure barium sulfate. The impurities vary depending upon the source of the ore, but are predominantly inert materials such as quartz. Pure barium sulfate is white, but impurities often cause Barite to be gray or pink. The color is not indicative of the purity or quality of the Barite. Of late due to scarcity in availability of 4.2 SG Barite, industry has started accepting 4.15 / 4.10 SG Barite also

Calcium Carbonate

With increased awareness of formation damage and simultaneous compulsion of keeping low mud weight (especially in shallow sections of the deep-water environment), Calcium Carbonate has emerged as a clear choice. It has a specific gravity is 2.7+/-0.1 which facilitates inclusion of adequate amount of weighing material to help have enough required solids in mud, which is not feasible with Barite or any other weighing material. The additional flexibility of sizing the Calcium Carbonate to desired micron level further favors the use of Calcium Carbonate in shallow sections of ultra-Deep-water / Deep-water environment. Thus, Calcium Carbonate although is not as popular weighing material as Barite, but has specific situation based application, where it is even preferred over Barite.

Hematite

Hematite is a naturally occurring iron oxide mineral. The product used as a weighting agent in mud is a micaceous hematite mined in Brazil. It is a very pure material having a specific gravity of approximately 5.1. It is harder than Barite and as a result, can be more abrasive under some conditions. To minimize the abrasive potential, it is ground so that no more than 0.5% will be retained on a 200-mesh screen.

CLAYS-

Wyoming Bentonite

It is a naturally occurring Bentonite which is mined Wyoming Bentonite in Wyoming. There are no other commercial sources of this mineral of comparable quality. It is predominately Sodium montmorillonite, but contains some Calcium Bentonite and sand or silt. As a result, it is somewhat variable in performance.

When the product fails to meet the API specs for viscosity yield, it is often treated with a polymer to increase the viscosity. The polymer causes a form of flocculation that increases the YP of the suspension at API test conditions. This may allow the product to pass the test, but does not improve its performance at high temperatures or in a treated mud. Its performance under most conditions is a function of the original quality of the clay. Filtration tests are probably the best way to evaluate the quality of the product.

Sub-Bentonite

The sub Bentonites are primarily Calcium Bentonites of poor quality. It was once popular to add these clays to muds to get a particle size mix. This is completely unnecessary and simply adds to the solids concentration of a mud. The reduction in fluid loss which was claimed for these products was a result of thicker filter cakes. These clays are like the clays contained in bentonitic shales and should not be used.

Beneficiated Bentonite

These products are Wyoming Bentonite treated with a polymer, usually a polyacrylate, to cause flocculation. The result is a product which produces suspensions of about twice the viscosity of ordinary Wyoming Bentonite. The same performance can be achieved at less cost by treating a Bentonite suspension with a flocculating polymer lime, or soda ash. The only advantage of this product is that it comes in one sack. It is ineffective in any environment where the polymer is ineffective such as salt water, Calcium contamination, and high temperature.

Attapulgite

It is a needle shaped non-swelling clay, which derives its viscosity building from its shape and surface charge. For this reason, it is unaffected by water salinity. It will produce as much viscosity in saturated salt water as it will in fresh water. An attapulgite particle is a bundle of fine individual needle like structured clay. High shear is required to break up these bundles and disperse the needles. As a result, the viscosity of an attapulgite suspension is dependent on achieving a high degree of shear.

Since the particles are needle shaped and non-swelling, attapulgite offers no fluid loss control. Its primary use is to build yield point and gel strength in a saltwater mud.

THINNERS-

Lignosulfonate

The Lignosulfonates used to treat drilling fluids are basically sulfonated lignin, complexed with chrome or iron and chrome. The starting raw material is spent sulfite liquor, which is a waste product from the paper industry. Since the basic raw material is a waste product, the finished product is cheap compared to the cost of a product tailor-made from pure raw materials. For this reason, it will be difficult to ever replace this product with anything comparable or better at the same price.

Lignosulfonate is essentially 100 percent soluble in water. It forms an acid solution with a pH of about 4.0. The Lignosulfonate molecule is quite complex. It is a large anionic molecule which adsorbs at the positive charge sites on clay platelets.

Lignosulfonate degrades at high temperatures. As it degrades, small amount of carbon dioxide is liberated. Over time, this can lead to a Carbonate or bicarbonate problem. At low pH, degradation may occur at temperatures as low as 250°F. At a pH of 10 or higher, degradation is retarded. The active Lignosulfonate content of a mud is reduced by about 2 lbs. /bbl in 16 hours at 300°F. Very rapid degradation occurs at temperatures above 350°F. At this temperature, hydrogen sulfide is liberated in small quantities.

Lignosulfonate works best at a pH of 10.0 to 11.0. At this level, the degradation is slow and its ability to de-flocculate is enhanced. About 2-4 lb./bbl active content is required in fresh-water mud to achieve maximum de-flocculation. Approximately twice this amount is required in saltwater mud.

Common Lignite

A lignite mineral known as Leonhardite is mined in North Dakota. This mineral is ground and sold as lignite to be used in treating muds. It is the Humic acid content of this material that functions as a thinner.

It has a very low solubility when added to water, but increases in solubility as pH is increased. It is about 75 percent soluble as a maximum. It is an acid material and forms solutions with a pH of about 4.5 when reacted with Calcium, it forms an insoluble precipitate. The fact that it is an acid and precipitates Calcium makes it a very good material to treat cement contamination. Concentrations above 10 lb./bbl should be avoided. Calcium added to such a solution will form a rigid gel.

Lignite is an adequate thinner at low temperatures, but at temperatures above 250°F it does not de-flocculate sufficiently well. It has been mistakenly considered a high temperature thinner simply because it has a thermal stability above 400°F.

The primary function of lignite is as a fluid loss-control additive. Lignite, Lignosulfonate, and Bentonite at equal concentrations will each reduce the fluid loss of a mud an equivalent amount.

Causticized Lignite

Causticized lignite is a blend of unit part caustic to six parts lignite. In solution, it has a pH of about 9.5. Its cost is approximately 40 percent higher than the combined costs of the lignite and caustic that it contains.

The performance of this product is no different from that of lignite and caustic added separately. Its only advantage comes from the fact that it contains enough caustic to solubilize when added to water. This is a disadvantage when treating cement contamination, since the acid behavior of common lignite helps to neutralize the hydroxyl effect.

Modified Lignite

There are many ways that lignite can be modified. Some products have been neutralized by the addition of potassium hydroxide. The potassium ion is supposed to aid in shale stabilization. Lignite can also be modified by complexing with a heavy metal. The heavy metals supposedly add temperature stability to Lignosulfonate treated muds. The addition of both potassium and the heavy metal is detrimental to fluid loss control.

Phosphate

Polyphosphates are good deflocculating agents in freshwater muds at low temperatures. Approximately 0.1 to 0.2 lb./bbl will completely de-flocculate a mud. There are several polyphosphates which are used in mud treating.

Sodium tetra-phosphate is the most common form used as a mud thinner. It will not function as a thinner in the presence of Calcium or high salt concentrations. It is a sequestering agent for Calcium and is often used for the purpose. It should not be used to treat for cement contamination since it is an alkaline material. At temperatures above 180°F, the polyphosphates are converted to orthophosphate which is a flocculating agent. This establishes the upper temperature limit for use of phosphates as thinners.

Another polyphosphate which is used as a thinner is tetra Sodium pyrophosphate. It is more alkaline than Sodium tetra-phosphate, but otherwise functions in much the same manner.

POLYMERS

There are basically two types of water soluble polymers used in muds. They are the natural polymers, polysaccharides, and the synthetic, vinyl polymers.

The natural polymers consist of starch, modified cellulose (CMC and HEC), gums, and Xanthan polymer. These materials increase the viscosity of water due to their molecular size, shape, and ability to absorb water.

The vinyl polymers consist of partially hydrolyzed polyacrylate and vinyl acetate Maleic Anhydride co-polymers. The performance of these polymers varies considerably due to difference in molecular weight and degree of hydrolysis. They have many similarities. They increase the yield point of clay suspensions by adsorbing on the clay particles and linking them together.

The synthetic polymers, which depend on adsorption, are inherently sensitive to their environment. Calcium is especially detrimental. They react with Calcium and become insoluble, causing high viscosities and gelation of the mud.

The natural polymers, which perform by thickening water, are essentially non-charged and are less sensitive to their environment. Starch and HEC will perform in almost any kind of brine, and Xanthan polymer will perform in salt water. CMC loses its effectiveness in salt water, above 50,000 mg/1 chloride. Gums such as guar gum are sensitive to Calcium and pH, and should not be used in Calcium concentrations above 250 mg/1 and pH above 9.5.

Starch, CMC, and HEC are primarily fluid loss control agents, but also may be used as viscosifiers. Guar gum is primarily used to viscosity brine work-over fluids. It imparts some fluid loss control. Xanthan polymer is a viscosifier and affords very little fluid loss control.

More information on the use of polymers is given in Section 4 under Treating Agents.

OIL MUD CHEMICALS

An oil mud is basically composed of a continuous oil phase with water emulsified into it as non-continuous internal phase and supplemental materials added to impart viscosity, fluid loss control and oil wetting of the solids.

A primary emulsifier is used to stabilize the emulsions and impart some viscosity and fluid loss control. If the mud is to be used at temperatures above 275-300°F or if Calcium chloride water is to be emulsified, a supplemental emulsifier is normally added.

All the solids in an oil mud must be wet with oil to prevent aggregation of the solids and weakening of the emulsion. An oil-wetting agent is used for this purpose.

To prevent settling of Barite and supply hole-cleaning capability, supplementary viscosifying agents are sometimes needed. Supplemental fluid loss control agents are also used in some cases, especially at very high temperatures.

All the oil mud product lines contain these basic materials. Individual products are often blends of more than one of these materials. The types of chemicals used to perform these various functions are often different in the various oil muds, leading to finished muds of different characteristics.

Emulsifiers

Emulsifiers are surfactants that reduce the surface tension between the water droplets and oil (or synthetic). They stabilize the mixture by being partially soluble in water and partially soluble in oil. They are usually long-chain alcohols, fatty acids or polymers and can be anionic, Cationic and non-ionic. One end of the emulsifier molecule has an affinity for water while the rest of the molecule has an affinity for oil or synthetic fluid. Emulsifier particles form a coating around the water droplets to prevent them from coalescing, as shown in the Figure.

Figure-Arrangement of Emulsifier around A Water Droplet

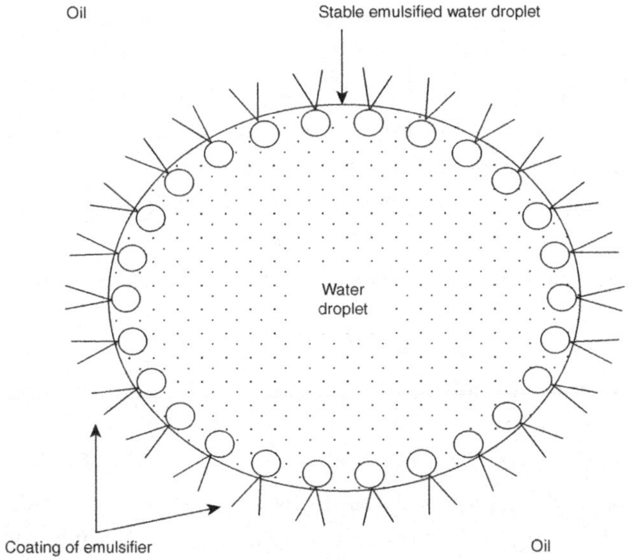

Wetting agents

A wetting agent is also a surface-active agent that reduces the interfacial tension and contact angle between a liquid and a solid. This causes the liquid to spread over the surface of the solid, as shown previously in Figure. Wetting agents have one end that is soluble in the continuous-phase liquid and the other that has a strong affinity for solid surfaces. In non-aqueous fluids, wetting agents cause solids to be wet with oil or synthetic liquid. The solids (Barite, clays and drill solids) must be wetted by the continuous liquid phase of the drilling fluid or they will aggregate and settle. Most solids naturally have a preferentially water wetting tendency and require wetting agents to wet them with either oil or synthetic fluid.

Viscosifiers

Although emulsified water increases viscosity, viscosifiers and gelling agents are also required.

Untreated clays cannot be used as viscosifiers because they do not hydrate and yield in oil or a synthetic fluid. If the clays are first coated with an amine, so that they are organophilic, then they will yield and viscosify in oil and synthetic fluids. These clays yield faster with increased shear and temperature. If the amine is not temperature-stable, the clay will become an inert solid. High-temperature organophilic clays may be needed for high-temperature applications.

Alternative non-clay viscosifiers are available to increase viscosity. They include asphaltic materials, fatty acid gellants and polymers. Developing viscosity is a problem when mixing new fluids in mud plant, where low shear mixing and low temperatures do not allow amine-treated clays to yield. Freshly prepared muds should not be treated with more organophilic clay than will be required when drilling. Overtreatment with clay leads to excessive viscosity once the fluid is exposed to shear and temperature in the well. Alternative viscosifiers such as oil-soluble polymer, should be used to provide additional suspension, while mixing and transporting new mud. Many polymer viscosifiers degrade with increased exposure to temperature and counteracted by the corresponding activation or yield of organophilic clay under these conditions and shear. In freshly prepared fluids this is counteracted by the corresponding activation or yield of organophilic clay under these conditions.

Filtration-control additives

HTHP filtration control of invert emulsion muds are affected by the viscosity of the continuous fluid phase, the oil or synthetic-to-water ratio, the tightness of the emulsion, water-wetting of the solids, the solids content, and the amount of amine-treated clay in the system. Many synthetic systems do not require special filtration-control additives in formulations designed for use at temperatures less than 300°F (149°C). At temperatures above this, most systems require a filtration-control additive. Gilsonite or asphalt, amine-treated lignite and polymers are the most common filtration-control additives.

Low end rheology boosters

Of late due to increase in activities in Deep-water and ultra-deep-water environment, mud engineers have come across a situation not occurring before. The temperature gradient reversal is a result of lower temp at sea bed and below and above this depth the temperature of the mud keeps increasing! This essentially means that the mud on the sea bed is comparatively the thickest and increased viscosity is not due to the presence of organophilic clay, but the contraction of continuous non-aqueous phase and dispersed droplets of brine in inner dispersed phase. Thus, the initial preparation of mud cannot have more clay and yet the mud should impart adequate rheology to the system to be able to bring cuttings to the surface, which becomes even more challenging due to the larger diameter of risers. Organo-phillic clays do not yield to their potential at lower temperature and the mud still needs viscosity imparting component to be able to perform the role of bringing the cuttings to the surface, which at times can be as deep as more than 3000 m, considering the 2000+m of water depth itself! This has led to the inclusion of a new class of polymers with inherent property of quickly yielding even at low temperature and thus impart the required lifting capacity to begin with. The fact that such boosters degrade soon when exposed to higher temperatures, favors the situation because by the time they degrade when exposing the mud to higher temperatures while continuing to drill below the sea-bed, the original organo-phillic clay starts yielding. Imagine this as a relay race situation, where the first lap is completed by low end rheology booster and they then pass on the baton to organophilic clay which then continues the race. Such boosters are available in market under varying names being marketed by each company as their own proprietary product under the banner of rheology boosters.

CHAPTER 8

IDENTIFICATION AND TREATMENT OF CONTAMINANTS

The most common chemical contaminants of water-base muds are:

1. Cement
2. Calcium or Magnesium
3. Salt
4. Carbonates and Bicarbonates
5. Water, especially in SOBM
6. Hydrogen sulfide
7. Carbon-di oxide

All these contaminants cause the mud to deteriorate and deviate in its desired properties, because of which it becomes incapable of performing the roles expected of it. Excess solids can also be termed as a contaminant, but excluded here as a separate chapter has been dedicated to address this problem.

CEMENT

The active part of the cement that causes problems in mud is Lime, $Ca(OH)_2$. As cement sets, the amount of active lime in the cement decreases. In green or unset cement, the active lime content is quite high while fully cured cement has very little active lime. This makes it impossible to predict the extent of contamination that will occur, even when the amount of cement to be drilled is known.

Lime has a relatively low solubility so that only a fraction of a pound per barrel will go into solution depending on the conditions of temperature, pH, and salinity. Both increased temperature and pH decrease the solubility of lime while increased salinity increases its solubility. Normally, only 0.25 to 0.5 lb./bbl of lime will go into solution in a mud. The remainder of the lime will remain in the mud as un-dissolved particles.

Diagnosis

The pH, Pf, and Pm will increase until the limit of lime solubility is reached. At this point, the pH and Pf will stop increasing since they are measuring only the alkaline material in solution. Pm will continue to increase as lime is added, because it is measuring the total alkaline material in the mud, both dissolved and un-dissolved. When the pH is high (approximately 12 or higher) and Pm>3Pf, an excess of cement or lime is indicated.

When the cement is green, it will disperse in the mud into fine particles. The greater surface area will allow more reaction with the mud and more severe flocculation problem

Effects on Mud Properties

The Cement contamination (or lime) in the mud will cause the mud to flocculate, resulting in increased YP, gel strength, and fluid loss. The degree of flocculation will depend on the solids content of the mud and the presence of enough thinner to offset the flocculating tendency of the lime. Since it is the flocculation of solids that causes the detrimental effects on mud properties, a high concentration of active solids will result in a more pronounced effect on the mud properties. The more dispersed clay solids that a mud contains, the more difficult it is for the Lignosulfonate to keep these solids from flocculating when exposed to cement or lime. If the mud contains no de-flocculant, then it will flocculate even with relatively low solids content. Thus, the effect of cement contamination will vary from mud to mud. In general, an untreated or lightly treated mud or one with high active solids content will react more violently than a highly treated, low solids mud.

Treatment

It is better to dump the most severely contaminated mud rather than try to restore it with chemical treatment. The mud cleaner and centrifuge should be used on a weighted mud to remove the fine particles and the De-sander and de-silter should be used on an un-weighted mud. After all the cement has been removed (that can be removed by the solids control equipment), chemical treatment can be used to restore the mud to its desired properties. A choice can be made whether to convert the mud to a lime, mud or to precipitate the lime.

The chemical treatment is based upon the concept that cement contamination causes an increase in Lime content which needs to be taken out of the system. One popular method is Precipitation of lime as well as neutralize the hydroxyl ions. Sodium bicarbonate and Sodium acid pyrophosphate (SAPP) are the two most commonly used additives for treating cement. To treat one pound of lime requires roughly almost same amount of Sodium bicarbonate or 1.5 pounds of SAPP. Both the materials will precipitate all the Calcium, but neutralizes only half of the hydroxyl ions. The remaining $_{OH}^{-}$ ions convert into caustic (Na OH). This means that the pH of the mud will increase when the un-dissolved lime is reacted with Sodium bicarbonate or SAPP. This makes it very difficult to treat out large amounts of lime or cement without causing a caustic reaction which will flocculate the mud. Lignite and Lignosulfonate are acid materials which can be used to neutralize the caustic which is formed and de-flocculate the mud. About 3 lbs./bbl will neutralize the caustic formed from 1 lb./bbl of lime.

Neither soda ash nor one of the alkaline forms of phosphate should be used to treat cement. These materials precipitate Calcium, but do not neutralize any hydroxyl ions, thus causing the pH to increase more rapidly. Causticized lignite should not be used for the same reason.

Precipitation reactions occur in the water phase. As the lime in solution reacts with the Sodium bicarbonate or SAPP, some of the un-dissolved lime goes into solution to replace it. The reaction will continue until the treating agent has been used up or until all the lime has been dissolved and reacted with the treating agents. Since the lime must first be dissolved before it reacts with the treating agent, the reaction goes slowly.

The decision of whether to use Sodium bicarbonate or SAPP will normally be made based on availability. The only difference in performance is that a slight overtreatment with SAPP will thin the mud since it also acts like a de-flocculant at temperatures less than 180°F.

Pretreatment

The question often arises as to whether a mud should be pretreated before drilling cement? If the pretreatment has the potential of damaging the mud properties, then it should not be used. Since the amount of active lime which will get into the mud is never known. It is not technically feasible to prescribe the correct amount of treating agent beforehand. An over treatment with either Sodium bicarbonate or SAPP is potentially as harmful to the mud as is the cement.

The best pretreatment is the elimination of unnecessary solids from the mud and treatment with Lignosulfonate and lignite. These additives will combat the tendency of the mud to flocculate and prevent the most severe reactions of cement on the mud.

CALCIUM AND MAGNESIUM

Calcium and Magnesium are closely related ions and have similar effects on mud properties. They replace Sodium ions on Bentonite, thereby reducing the degree of hydration, and promoting flocculation and the aggregation of the Bentonite platelets. This leads to increases in yield point, gel strength, and fluid loss of the mud.

Calcium and Magnesium may enter the mud system from several different sources such as make-up water, saltwater flows, contaminated mud products, and drilling of gypsum. Seawater, for instance, contains about 1250 mg/l of Magnesium and 425 mg/l Calcium. If well water is used for make-up water, it may contain these ions and should be checked periodically. Formation salt water may also contain various amounts of

Calcium and Magnesium. Some of the mud products such as Barite may contain Calcium. The API specifications for Barite allow up to 250 mg/l Calcium in the Barite. This is equivalent to approximately 0.11 pound of gypsum per sack of Barite. Gypsum or anhydrite beds are also a common source of Calcium. Instances are not uncommon when the line transferring Barite and cement to rig from boat get contaminated with each other! This can cause a problem with cause not identified soon. The Barite passing through cement will severely contaminate the Barite which then unknowingly enters into the circulatory system while mixing the line contaminated Barite

Diagnosis

Increase in Calcium or Magnesium concentration will cause several changes in the mud system in addition to an increase in the total hardness. A reduction in pH will normally occur. This is much more pronounced from Magnesium additions than from Calcium due to the reaction of Magnesium with hydroxyl ions to form the insoluble Magnesium hydroxide, $Mg(OH)_2$. Lime is much more soluble than Magnesium hydroxide so that the resulting decrease in pH will be less.

Calcium and Magnesium will also cause an increase in fluid loss due to the strong dehydrating and the flocculating effect of these ions on Bentonite. Even in cases where the effect of flocculation is counteracted by Lignosulfonate, the fluid loss will be increased due to dehydration of the Bentonite.

Flocculation of the mud is the most serious adverse effect of Calcium or Magnesium contamination. This causes an increase in yield point and gel strength of the mud. The severity of the problem is a function of the concentration of the active clay solids, the amount of Calcium or Magnesium introduced into the mud, and ability of the thinner to counteract flocculation. Flocculation from any cause is always aggravated by high solids content. If the concentration of Calcium or Magnesium in a mud gets high enough, it will overwhelm any thinner and cause flocculation. Temperature is also important because the increased temperature also promotes flocculation and will increase the severity of a contamination induced problem.

In summary, an increasing trend of total hardness or a sudden increase to values above 200 mg/l signals the presence of Calcium or Magnesium in the filtrate. The harmful effects on mud properties will be increased fluid loss, yield point, and gel strength. The pH will decrease, especially due to the addition of Magnesium, and the plastic viscosity will tend to decrease as the hardness ions reduce the hydration of the Bentonite platelets.

Effects on Mud Properties

In addition to the effects of dehydration and flocculation of Bentonite, Calcium and Magnesium also affect other mud additives. Many of the polymers and lubricants are sensitive to water hardness.

The synthetic vinyl polymers (poly-acrylates etc.) react with Calcium and become insoluble. Gelation of the mud and high viscosities that cannot be controlled will develop at Calcium concentrations above 200 mg/l. Many of the natural polymers are also sensitive to Calcium. CMC loses its ability to hydrate as Calcium concentration is increased. At concentrations above 1000 mg/l, it becomes ineffective. Gums such as Guar, are sensitive to Calcium and should not be used where Calcium concentrations are above 250 mg/l. Of the commonly used polymers, only starch and HEC perform well at high Calcium concentrations.

Many other mud additives "salt-out" or become ineffective in the presence of high Calcium or Magnesium concentrations. Lignite is precipitated by hardness ions. The precipitant appears to be hydrated, thus building viscosity. High viscosities may develop if more than 6 to 10 lbs./bbl of lignite is precipitated in the mud. Lignosulfonate also becomes ineffective at high hardness concentrations.

Treatment

The treatments for Calcium and Magnesium are different and must be considered separately. The first line of defense against any type of contaminant is to maintain a mud with minimum clay solids content. This is true because the effect of the contaminant on the clay solids is responsible for the changes in mud properties that occur. The source of the contaminant is also important in selecting the type of treatment to use. If the amount of contaminant is small or non-recurring, the decision will normally be to precipitate it from solution. Soda ash is normally used to precipitate Calcium. One lb./bbl of soda ash will precipitate 1078 mg/l of Calcium. Care should be taken not to over treat with soda ash. This could result in the development of a Carbonate problem.

Treatment with soda ash should be calculated to leave approximately 100 mg/l of Calcium in solution.

Caustic should be added to a mud to treat out Magnesium. Precipitation will begin when the pH reaches 10 and will go to completion at a pH of 10.5. One lb. /bbl of caustic will precipitate 864 mg/l of Magnesium. Magnesium precipitates as a very fine particle, and at high concentrations will gel water. Consequently, no more than a few thousand mg/l can be precipitated without causing severe increases in the mud viscosity.

If the extent of Calcium and Magnesium contaminant is quite large or a continuing feed-in, the decision may be to treat or convert the mud to a type that will tolerate the contaminant.

For instance, if a long interval of gypsum is to be drilled, the mud may be converted to a gyp mud rather than continually treating the mud to precipitate the Calcium that goes into solution.

To convert to gyp mud, some dilution may be necessary if the active solids content is too high. Approximately 2 to 4 lbs./bbl of Lignosulfonate should be added to de-flocculate the mud and caustic added to increase the pH to at least 10. The Calcium content in the filtrate will range from 300 to 1200 mg/l depending on the pH of the mud. A high pH decreases the solubility of Calcium and limits the Calcium content in the filtrate.

Since gypsum has a limited solubility, the filtrate analysis for Calcium may not indicate the total amount of Calcium in the mud. As Calcium is precipitated from the water, additional gypsum will go into solution. Repeated treatments with soda ash will be required until the filtrate Calcium has been reduced to the desired level.

SALT

The salinity of water affects the hydration, dispersion, and flocculation behavior of clay particles. As salinity is increased, the degree of hydration is decreased and the tendency to flocculate is increased. This has pronounced effects on both the filtration and flow properties of a mud.

The most common sources of salt are make-up water, saltwater flow, and drilling of salt beds. It is important to determine the source of a salinity increase, so that proper remedial action can be taken. The salinity of the make-up water should be frequently checked to determine if a change in mud salinity is due to the make-up water or from a down-hole source.

Diagnosis

Any sudden increase in chloride concentration of more than a few hundred mg/1 indicates that salt or saltwater from some source is entering the mud system.

A significant increase in salinity will cause the mud to flocculate. The yield point, gel strength, and fluid loss will all tend to increase. The amount of increase will depend on the amount of active solids in the mud, the amount of Lignosulfonate, and the amount of increase in salinity. The pH of the mud will also be decreased and a slight increase in total hardness may occur. The increase in Sodium ions from the salt may exchange some of the Calcium from the clay solids and cause the total hardness of the filtrate to increase by 100 to 200 mg/l.

To differentiate between a saltwater flow and drilled salt, it may be necessary to estimate the volume of formation water influx that would be required to cause the observed increase in salinity. If an increase in much volume of this amount cannot be accounted for, the increase in salinity must have been caused by drilled salt.

Formation waters may contain significant amounts of Calcium and Magnesium chloride as well as Sodium chloride. In these cases, large increases in total hardness will occur. In pure Calcium chloride water, Calcium to chloride ratio of 40 to 71 will exist. If the formation water contains Magnesium, the pH of the mud will rapidly decrease to 9.5 or slightly below due to precipitation of Magnesium hydroxide.

Effects on Mud Properties

A chloride increase of a few thousand mg/l will normally cause a mud to flocculate, resulting in increased YP, gel strength, and fluid loss. The degree of flocculation will depend on the active solids content of the mud and whether enough thinner is present to minimize the tendency to flocculate. The salt will also cause dehydration of the clay particles. Dehydration is slower than flocculation and may take a day or more before the clay particles completely shrink to a new equilibrium degree of hydration. The result of dehydration is a reduction in plastic viscosity and an increase in fluid loss. If a mud has been treated with Lignosulfonate before the addition of the salt, the clay platelets are protected from the dehydrating effects of the salt and very little increase in fluid loss will occur.

In summary, the effects of increased salinity on mud properties are largely due to flocculation and dehydration of the active clay solids. These effects are decreased by increased concentration of Lignosulfonate in the mud. They are also decreased by decreased active clay solids content. A mud containing a minimum amount of clay solids and 4 to 8 lbs./bbl of Lignosulfonate will tolerate an increase in chlorides of several thousand ml with minimal effects on the mud properties.

Treatment

Unlike the other contaminants, salt cannot be chemically precipitated. The only way the salinity can be reduced is by dilution with fresh water. This means that the mud treatment must be designed to treat the effects rather than eliminate the presence of the salt.

If a large amount of salt is to be drilled, the mud will continue to increase in salinity and keep the mud properties in a state of continual change. To avoid this, the mud should be converted to a salt mud. The dilution will often be needed to reduce the active clay solids content. Then the mud should be treated with 4 to 8 lbs. /bbl of Lignosulfonate and salt added, until the mud is nearly saturated with salt (chlorides 150,000 to 180,000 mg/l). Pre-hydrated Bentonite can be added as needed to increase the yield point and decrease the fluid loss. If the bottom hole temperature is less than 250°, starch may be added for additional fluid loss control.

In a salt-saturated mud, maintaining stable mud properties may become difficult at temperatures above 300°F. This is due to precipitation of the Lignosulfonate by the salt. Even small amounts of Calcium will compound the problem. In case Calcium is present in the mud, it should be removed and fresh Lignosulfonate added. Extra effort should be made to maintain a low solids mud.

CARBONATES

The Carbonate family consists of carbon dioxide ($C0_2$), bicarbonate (HCO_3^-), and Carbonate (CO_3^{--}). Equilibrium exists in solution among these three components depending on the pH of the solution.

If any one of these three components are added to a water solution, it will be partially or completely converted into the other two components, depending on the final pH of the solution. For example, as $C0_2$

is dissolved in water, it begins to convert into the HCO_3^-. If the pH of the water increases to 8.3 by the addition of hydroxyl ions (OH^-), the CO_2 will be completely converted to HCO_3^- Further addition of OH^-, increasing the pH, will begin to convert HCO_3^- to CO_3^{--} At a pH of 10.3, there will be approximately equal concentrations of HCO_3^- and CO_3^{--} and at pH 11.3, there will be about a 10/90 ratio of HCO_3^- to CO_3^{--}. In this manner, the concentrations of HCO_3^- and CO_3^{--} can be increased as more CO_2 is dissolved in the water. Carbon dioxide can enter the mud from the atmosphere (aeration), from the formation, or from thermal degradation of organic mud additives such as Lignosulfonate or lignite. Carbonates can also enter the mud because of direct additions of soda ash (Sodium Carbonate) or Sodium bicarbonate or by caustic extraction of some Carbonate minerals that are present in some Barites.

Regardless of how these ions enter a mud, they can accumulate to a concentration that will cause the mud to flocculate. The reaction with clay solids that causes them to flocculate is not completely understood. The HCO_3^- seems to be the primary cause of the flocculation. The concentration that will cause flocculation is dependent on the amount of active solids in the mud and the temperature. A high active solids content and high temperature both promote flocculation and will increase the degree of flocculation induced by any contaminant. In some situations, the first sign of flocculation may occur at HCO_3^- concentrations as low as ten milli-moles per liter (610 mg/l).

Unlike others chemical contaminants, the flocculation caused by Carbonates is irreversible by chemical treatment with Lignosulfonate. If caustic is not added with the Lignosulfonate, a reduction in pH will occur, causing more HCO_3^- to convert from CO_3^{--}, more severe flocculation may occur.

Diagnosis

The diagnosis of a soluble Carbonate problem involves either evaluation of alkalinity test results or a direct measurement of Carbonates using the Garrett Gas Train (GGT). Often the GGT measurement is not available and an evaluation of alkalinity data must be used to identify the Carbonate problem.

The Pf alkalinity is the ccs of 0.02 normal sulfuric acid required to reduce the pH of 1 cc of filtrate to 8.3. In the titration, the acid is consumed in neutralizing the hydroxyl ions (OH^-) and in converting the CO_3^{--} ions to HCO_3^- ions. As a result, Pf alkalinity is a measure of the combined OH^- and CO_3^{--} concentrations in the filtrate. The pH is a measure of the OH^- ion concentration.

Since all the methods of Carbonate analysis using alkalinity measurements have certain limitations, it would be wise to cross check the results using at least two of the methods. Large discrepancies may mean inaccurate pH and alkalinity measurements and require rechecking these properties. In general, the pH, Pf Method will be the most accurate followed by the P1-P2

The most accurate method for determining Carbonate content of a mud is with the Garrett Gas Train. The GGT was originally developed to determine sulfide content of a mud. By using a Drager tube designed to measure CO_2, it can be used to determine Carbonate content. In the test, acid is added to a filtrate sample, converting all the Carbonates to CO_2. The CO_2 is collected and metered through the Drager tube. From the darkened length of the tube, the total Carbonates (CO_3^{--} + HCO_3^-) can be calculated.

The advantage of this method is that it is a direct measurement of Carbonates and does not require pH and alkalinity measures and interpretations. It does not differentiate between CO_3^{--} and HCO_3^-. Since HCO_3^- appears to be more harmful to mud properties than CO_3^{--}, it may be desirable to estimate the HCO_3^- concentration by use of the theoretical equilibrium that exists between CO_3^{--} and HCO_3^- at the pH of the mud.

Effect on Mud Properties

The effects of excessive soluble Carbonates on mud properties are basically due to flocculation of the mud. This causes increases in the yield point, gel strength, and fluid loss. These effects are normally more pronounced, down-hole at elevated temperatures than at the surface. Consequently, the flow properties of the mud at the flow line will be increased over those at the suction. The first indication of a problem at the surface is usually an increase in the ten-minute gel strength. The mud will look good, as-long as it is stirring, but will quickly form a fragile gel when sitting still. As the problem develops, the HTHP fluid loss will increase and eventually the yield point, initial gel strength, and API fluid loss will all increase. It should be remembered that the mud may be quite flocculated down the hole before severe flocculation occurs at the surface. This can be observed in increased pressure required to initiate circulation.

In addition to causing flocculation, the Carbonates may also convert Calcium-base clays to Sodium-base and thereby increase their yield. This will have the same effect on the mud properties as the addition of more Bentonite, causing an overall increase in the thixotropic nature. This effect will be more pronounced in the mud having high concentrations of active drilled solids.

A Carbonate problem is often accompanied by a reduction in pH or an increased consumption of caustic to maintain a constant pH. This is due to CO_2 being dissolved in the mud or from the process of extracting soluble Carbonates from certain Carbonate minerals in the Barite. This effect will be most pronounced in the "bottoms-up" mud due to the increased temperature down hole.

The Calcium content of a mud can decrease to zero as Carbonates enter the mud due to precipitation as Calcium Carbonate. Some amount of Calcium is usually indicated by the total hardness titration even in the presence of high Carbonate concentrations. A total hardness as high as 150 ppm may be measured in muds containing Carbonates. This is probably due to Calcium carried by the lignite and Lignosulfonate as a complex ion or compound and not Calcium ions in the filtrate that are free to react with the Carbonate.

Treatment

The effects of flocculation caused by Carbonates cannot be decreased by treatment with Lignosulfonate or other de-flocculants. Dilution is also usually of little or no benefit, especially if contaminated Barite is the source of the problem. This necessitates treating the actual problem rather than its effects. Removal of Carbonates involves the addition of a soluble form of Calcium to precipitate Calcium Carbonate. Usually, lime or gypsum is used for this purpose.

Calcium precipitates CO_3^{--} as Calcium Carbonate (Ca $C0_3$). Calcium will not precipitate HCO_3^- unless it is first converted to CO_3^{--}. If gypsum ($CaSO_4 \cdot 2\,H_2O$) is added to a mud that contains both CO_3^{--} and HCO_3^-, it will precipitate only the CO_3^{--}, causing a reduction in pH and leaving in solution all the original HCO3. Enough caustic must be added with the gypsum to convert all the HCO_3^- to CO_3^{--} to precipitate both the CO_3^{--} and HCO_3^-.

$CO_3^{--} + HCO_3^- + OH^- + 2Ca^{++} \rightarrow 2Ca\,CO_3 \downarrow + H_2O$

If a mud containing Carbonates has a pH that is lower than desired, lime can be added to precipitate Carbonates and increase the pH.

Even at this treatment level, the pH of the mud will increase as lime is added. Reaction of one lb./bbl of lime with HCO_3^- will produce ~ 0.5 lb./bbl of caustic. Normally, fresh Lignosulfonate treatment is required after any precipitation takes place due to physical extraction of Lignosulfonate from solution by the precipitant. Since Lignosulfonate is an acid material, it will help in preventing the pH from increasing. Approximately 3 lbs. /bbl of Lignosulfonate is required to neutralize the 0.5 lb. /bbl of caustic resulting from 1 ppb of lime reacting with HCO_3^-. Obviously, large amounts of Lignosulfonate will be required to keep the pH from rising to undesirable levels if a large amount of lime is required. In this case, it would be better to switch to gypsum additions or to a combination of gypsum and lime

Addition of lime or gypsum to a mud should happen slowly. When additions are made rapidly, the lime will react with the clay solids before they are used up in the Carbonate reaction. The total calculated amounts should be added over several circulation cycles when large amounts are being added. Mud checks should be made between incremental additions and a new calculation of concentration of Carbonates made as a safeguard against miscalculation.

There is no need to add Lignite while treating a Carbonate problem since lignite will also react with the lime or gypsum and interfere with the treatment.

It is certainly advantageous to detect and treat a Carbonate problem in the early stages, even before significant increases in flow properties occur. When 10 millimoles/l of HCO_3^- are indicated,

0.25 lb./bbl of lime can be added without risk of harming the mud. If a Carbonate problem is being caused by CO_2 that is generated by thermal degradation of Lignosulfonate and lignite, small daily additions of lime will be required. This is very likely to occur in wells that have a bottom-hole temperature of over 300°F. The pH of the mud should be maintained between 11.0 and 11.5 so that most of the Carbonates that do exist in the mud will be in the form of CO_3^{--}, rather than the more harmful HCO_3^-. By maintaining the pH with lime and caustic, the total concentration of Carbonates can be minimized. Overtreatment with lime will not occur, if the pH does not exceed 11.5 since excess lime cannot exist at a pH of 11.5 or below. Under treatment with lime will result in an increasing trend in Mf-Pf. This will call for an increase in lime additions. Careful attention to prevent a Carbonate problem is easier and considerably cheaper than the difficult task of eliminating a severe Carbonate problem.

CONTAMINANT: WATER FLOW IN SOBM

Indicator	Treatment
Decrease in O/W ratio (water intrusion)	Increase mud weight
Decrease in mud weight	Increase mud weight
	Add primary emulsifier and wetting agent
	Add lime (conventional system)
	Add oil to adjust O/W ratio
	Add salt to adjust brine salinity

In extreme cases, the contamination of water in an Oil base mud can lead to complete breakage of emulsion thereby causing separation of solids and conversion of Oil base mud into a thick slurry. This thick slurry can sometimes be so viscous that it can cause blockage of circulatory passage. This is one of the limitations of the use of oil base mud in rank exploratory wells where such situation can occur. The table above describes the problem of water flow in oil base mud and the suggested remedial measures / preventive measures to be taken to mitigate/prevent this problem. Such occurrences are not common and in isolation, they do not qualify for not using SOBM.

Preventing SOBM from adverse effect of water intrusion is to over-engineer the system to higher emulsion strength to withstand minor contaminations without too much deterioration. While drilling in areas notorious for water flow, the ES should be maintained ~ 1000. In addition to it starting with higher OWR also minimizes the re-conditioning process, should there be a minor intrusion of water.

CO2 CONTAMINATION IN SOBM

Indicator	Treatment
POM decrease	Increase mud weight
	Add lime
Rheology increase	Add primary emulsifier and wetting agent
	Add oil for dilution
	Increase mud weight
	Add lime

H2S (Hydrogen Sulfide) CONTAMINATION

Hydrogen sulfide is by far the most severe contaminant due to its potential unsafe nature, leading to fatality in extreme cases. That is the principle cause of dedicating an exclusive discussion on Hydrogen sulfide management, as below.

Managing Mud Engineering Operations While Exploring / Exploiting Hydrocarbons in H2S Environment

Hydrogen sulfide is a naturally occurring gas that is produced along with natural gas and crude oil. It is a colorless gas with a composition of two parts Hydrogen and one-part Sulfur. H_2S can also be produced by decomposition of organic matter in drilling fluid or water.

H_2S forms during sulfate reduction via the generalized formula

$$2CH_2O + SO_4^{2-} \longrightarrow H_2S + 2HCO_3^-$$

Where,

CH_2O = Organic matter,

SO_4^{2-} = Sulfate, and

HCO_3^- = Bicarbonate or alkalinity.

Despite their common presence in marine sediments, H_2S concentrations do not usually reach very high levels. Relatively low concentrations of H_2S are found primarily because either the reactants (organic matter and/or sulfate) may be limited or the iron present in marine sediment combines with the sulfide to form iron sulfide minerals (e.g., Pyrite).

Properties

i. Slightly heavier (1.2 times) than air; it seeks lower areas, especially pits and cellars.
ii. Odor of rotten eggs in small doses. Higher concentrations cause paralysis of the olfactory nerve within 60 seconds, because of which, odor cannot be detected.
iii. Extremely toxic (poisonous).
iv. Explosive in air (flammable limits 4.3 – 45% by vol.)
v. Soluble in water (10.672 g/100ml H_2O)
vi. Boiling point -75° F
vii. Melting point -119°F
viii. Density of liquid 0.790 @ 60°F
ix. API gravity 47.6
x. Reacts with most metals to form sulfides, which are usually insoluble precipitates.
xi. Dissolves in water to form a weak hydro-sulfurous acid.
xii. It extinguishes the original flames, but burns itself with blue flame.

1. Health Effects Of Hydrogen Sulfide Overexposure

1. If steps are not taken to control exposure, continued inhalation of Hydrogen Sulfide, Hydrogen Sulfide could result in:
   i. Loss of the sense of smell
   ii. Dizziness, unconsciousness leading to death if exposed for a long time
2. Low concentration exposures (under 10 PPM)
   i. In low concentrations, Hydrogen Sulfide can be detected by its odor; the smell cannot be relied upon to forewarn of dangerous concentrations, because it rapidly paralyzes the sense of smell. A longer exposure to the lower concentrations may result in the loss of the sense of smell.
   ii. Symptoms from repeated exposure to low concentrations usually disappear after being removed from the exposure.
3. Higher concentration exposures (10 PPM and above)

i. Concentrations that are prolonged (and continuous) or of high concentrations may lead to death.
ii. It should be well understood that the sense of smell will be rendered ineffective by hydrogen sulfide, which can result in an individual failing to recognize the presence of dangerously high concentrations. Exposure to hydrogen sulfide causes death by paralyzing the respiratory system.

2. Preventing Exposure

All the control methods discussed below are essential to minimize additional sources of Hydrogen Sulfide absorption from inhalation. Strict compliance with these provisions can virtually eliminate several sources of Hydrogen Sulfide exposure that significantly contribute to excessive Hydrogen Sulfide absorption.

1. Review the site-specific safety programs as well as the site emergency action plan.
2. Ventilation systems may provide for venting of the Hydrogen Sulfide vapor prior to entrance into the area.
3. Confined Space Entry Procedures will greatly reduce the hazards to the crew and should be followed whenever an entry into a confined space is required.
   - The area is to be evacuated and reevaluated prior to reentry.
4. Gas detection equipment shall be used whenever an entry into an area which may contain hydrogen Sulfide vapor.
   - Equipment to be operated as per the manufacturer's instructions.
   - Detection equipment to be calibrated prior to use as per the manufacturer's instructions.
   - Continuous monitoring to be exercised when Hydrogen Sulfide has been detected.

3. Detection of H2S in Mud and Ambient Air

It is not professionally a sound practice to depend upon smell to establish presence/absence of H_2S. While drilling in H_2S prone areas, it is a good practice to use Lead acetate paper to check the qualitative presence / absence of H_2S by soaking the lead acetate paper with filtrate of the mud. If the same turns black, presence of H_2S is indicated. Lead acetate paper turns black even if it is kept exposed to H_2S environment, but this can take longer than desired time to detect H_2S.

4. Measurement of H2S

From a safety standpoint, any rig drilling in the potential H2S zone should be equipped with H_2S monitors and alarms. The most reliable and accurate means of measuring the sulfide content in solution in the fluid is with a Garret gas train apparatus.

GGT Procedure for measuring sulfide content:

GGT apparatus is proprietary product of a specific company and details of the same can be found on the company's website.

5. Adverse Effect of H2S on Mud Properties and Parameters

Since H_2S is a charged molecule its impact on the mud will depend upon the type of mud which is contaminated with H_2S. In SOBM the inherent component Lime has the additional capability to dissolve H_2S and thus SOBM is by default H_2S proof to some extent. Nevertheless, it still needs additional scavenger because with a continuous influx of H_2S ingression from well-bore, the lime content keeps depleting and a stage is reached where there is no more free lime available in SOBM.

H_2S is also soluble in continuous phase, but the degree of solubility changes in temperature and pressure. Although intrusion of H_2S is not going to pose any immediate threat to safety, but if left un-attended, it can cause serious damage to personnel on board the rig. The most commonly used H_2S Scavenger is Zinc Carbonate. The dose of scavenger depends upon the severity of influx and could range between 1 to 3 ppb. The reaction rate with zinc Carbonate depends on the mechanical shearing; therefore, shearing should be imparted while mixing this product to maximize the coverage area of reactant.

6. Mitigating adverse effect of H2S in mud

The standard procedure followed on rig during H_2S show is as follows-

While drilling in a known H_2S prone area as a measure of precaution the lime content of the SOBM is purposely kept in the range of 8 to 10 ppb and the plan is – at the first sign of H_2S show the mud shall be treated with lime to maintain at least 7 ppb dose at any given time. With the dissolution of H_2S in Lime the available lime keeps dropping as the Lime is converted into Calcium sulfide. The Calcium sulfide remains in the system and can convert back into H_2S and Lime depending on the alkalinity of the media, as this is a reversible reaction. This will become clear with the following equation

Ca O (Lime) + H_2S = Ca S (Calcium Sulfide) + H_2O (Water)

Ca S + H_2O = H_2S + Ca O

Therefore, it is not a correct practice to exclusively depend upon Lime only. Additive Zinc Carbonate reacts with H_2S and forms Zinc sulfide. This reaction is irreversible and hence there is no threat of Zn S converting back to H_2S regardless of changes in physical conditions of mud. This reaction is represented by the following equation

Zn CO_3 + H_2S = Zn S + H_2O + CO_2

Carbon di-oxide gets diffused and hence the reaction is irreversible in nature. Therefore, the H_2S is destroyed. The dose of H_2S scavenger is dependent upon the ppm level of H_2S. Field practices have shown that for every ppb of $ZnCO_3$ can handle 500 ppm H_2S safely.

A point to remember: if the mud is pre-treated in anticipation of the possible influx of H_2S from the hole, then we run into the risk of not being able to detect H_2S show on the surface. Conversely, if we do not treat the mud with H_2S while drilling in an area where the possibility exists of the presence of H_2S then we are exposed to the risk of H_2S show at the surface. Therefore, to strike a balance between both the situations, it is preferred to keep pH on the higher side as a first defense against H_2S and keep $ZnCO_3$ handy near the hopper for immediate mixing, should there be a drop in pH or any other sign indicating show of H_2S on the surface.

Normally the pH to be maintained during drilling in the H2S prone area is 10.5
Recommended Treatment: Basic ZnCO3 – 2 to 3 ppb (1ppb = 500mg/L)

Touching upon WBM the best defense is to keep the pH of mud > 10. This ensures that H_2S will not be present in the mud in the Free State. However, H_2S being acidic in nature, will drop the pH of WBM and the same be maintained with the dose of NaOH and/or Lime to a value of > 10 always. In addition to maintaining pH, >10 additive Zinc Carbonate should also be used.

Maintaining high pH with the combination of Lime and caustic soda is a preferred option. H_2S contamination treatment should be planned and materials procured in time to ensure that rig is equipped with all contingency measures to manage continuity of operations in H_2S environment, as far as managing mud is concerned. The lime reacts with H_2S to form Calcium sulfide, which is only partly soluble whereas caustic soda reacts with H_2S to form Sodium sulfide, which is completely soluble in water, causing it to release H_2S when pH drops to <8.0

When testing for presence of H_2S, it is extremely important that the filtrate used is as fresh as possible. In addition, the filtrate should be taken from a fluid sample freshly collected from flow line. Minimize exposure of fluid/filtrate to atmosphere as this reduces the accuracy of the measurement

Regardless of the type of mud if hole conditions permit, it is a preferred option to raise the mud weight to stop the intrusion of H2S in mud. Apart from it the tubulars also need to be coated with suitable corrosion inhibitors @ 3-4 gal/bbl to prevent them from corrosion.

Managing H2S in Brine:

By and large, high pH always helps in negating the impact of H_2S intrusion to an extent and same holds good for brine also. Apart from it every mud company has their own package of H_2S scavenger and corrosion inhibitor specially designed for brine. Since brine is used during completion/well testing, by the time we reach this stage we already get enough indication about the presence of H_2S and hence brine can be pre-treated with H_2S scavenger if H_2S presence is confirmed.

To safeguard the tubular from H_2S corrosion, it is always a safe practice to select the tubing of that metallurgy which resists H_2S corrosion. In addition to it, the brine can also be treated with suitable corrosion inhibitor to further prevent the tubular from probable attack of H_2S

under bottom hole conditions. While selecting suitable corrosion inhibitor, caution should be exercised to select the type of inhibitor which not only negates the impact of H_2S, but also does not adversely affect any other component of the rig.

Managing H2S concentration below the safety limit during well testing:

This is relatively less spoken part of H_2S management. The crude oil, which flows to surface during well test operations may contain very high concentrations of H_2S and the same cannot be transported by road to other locations unless the H_2S dissolved in crude is brought down to a limit considered safe by local regulatory authorities for transportation. Since there are not adequate resources /equipment of such type (which can treat outflowing crude for eliminating H_2S contamination) available on rig during flowing of well for testing purposes, it is often not a simple exercise. The selection of the right type of H_2S scavenger is of utmost importance as the crude flowing during well test is also associated with water which contains dissolved H_2S. A simple approach is to administer the dose of caustic soda at the choke manifold with the help of a chemical injection pump before the crude enters separator. However, Caustic alone may not be sufficient and either an amine type of scavenger or a combination of both is a much preferred option. The toughest challenge in such cases is to match the dosing rate of chemical with the flow rate in such a way that the chemical reaction is uniformly effective throughout the tenure of the flow. In combination with chemical dosing, increased temperature and additional settling time in storage tanks before loading in tankers for onward transportation also helps in achieving the objective of <20 ppm dissolved H_2S in crude, considered safe for transportation in tankers. This has been achieved on one of the rigs during well testing. The challenge was to devise a treatment tank along with mixing system and hooking up the same with injection pump on choke manifold to facilitate uninterrupted injection of the chemical. The diagram below shows a schematic of the treatment modalities devised on the rig to continuously treat the outflowing fluid, (including oil and water) to bring down the H_2S concentration in the desired range of < 20 ppm. The system was so efficient that the final level of H_2S came down to close to Zero ppm at the loading point of tanker for onward transportation.

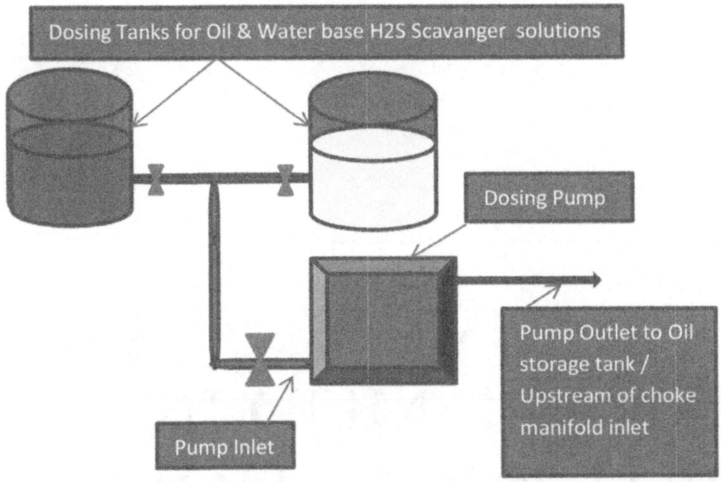

SCHEMATIC TO DISPLAY THE TREATMENT ARRANGEMENT FOR ELIMINATING DISSOLVED HYROGEN SULFIDE FROM THE FLUID FLOWING TO SURFACE DURING WELL TESTING.

Corrosion

H_2S can initiate metal corrosion by reacting the with metal. H_2S can also cause metal embrittlement caused by penetration of H_2 in the metal. The attack mechanism is complex. H_2S is a weak acid when dissolved in water and can act as a catalyst in the absorption of atomic hydrogen in steel, promoting sulfide stress cracking (SSC) in high strength steel.

In an environment with H_2S corrosion the most common types of corrosion include uniform corrosion, corrosion fatigue, sulfide stress cracking, hydrogen blistering, hydrogen embrittlement etc.

The literature survey reveals that in sporadic instances, tubulars were externally coated with corrosion inhibitor. It is not clear whether this alone is enough - because invariably on all such occasions, mud/brine is also treated with additional doses of corrosion inhibitors. The treatment level and various chemicals are already discussed in preceding chapters.

As a guiding rule, maintaining a high pH and overbalance on rock not only helps to minimize the probability of adverse effects of H_2S but also minimizes the show of H_2S on the surface. For various technical reasons, it may not be possible to have enough mud weight to eliminate show of H_2S on the surface and hence taking all other precautions is a better way of managing H_2S corrosion.

CHAPTER 9

EFFECTIVE WAYS TO MANAGE MUD IN HIGH ANGLE HOLE

INTRODUCTION

The successful completion of the deviated or horizontal wells is partly dependent on the ability of the drilling fluid to maintain stable hole-condition and remove the cuttings from the well bore at the high angle.

The functions a drilling fluid performs, which are severely compromised by drilling highly deviated and horizontal wells are:

1. Hole-cleaning
2. Bore hole stabilization
3. Prevention of stuck pipe

The most important of these is hole-cleaning, which has a direct effect on the other two.

The fluid requirements for effective hole-cleaning in high angle and horizontal wells differ from those in a vertical well.

Hole-cleaning efficiency in a vertical well is related to the particle slip velocity versus the annular velocity. This is a very simplified approach, not considering particle interaction, hole geometry or particle geometry. The results obtained are qualitative and as a thumb rule, hole-cleaning in vertical wells is not a problem.

In high angle and horizontal drilling, the direction of cutting's slippage is perpendicular to the direction of fluid flow. Various other forces also start to come into effect such as that due to pipe rotation. What is more important is - we now have an eccentric annulus due to the pipe no longer being in the center of the well. These factors can adversely affect hole cleaning efficiency of the mud, if not taken into consideration while drilling an inclined hole.

The most important factor which complicates hole-cleaning in deviated wells is the formation of a cutting bed on the lower side of the well. Obviously, as the well becomes more deviated, the tendency to form this cutting bed also increases considerably.

The dynamics attached to this cutting bed are related to the flow regime and deviation. Essentially problems develop with the cutting bed because:

A. It becomes too thick due to poor hole-cleaning. During a survey or connection drill-pipe lying on it can get stuck.
B. The cutting bed formed in the hole or casing section with an angle of between 40 (to 65) slips when the pumps are turned off. This results in deviation increased drag on trips or connections and can even lead to being stuck in the casing.

Obviously, reducing cutting bed size by improving the hole-cleaning should be considered to avoid the related problems.

FLUID DESIGN

Before selection of a fluid system, it is always preferred to list the drilling problems generally encountered in this field and the objectives we are trying to achieve.

We know that fluid viscosity has little effect on cuttings removal in turbulent flow regimes, so, no optimization should be done here. However, with annular velocity in the range of 65 m/min +/- and / or Reynolds number ~ 1900+/100, it is possible to design a fluid which will exhibit good viscosity in laminar flow and tend to move into turbulence quite readily. Such a fluid is a shear thinning fluid.

The associated problem of solids settling, while and after the pumps are shut down, is particularly significant at high angles of inclination. The Driller often removes moderate accumulations of settled solids by washing and reaming to bottom after trips.

Clearly, one function which the drilling fluid must fulfill is that of suspending both drill solids and weight material at low and zero shear rates. Development of gel structures which can lead to surge and swab pressure should nevertheless be avoided.

Chemical stabilization (or more correctly, avoidance of destabilization) of the well bore cannot be ignored. As in the case of settling, the consequences of failure to minimize destructive chemical interaction between the drilling fluid and the formation become more severe as deviation becomes high.

The suspension should not be achieved at the cost of developing the gel structure of a type likely to induce surges and swab problems. This essentially means that the fluid should can provide higher fragile gels.

Field experience has led to the following empirically derived, generalized scheme for the design and application of drilling fluids for use in high angle well bores.

1. Fluids should show high viscosity at low shear rates. This behavior should persist under the most adverse conditions of temperature and pressure applying in the well. **(Measurement of the rheological performance of the fluid under various conditions of temperature and pressure, culminating in bottom-hole conditions, is recommended)**
2. Fluids should not produce troublesome gels. It is important to ensure that gels are fragile, so that surge and swab problems are not introduced.
3. Fluids should show inhibitive, neutral or stabilizing interaction with the formation.
4. Maximum pump rates should be used. This is not specifically an attempt to induce turbulence, which is not felt to be a necessary element in cleaning even lateral sections. This point is raised

rather, in recognition of the fact that annular velocity plays a role in hole-cleaning.

5. The possible formation of cuttings beds should never be discounted. If the cutting bed does form, a change in approach will almost certainly be required to remove the bed. This is one instance where the use of turbulent flow is recommended. A brine pill, pumped as fast as possible, should be used to stir the bed up and bring it out into the general fluid flow. A plug flow (or very highly viscous) pill may be pumped immediately behind the turbulent pill to carry the new mobile cuttings to the surface. The pipe should be worked and reciprocated throughout and, in all cases which warrant, jarring may be used. Since mud weight also has a contributing positive role in hole-cleaning, it is desirable that the weight of the low vis pill (brine pill in this case) is same as circulating mud.
6. The engineer should not be guided by quantitative calculations of Reynolds numbers nor of critical velocities. Poor hole-cleaning of the fluid should direct the engineer to work on improving the low shear rate rheology of the fluid (in conjunction with rising the pump rate, if possible).

FLOW RATE

Increasing the flow rate improves the hole-cleaning, be it laminar or turbulent flow. Improving the hole-cleaning obviously lessens the cutting bed. Laminar flow, by its definition, implies that the fluid velocity at the well bore tends to zero. The velocity increases towards the center of the annulus. The shape of this profile is dependent upon the flow rate and rheological properties of the fluid. As the velocity of the fluid at the well bore tends to zero, particle settlement takes place, a cutting bed builds up, eventually leading to stuck pipe.

Turbulent flow is by far the best means of hole-cleaning. However, its use can cause hole washouts.

The restrictions on flow rate are obviously pump capacity, pressure limitations and ECD. It must be remembered that ECD will increase when drilling horizontally whereas formation strength will not!

Rheology

Increasing the yield point increases the hole-cleaning potential of a fluid in vertical well sections. In a deviated well, this may not be the case. Once the angle of deviation is over 40° Higher YP may cause increased thickness cutting bed. Decreasing the YP to allow turbulent flow to take place is the way to improve hole-cleaning. However, as said previously hole washouts may occur, also larger annulus may not have turbulent flow.

A successfully field-tested solution to this problem is to pump a 50 to 70 bbl low viscosity pill. These can either be weighted up with salt or have spacers to ensure the correct hydrostatic head is maintained. Localized turbulent flow is thus generated which stirs and removes the cutting bed build up.

DECREASED RATE OF PENETRATION

This causes a lower concentration of cutting to be in the fluid, thereby reducing inter - particle reaction and collisions hence improving hole-cleaning.

CUTTING SIZE

The larger the cutting size, the more difficult to remove. Cutting size is dependent upon bit size, fluid viscosity and formations drilled. These should be considered in well hydraulic plans.

FLUID DENSITY

The mud weight required to drill a hole is selected to give a hydrostatic pressure adequate to prevent collapse of the well bore due to mechanical stability problems caused by the large imbalances between the rock stress and the pressure exerted by the mud on the walls of the well bore.

With inclination, the pressure required to prevent bore hole-collapse is increased and consequently a higher mud weight will be required. The difference between the pressure required to prevent collapse and the fracture pressure is reduced and the correct selection of mud weight is therefore critical.

DEVIATION

Bore-hole geometry can also play a big role in reducing or increasing the problems associated with drilling high angle wells. The greater the deviation, the smaller the vertical vector of fluid flow, thereby reducing hole-cleaning. From this aspect, it is advisable to build angle as rapidly as possible and as late as possible.

From a mud engineering standpoint, it is obvious that significant improvement in fluid engineering cannot provide the same overall gains in cutting removal, that is easily achieved with variation in directional drilling programs.

Effect of Inclination on Particle Transport

Studies have shown that the inclination of the well has significant influence on the motion and distribution of drill cutting in the hole. In addition to the hole angle the annular velocity and rheological properties of the drilling fluid also affect the behaviour of the cuttings.

As the inclination increases, there is a tendency for cuttings to accumulate a low side of the annulus and this eventually leads to the formation of cuttings' beds.

Four regimes have been identified as shown in Table 1.0. The exact angle at which the cuttings bed form, depends primarily on the annular velocity as shown in the figure below

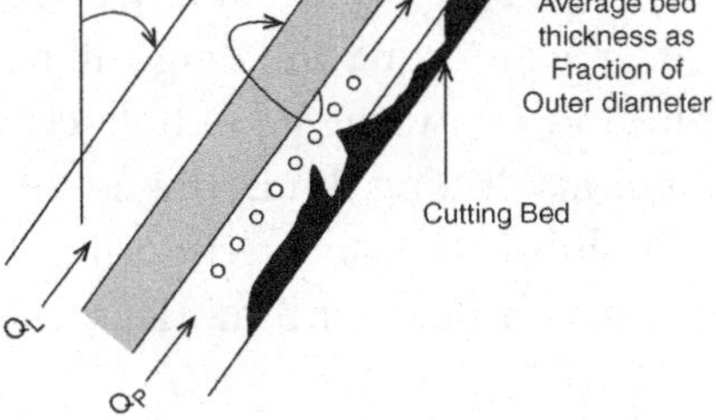

Representation of inclined annulus with cutting bed

TABLE I

S. No	Hole angle	Comments
1	10 deg or below	No issue, It's a vertical well
2	11 to 35	Cutting bed starts building
3	36 to 60	Very critical. Sever cutting bed is formed which slips when pumps are shut
4	61 to 90	Cuttings bed present but doesn't slip as much as above

FLUID LOSS CONTROL

In deviated holes, factors that indicate that a tight fluid loss should be maintained with a lubricious filter cake are:

1. The drill string lies against the low side of the hole; hence the chance of differential sticking, excessive torque and drag are increased.
2. High angle wells tend to require a higher mud weight, further increasing the chance of differential sticking.
3. A high fluid loss can result in lubrication of the fracture planes in the near well bore and consequent collapse of the well.
4. In permeable sandstone, it is possible for the mud filtrate to leak-off into the formation and increases the pore pressure. This results in a decrease in fracture gradient which in extreme cases can lead to whole mud losses.

LUBRICITY

The high drill string - well bore contact and the build-up of cuttings bed means that a high degree of torque and drag will be experienced in high angle wells. In addition to ensuring good hole-cleaning it is important to use a highly lubricious drilling fluid. The use of an oil based drilling fluid will provide the lubricity characteristics necessary. If water based fluid is to be used, select a fluid with lubricity characters, comparable with SOBM.

Glycols, a special class of poly-ol group of compounds often provide additional lubricity and aid in shale inhibition by minimizing pore-pressure transmission both, by clouding and non-clouding mechanism.

INHIBITION

In long deviated wells, the inhibiting nature of the fluid becomes relatively more important for several reasons:

1. The hole is open for longer periods. Hole-stability has consistently been shown to be time dependent. The combination of poor hole-cleaning and sloughing or swelling shale caused by insufficient inhibition may result in the hole being lost.
2. Gauge hole is important in maintaining directional control; a fluid without sufficient inhibition can result in over gauge hole due to dispersion.
3. Insufficient inhibition will result in a build-up in solids in the fluid requiring high dilution rates to control. A build-up of fines can result in problems controlling the fluid loss and will adversely affect the rheological properties of the fluid resulting in a less shear thinning fluid with a poorer hole-cleaning capacity and higher circulating pressure losses.

Barite sagging:

It is a special type of situation which is invariably found only in deviated holes and hence it deserves a special mention here.

Sagging of Barite or other solids of mud is out of the result of the poor thixotropic feature of mud because of which, the solid particles tend to settle at the bottom part of the mud column there by creating dissimilar patches of mud with varying densities. The top most part of mud is the lightest and the bottom-most part weighing the highest.

Due to higher lubricity, this problem is more common in oil base mud and is further compounded in inclined holes. If left unattended, it can result into severe well-bore instability, which can cause rig NPT.

Although it is not a rule, normally sagging is observed in the mud which has weight ranging between 10 to 14 ppg. This is, because these mud weights have enough solids to initiate sagging and yet they also have enough void space in the mud itself to allow the solids to sag. For mud weights beyond 14 ppg. Normally the amount of solids is so high that even if mud does not have adequate thixotropy it is so tightly packed with solids (~ 500 ppb) that there is hardly any additional void space available for solids to settle down. Similarly, for mud weights below 10 ppg, the amount of solids is not so high that settlement of part of these solids can cause a density difference of that order which can be noticed with the help of an ordinary mud balance.

However, the probability of sagging cannot be completely ruled out in these ranges also and hence adequate measures of imparting minimum low-end rheology in mud must be taken always to eliminate sagging. Needless to mention sagging can often result in promoting stuck pipe when in combination with sudden release of additional cavings or when cuttings do not get dislodged from the hole soon after their generation.

A systematic approach to address / resolve sagging is to first evaluate the extent of sagging for which following procedure is being described below.

Procedure for testing the sagging tendency of mud

Following is the general description of the procedure. For detailed and more precise information, readers are suggested to visit the web-site of the manufacturer.

The procedure of testing sagging tendency of mud is based upon a simple, yet effective procedure involving the use of a Sag shoe. It is a cylindrical disc with a notch at one end, which uniformly descends in thickness to provide conditions which can induce sagging of Barite in mud.

This sag shoe is placed inside the heating cup of the rotational viscometer which is then filled with mud under question and allowed to equilibrate with the heating cup temperature which has been preheated to 125 Deg F. After uniform shearing with the help of selected viscometer speed, the bottom most part of mud near the descending notch is extracted with the help of a syringe and weighed after transferring in a 20-ml retort cup (which has been pre-weighed empty along with lid). This procedure is repeated twice and the weight difference between empty, the empty mud and weighed mud is compared. The difference in both the numbers is taken as sag number.

Lower the value of sag factor, better is the quality of mud in terms of resistance to Barite sagging. This is an empirical number derived as a ration of two identical unit values, it is dimensionless.

SUMMARY

The drilling fluid plays a crucial role in the drilling of high angle and horizontal wells. Considering offset data and shale analysis studies a suitable drilling fluid and the mud density profile can be selected to ensure stable hole conditions.

Hole-cleaning is more dependent on flow rate than viscosity at high angles. Rheological properties of the mud should be selected to give a flat velocity profile. Turbulent flow will give a flat velocity profile and should be selected where hole washout is not likely to occur. Should turbulent flow be undesirable or unachievable, a highly non-Newtonian fluid should be used to ensure a flat velocity profile. Turbulent flow should be induced by either pumping low viscosity pills or increasing the pump rate if possible to minimize build-up of cuttings' beds.

On deviated wells, the following techniques are advised.

1. If the turbulent flow regime cannot be maintained / achieved pump in laminar regime as fast as possible.
2. Do pump tandem sweep of LO and HI VIS after 40° as this works better than hi-vis alone
3. Build angle as late as possible and land casing as quickly as possible.
4. Prior to logging or running casing thicken mud up to prevent slow down settlement in highly deviated section.
5. Use as low solids mud as possible.
6. Use lubricants if necessary, but be aware of the possible problem of lubricants increasing cuttings bed slip.
7. Be aware of different assemblies being used in deviated section.
8. The use of top drive that allows stands to be pumped and back reamed out of the hole considerable reduces problems that develop due to cuttings bed build-up.
9. Fluids should show high viscosity at low shear rates. This behavior should persist under the most adverse conditions of temperature and pressure applying in the well. **(Measurement of the rheological performance of the fluid under various**

conditions of temperature and pressure, culminating in bottom-hole conditions, is recommended)

10. Fluids should not produce troublesome gels. It is important to ensure that gels are fragile, so that surge and swab problems are not introduced.

It should be stressed that recommendations are made from a drilling fluid standpoint only. There are obviously other aspects to consider e.g. Drilling at a controlled rate, rotating and reciprocating the pipe and all such established good drilling practices, which contribute to the goal of reaching to the desired TD with selected trajectory.

CHAPTER 10

OIL / SYNTHETIC BASED FLUIDS

INTRODUCTION

An oil base fluid can be described as a drilling fluid which has oil as its continuous or external phase and the water or brine, if present, as the dispersed or internal phase. The solids in an oil base fluid are oil wet. All additives are oil dispersible and the filtrate of the mud is oil. The water is in the form of emulsified internal phase.

Invert emulsions are oil muds that are formulated to contain moderate to high concentrations of water/Brine. Water/Brine is an integral part of the invert emulsion and normally contains a salt such as Calcium or Sodium chloride. An invert emulsion can contain as high as 60% of the liquid phase as water/brine. Special emulsifiers are added to tightly emulsify the water as the internal phase and prevent the water/brine from breaking out and coalescing into larger water droplets. These water droplets if not tightly emulsified, can water wet the already oil wet solids and seriously affect the emulsion stability. Invert emulsions are usually tightly emulsified, low fluid loss oil muds. An improvement in drilling rates has been seen when the fluid loss control of the system is relaxed, thus the name "relaxed" invert emulsion. Also, the relaxed invert emulsion fluids do not use as much emulsifier as the regular invert emulsion systems.

OIL-BASE FLUID APPLICATIONS

Drilling

Oil-base fluids allow the drilling of reactive formations more efficiently than water-base fluids.

The all-oil filtrate provides an environment in which water-sensitive clays do not hydrate. More recently, the advent of relaxed filtrate, low-solids, oil-base fluid has resulted in an exceptional drilling rate.

Perforating

Oil-base fluid will not seal off perforations due to non-reactive nature of the continuous phase. Therefore, bullets and jet charges shot in such environment do not form a mushy/lumpy structure, thereby allowing un-interrupted passage to hydrocarbon to flow through. Job.

Work over and Completion

Using oil-base fluid in the open-hole section of wells simplifies operations as the fluid provides maximum lubrication to the borehole, reduces torque and reduces the chances of differential sticking. The procedure aids the operator in landing liners in troublesome holes. The same benefits can also be seen, when running tubing and setting packers.

Coring

Maximum recovery is normally achieved because formation clays do not react with oil-base fluids. The operation is made easier by elimination of core swelling and jamming of the core barrel and differential sticking.

High-Temperature Drilling

Wells with temperatures above 400°F (204°C) frequently use oil-base fluids. The solids/chemical tolerance of oil-base fluids is much higher than that of most water-base fluids. Also, there is less oxidation reduction taking place in an oil-base fluid as the oxygen content is negligible, compared to the water base system. Consequently, certain chemicals/solids are not highly affected by temperature.

Wellbore Stability

Soft shale can be controlled or hardened by exposure to oil-base fluids which have high-salinity water in the emulsified phase. Transfer of water from the shale to the oil-base fluid is attributed to osmotic forces across the semi-permeable membrane around the emulsified water droplets. Care needs to be exercised here by ensuring that the salinity of the rock and mud is not too widely different.

Corrosive Gases

Hydrogen sulfide and carbon dioxide bearing zones, which are hazardous to metal goods and damaging to water-base fluid rheology, are better controlled in an oil environment due to absence of electro-chemical interaction. The oil coating protects the metal from corrosion, and lime or scavengers can be added to control gases, if necessary.

Salt and Contaminating Formations

At temperatures above 250°F (121°C) and depths below 12,000 ft (3658 m), plastic flow of the salt was encountered in several wells. High-weight, oil-base fluids were used to overcome these problems. The KCl and $MgCl_2$ salt sections are not washed out when drilled with an oil-base fluid.

Differential Sticking

Earlier, the application of an oil-base fluid for freeing stuck pipe was discussed. The chance of becoming differentially stuck while drilling with an oil-base fluid system is minimal. The oil filtrate, lubricity and low fluid loss are ideal for preventing wall sticking.

Drilling Directional Holes

Oil-base fluid is more suited for directional work because it helps in maintaining a gauge hole and reduce torque.

Packer Fluids

Oil-base fluids are used as packer fluids because they serve as a stable non-corrosive barrier against corrosive formation fluids. In certain areas with deposits of natural Gas Hydrates, oil-base fluid casing packs are utilized between the strings of casing that pass through the permafrost zone. This provides insulation to the frozen ground from the warm oil being produced.

High Density Fluids

Oil based fluids have a higher tolerance to active and inert solids. Because of this inherent characteristic in combination with surface wetting agents, high density and drill solids tolerant fluids are obtainable with adequate rheological properties.

OIL-BASED FLUID CHEMISTRY

Emulsions

Liquids which are soluble in one another are said to be *miscible.* An example of soluble liquids is water and alcohol. Liquids which are not soluble in each other are *immiscible.* An example would be water and oil.

When immiscible liquids are mixed vigorously, one liquid will be suspended or dispersed as droplets in the other liquid. Such two-phase systems are called *emulsions* (see Figure 1). The dispersed or suspended liquid is called the internal, or dispersed, phase. The other liquid is called the external, or *continuous,* phase.

The most common types of emulsions are oil in water and water in oil. An example of an oil-in-water emulsion is milk. Examples of water in oil emulsions include grease, margarine, and oil-base fluids. The first and most common emulsions studied were oil in water. Water-in-oil emulsions were studied later and sometimes are called *invert* emulsions.

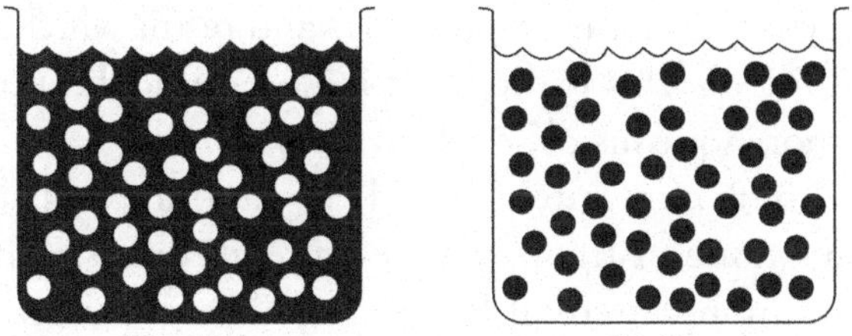

Figure 1 Emulsions

Dispersing one liquid as small droplets in another liquid increases the surface area between the two liquids. The surface energy/tension that exists at the interface between two liquids is called the *interfacial* energy or interfacial tension. This interfacial energy/tension tends to minimize the surface area and work must be performed (energy expended) to increase interfacial surface area and overcome these forces. A mixer can form a mechanical emulsion between two pure liquids, such as water and hydrocarbon, but the emulsion breaks and the two phases separate when the agitation ceases. This de-emulsification is due to interfacial energy/tension minimizing interfacial surface area. To form a stable emulsion, a third component is necessary, that can decrease inter-facial tension between the two liquid phases.

NON-AQUEOUS EXTERNAL PHASE

Base Oils

Hydrocarbon oils are the continuous phase in oil-base fluids. They are non-polar, low-surface energy/tension liquids and interact only weakly with mineral solids. This characteristic is the basis for the use of oil-base fluids as non-reactive, inert drilling fluids. Hydrocarbon oils will not allow clays to swell, which makes them ideal for drilling through hydratable shale.

The most commonly used oils today are synthetics where certain environmental regulations prevail, low-aromatic-content, low-toxicity mineral oils, and diesel oil. Crude oil has been used in the past, but finds little application in today's oil-mud drilling fluids. Crude is relatively cheap, often available, but may need topping to minimize flammability since a flash point greater than 180°F (82°C) is advised.

Crude oil contains native asphaltenes and resins which can interfere with other additives. This is the main reason crude did not gain much popularity as non-aqueous continuous phase.

Diesel oil is a moderately-priced, commonly available distillate which contains none of the native asphaltenes or resins in crude and is the most commonly used oil for mixing oil-base fluids. Diesel based systems are generally used today only in land drilling operations. Offshore drilling operations limit the use of diesel base systems due to the toxicity of the base oil. The aromatics in diesel oil can swell rubber gaskets, seals, and pipe rubbers, an aniline point greater than 140°F (60°C) (the higher the aniline point, in general, the lower the concentration of aromatic) are recommended. Mineral oils are recommended due to lower toxicity relative to diesel oil. Most of the toxicity of diesel oil is associated with the aromatic, particularly the multi-ring or poly-nuclear aromatic. Low-toxicity mineral oils have low to very low concentrations of aromatic and those present are predominantly single-ring. Synthetics are the preferred oil in offshore drilling operations where environmental regulations prohibit the discharge of cuttings and/or whole mud to the sea.

A summary of the ADVANTAGES and the DISADVANTAGES of oil-base (non-aqueous) drilling fluids follows. Unless specifically indicated otherwise, the information contained in this chapter applies to both oil and synthetic base fluids.

ADVANTAGES OF OIL-BASE FLUIDS

1. Shale Stability and Inhibition

Properly conditioned oil-base fluid should have a minimal effect on any shale formation. With correct internal phase salinities, oil fluids provide gauge holes through water sensitive shale combined with long-term borehole stability. This statement is true if tectonic stresses are not a factor.

2. Temperature Stability

Oil-base fluids have application in wells with high bottom-hole temperatures. Oil-base fluids have shown stability in wells with logged bottom-hole temperatures more than 500°F.

3. Lubricity

Torque and drag problems related to deviated well-bore, crooked hole, and sidetracks can be considerably reduced with the use of oil-base fluids. The lubricating properties of an oil-base fluid are a result of the chemical nature of the oil-wetting surfactants and the inherent lubricity of the oil itself.

4. Resistance to Chemical Contamination

Carbonates and salt formations do not adversely affect the properties of an oil-base fluid. CO_2 and H_2S can be easily removed with lime.

5. Gauge Hole in Evaporite Formations

Water based mud are invariably electrically charged entities and have an affinity for reaction/dissolution of inorganic salts like Sodium and Calcium chloride. This results in huge washouts and hole becomes practically unmanageable in terms of being able to case and cement. Due to non-reactivity of continuous phase of SOBM for all such inorganic salts, Oil mud is by default an ideal choice for drilling through such formations.

6. Solids Tolerance

Drilled solids are inert in oil-base fluids and less dispersible other than from that resulting from mechanical shear; therefore, the fluids have a much higher solids tolerance. The key to solids tolerance is less due to volume than it is to particle size and surface area of the solids. Therefore oil-base systems are more tolerant of solids resulting in reduced dilution requirements for solids control.

7. Reduced filtrate invasion

Oil-base fluids normally have a low fluid loss and the filtrate from a stable mud is generally all oil. The filtrate penetration into the formation would be less than that of filter paper as the well-bore surface area increases incrementally with distance from the well-bore. The filtrate from an oil-mud invades only a short distance into the formation production zone and therefore minimizes damage caused by invasion leading to flushing of hydrocarbons away from the well bore.

How much water-free is an all oil HTHP filtrate? We almost always take it for granted that an all oil HTHP filtrate is 100% oil. However, need not and is not always true. Karl-Fischer titration has shown that an apparently looking all oil filtrate can contain water in the form of extremely fine droplets completely inseparable from oil. On two such separate instances the water content of the all oil HTYHP filtrate has been found close to 1000 / 1500 ppm. Not many might have probably noticed this feature and kept wondering why the all oil mechanically stable rock also fails after a finite time interval. Another equally interesting and unusual character of oil base mud has been un-earthed when an expandable Bentonite remained in contact with non-polar base oil for over a month in an experiment conducted more out of curiosity than a disciplined scientific finding. The swelling of clay would create pressure, which we normally call Osmotic pressure. However, in a non-polar environment can we still call it as osmotic pressure or better refer as Capillary pressure is open to debate. The curiosity was to know the upper limit of the so called non-reactivity of base oil in shale. After a month, it was noticed that there had been a slight increase in volume of the base oil / bentonite mixture which could be quantified by the presence of pressure which was applied to bring the volume level same as it was at the start of the experiment. Thus, despite the absence of any aqueous environment and mechanical forces there had been an interaction between bentonite and base oil, whose mechanism can have more to do with the intrusion of base oil in between the layers of clay due to reasons other than chemical. This requires more detailed investigation to arrive at the real scientific cause than a plausible theoretical explanation.

8. Reduced Tendency for Differential Sticking

Oil base mud filter cake can be crudely split between a fluffy part and a thin slick filter cake.

A thin, slick filter cake minimizes the chance of differential sticking. Differential sticking is still possible if drilling highly overbalanced, especially with a high fluid loss system or due to the torturous geometry of the hole.

9. Drilling Under balanced

The wellbore stability resulting from the use of oil-base drilling fluids allows, in some instances, drilling with under balanced pressure, thereby increasing the rate of penetration (ROP).

10. Reuse

The excellent stability and solids tolerance of a well-conditioned oil-base fluid allows its use in multiple wells. Although the initial make-up cost can be expensive, the re-use of oil-base fluids can result in less expense than water-base fluids since the ability to recondition for multiple applications exists.

11. High Penetration Rate

The use of lower solids content oil-base drilling fluids, combined with the advent of polycrystalline diamond cutter (PDC) bits, has resulted in the ability to drill with high penetration rates throughout the world. The modification of drilling fluid properties and bit design allows for increased penetration rates in all types of shale, from soft gumbo to the older hard types. However, a word of caution here is worth mentioning – oil base muds facilitate faster ROP and not equally faster hole-cleaning!

12. Flexibility

With the advent of both "relaxed" oil-base fluids and low toxicity oil-base fluids, any size hole, any environment, and any lithology, can be drilled with a specially tailored oil-base fluid.

13. Reduction of Stress-fatigue

Stress-fatigue of tubular is reduced considerably when using oil-base fluids, i.e. less torque and corrosive attack from formation acid gases.

14. Reduced Corrosion

Oil fluids are not corrosive since oil, instead of water, is in contact with all the tubulars.

DISADVANTAGES OF OIL-BASE FLUIDS

1. High Initial Cost per Barrel

Initial make-up costs can be high. A barrel of oil-base fluid has a high initial cost and maintenance costs can climb substantially depending on the environments encountered. These costs fluctuate with the price of the oil. Environmental restrictions on discharge of oil mud cuttings exponentially raise the cost to a very high level, which can substantially reduce the over-all benefit of the use of SOBM.

2. Mechanical Shear Required

To achieve the emulsion stability and viscosity required, mechanical shearing is necessary to form a tight emulsion and disperse the organophilic platelets. This can be accomplished by using a high-pressure pump in conjunction with special shearing devices or shearing through the drill bit.

3. Reduced Kick Detection Ability

H_2S, CO_2, and hydrocarbon gases such as methane (CH_4) are all soluble in oil-base fluids. If gas enters the well bore, it can go into solution under pressure and as the gas moves up the hole it can break out of solution and expand at a very rapid rate, thus emptying the surface wellbore and reducing the hydrostatic pressure allowing additional kicks to be taken. If this occurs at a shallow depth, it may be too late to prevent a blowout. All kicks in oil-base fluids must be treated as a gas kick and circulated through the choke.

4. Pollution Control Required

Most areas where oil-base fluid is used have environmental restrictions. Rig modifications may be necessary to contain possible spills, collect or clean cuttings, and handle whole fluid without dumping. Environmentally acceptable disposal procedures must be followed when discarding cuttings and whole mud.

5. High Cost of Lost Circulation

Lost circulation with SOBM is not only very expensive due to the mud's cost per barrel, but the probability of encountering loss is also higher with the use of Oil base mud. The problem may not often be completely curable. Therefore, it may not be practical to use oil-base fluids where lost circulation is anticipated. Seepage losses are a common occurrence with oil-base fluids. Formations with low permeability and porosity may exhibit no losses when drilled with a water-base fluid, a partial or full loss of returns may occur when drilled with an oil-base fluid. Oil-base fluids, because of their low leak-off values, can act as efficient fracturing fluids.

6 High cost of handling oil base mud

Boats carrying oil base mud and the tanks holding oil base mud need frequent cleaning, the overall cost of using oil base mud is invariably high, as many cost components associated with the use of oil base do not reflect in the cost of barrel of mud.

7. Disposal Problems

Oil-base fluid cuttings may have to be cleaned before dumping. Some environments require that cuttings be sent to proper disposal areas. Also, whole fluid cannot be dumped. Therefore, if solids need to be reduced, dilution is the only answer. Oil-base fluid volumes only increase. Eventually the oil-base fluid must be disposed of, possibly at considerable cost. No one would ever admit, but at times when the project is nearing an end, few would pray for the losses to happen in unproductive zones to decrease their inventory of SOBM and save all costs associated with the storage and / or disposal of SOBM!

9. Solids Control Equipment

Oil-base fluids have inherently higher viscosities than water-base fluids. This increased viscosity, combined with the need for oil to dilute during the centrifugation process, results in less than effective solids removal.

10. Hole-cleaning

Hole-cleaning is more difficult with oil-base fluids. Some reasons for this are – (1) Cuttings do not disperse into the fluid as with water-base fluid, requiring larger cuttings to be removed; (2) oil fluids are more Newtonian in behavior than water-base fluids; and (3) oil-base fluids have less thixotropic behavior than water-base fluids.

11. Rig Cleanliness

Extra effort is required to keep a rig clean when oil-base fluids are used. Special dressing areas, steam cleaners on the rig floor, etc., are required.

12. Special Skin Care for Personnel May be required

Prolonged skin exposure to OBM and SOBM results in oils being removed from the body, resulting in dry and cracked skin. Proper hygiene and skin moisturizers are required to alleviate this situation.

13. Hazardous Vapors

As hot oil-base fluid flows across high-speed shaker screens and circulates in fluid pits, toxic hydrocarbons may be released. Therefore, shaker and fluid pit areas must be adequately ventilated.

14. Effect on Rubber

Oil-base fluids can cause either shrinkage or swelling of normal rubber parts. Therefore, it may be necessary to change out blowout preventer (BOP) rubber components, and the Hydril, etc., to oil-resistant rubber such as Nitrile. Diesel is worse than low-aromatic oils in this respect. Drill-pipe rubbers will also need to be replaced. Depending on the base fluid used, shaker bonding polymers may also be negatively affected.

15. Fire Hazard

Since the oil used in oil-base fluid can burn, extra precautions are required to prevent a fire. Additional gas, smoke, and fire detector and protection devices are necessary, including sealed electric motors and spark-less switches and controls.

16. Gas Stripping

Intrusion of gas into an oil-base fluid can cause the weighty material to settle. This is referred to as *gas stripping* of Barite, whereby the intruding gas changes the wetting characteristics of the Barite surface area. This can be a problem when using oil-base fluids on gas wells.

17. Aqueous Contamination

An influx of water due to surface leaks, rain, or formation has a significant impact on the properties of the system. Dilution is the only recourse to correct the condition and one that is not inexpensive to treat.

BASIC COMPONENTS OF AN OIL BASED MUD

The composition, number of additives and doses of additives shown here are only for illustration purpose. Real case may require deliberation and optimization to finalize the formulation.

SOBM: A typical barrel of SOBM has the following components

- Continuous external phase of oil
- Internal emulsified phase of Brine
- Emulsifier to keep both the phases as a single emulsified phase
- Organo-phillic clay to impart body / viscosity / thixotropy to the fluid
- Fluid loss control additive(s) to control the filtration loss
- Rheological modifiers to boost low end rheology

- Bridging and sealing agents to make the fluid loss resistant
- Wetting agents, to oil-wet the solids of mud
- Weighing agents to impart desired weight

The proportion of blending of each component to finally formulate a typical barrel SOBM is as follows:

• Continuous oil phase	70 % of the liquid phase
• Emulsified brine phase	30 % of liquid phase (10.3 ppg CaCl2)
• Emulsifier	~ 12 ppb
• Viscosifier clay	~ 6 to 8 ppb
• Wetting agent	1 to 1.5 ppb
• HTHP F/L control agent	~ 5 ppb
• Sealing agent	8 ppb
• Bridging agent	+/- 35 ppb (sized Calcium Carbonate)
• Weighing agent	Depending upon final desired weight
• Rheological modifiers	~ 1 ppb

SOBM whose formulation has been shown above, shall give mud of following properties.

• Mud weight	As desired
• Yp	+/- 24 lbs/100 ft2
• 6 RPM	~ 18
• Fluid loss API	< 4 cc (For WBM)
• Salinity	As chosen for hyrdate suppression
• Gels	8 / 10 / 12
• HTHP F/L	< 4 cc (For SOBM)
• Oil water ratio	70:30
• Activity of water phase	0.75(+/-)
• Emulsion stability	> 400 V

Carbon distribution pattern

Almost all the base oils have a fixed Carbon distribution pattern by which, their IBP and FBP ranges come into force. The pattern may start from as low as C-8 and range to as high as C=40, by and large. There is always a specific peak in the finger-print of every base oil, which represent a Carbon chain number representing the highest volume of the content of the base oil. For example, if a base oil contains 64 % composition of the liquid fraction comprising of 14 Carbon length chain, it is popularly called as C-14 peak base oil. This description has been purposely kept a bit like a common man's language than making it technically complicated enough to the extent that beginners cannot understand, leading to loss of the purpose of the book.

Aniline Point

The Aniline Point is an indicator of the detrimental effects (damage) oil will inflict on the elastomers, (rubber compounds) that encounter oil-based (non-aqueous) drilling fluids. It is commonly referred to as a test to evaluate base oils that are used in oil muds. The test indicates if an oil is likely to damage those parts of the drilling fluid system which are constructed from rubber materials.

The aniline point is called the "aniline point temperature," which is the lowest temperature (°F or °C) at which equal volumes of aniline ($C_6H_5NH_2$) and the oil, form a single phase. The aniline point (AP) correlates roughly with the amount and type of aromatic hydrocarbons in an oil sample. A low Aniline Point is indicative of higher aromatic, while a high Aniline Point is indicative of lower aromatic content.

It has been established through field experience that oils exhibiting an Aniline Point of less than 140° F (60° C) will be damaging to the elastomers (rubber parts) of a drilling fluid circulating system. The API has developed test procedures that are the standard for the industry.

Synthetics

Synthetic-base drilling fluids (SBF) use a synthetic type material as the continuous phase. The first SBF was used offshore Norway in 1990. Since that time, due to the ever-changing environmental status of SBFs, several types of synthetic materials have been used to formulate the synthetic-base fluid.

Unlike mineral and diesel oils which are distilled from crude oils, synthetic type materials are usually polymerized from ethylene. Since the synthetics are pure products made from ethylene, they contain no aromatics, thereby lowering the toxicity level normally associated with aromatic compounds. Like hydrocarbon oils, the synthetic type materials are more viscous than water. See from the properties of several synthetic materials.

Aqueous (internal) phase

Water contamination is unavoidable in oil-base fluids. Water is picked up at the change out from water-base fluid, from drilling water-bearing sands, and from drilling hydratable shales. Oil-base fluids can be formulated to accommodate additional water. The salinity of the water phase can be controlled with the use of dissolved salts. Calcium chloride ($CaCl_2$) is the most common salt used, although Sodium chloride (Na Cl) or drillers' salt, seawater, and various brines have been used. The choice is dictated by both the formation and economics.

Aw and Osmosis

Consider the hydrated Sodium chloride as shown in Figure 2. In addition to the shell of tightly adsorbed water molecules adjacent to the ions, there are more water molecules in the next layer, only moderately less adsorbed. Water molecules in the immediate vicinity of such ions are tied up in hydration and not entirely free to hydrate other ions. This is defined as *bound* water.

As the concentration of ions increases, so does the concentration of water molecules tie up in hydration. As the concentration of water

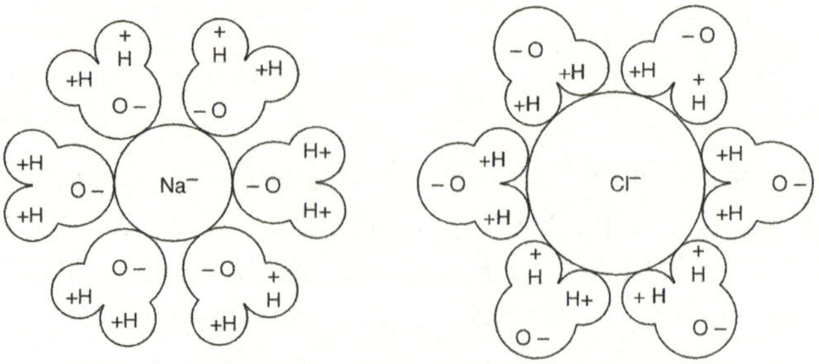

Figure 2 Hydrated Sodium and Chloride Ions

molecules tied up in hydration increases, the concentration of *free* water molecules (not involved in hydration) decreases. The availability of the water molecules in a solution to hydrate additional solutes can be quantified by the "activity" scale. All the water molecules in fresh water are available for hydration and the activity (Aw) is defined as "1". As salt, or any substance that hydrates, is added, the concentration of hydrated ions or molecules increases and the availability of the brine molecules in the solution to hydrate any additional material decreases the activity (Aw).

If the two solutions of different Aw, are poured together, diffusional mixing will result in a uniform Aw intermediately between the two. If two solutions of different Aw, are separated by a semi-permeable membrane, an unusual situation occurs, as shown in Figure 3. A semi-permeable membrane is defined as a thin membrane which will allow water molecules to freely pass, but not salt ions.

Water molecules in the solution of higher Aw will diffuse into the solution with lower Aw, as before, but not the ions. If the chamber with the solution of lower Aw is sealed, then the water diffusing through the membrane increases the pressure in the chamber. This is defined as *osmotic* pressure (Figure 3)

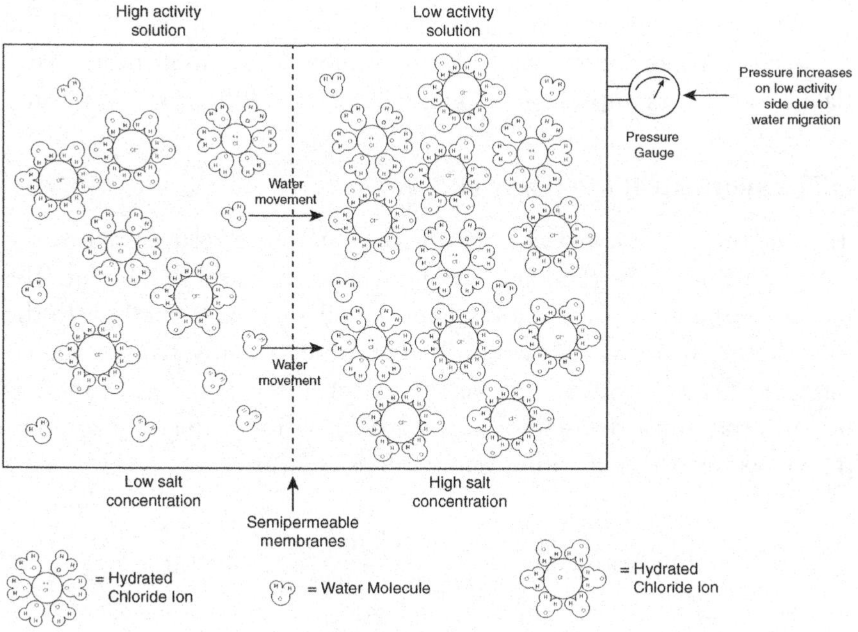

Figure 3 Osmosis

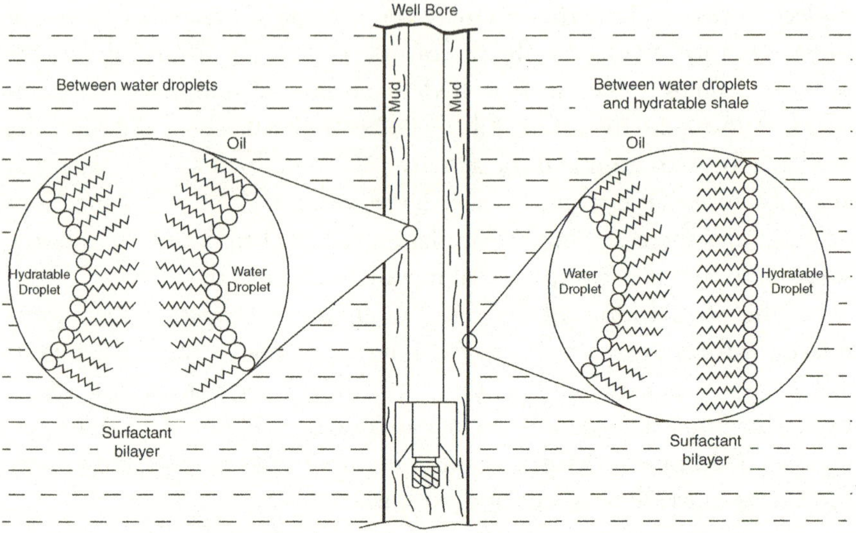

Figure 4 Semi Permeable Membrane

Semipermeable membranes are created between emulsified water droplets in water in oil emulsions and between emulsified water droplets and the formation drilled (see Figure 4). The layer of surfactant-rich oil will allow water molecules to pass, but not salt. Unless the Aw values are equal, an osmotic imbalance, or potential.

As the concentration of salt in the aqueous phase increases and the Aw decreases, less water is available to solubilize and hydrate other materials. Lime, for example, is less soluble in $CaCl_2$ solutions than in water.

Oil or Synthetic Ratio (O/W or S/W)

The oil- or synthetic-to-water (O/W or S/W) ratio relates only to the liquid portion of the mud and is not affected by the solids content. The oil- or synthetic to- water ratio relates the oil and water fractions to the total liquid fraction. Generally, higher mud weights require higher ratios. Different conditions favor the use of different ratios, so there is no ratio that must be used for any set of conditions. The calculation of the oil-to-water ratio requires retort values as follows:

$$\text{Oil ratio (O)} = \left(\frac{\text{vol\% oil}}{\text{vol\% oil} + \text{vol\% water}} \right) \times 100$$

$$\text{Water ratio (W)} = 100 \quad \text{oil ratio}$$

ELECTRICAL STABILITY

The electrical stability is an indication of how well (or tightly) the water is emulsified in the oil or synthetic phase. Higher values indicate a stronger emulsion and more stable fluid. Oil and the synthetic fluids do not conduct electricity. In the electrical stability test, the voltage (electrical

Potential) is increased across electrodes on a fixed-width probe until the emulsified water droplets connect (i.e., coalesce) to form a continuous bridge or circuit. The stronger the emulsion, the higher the voltage required to break down the emulsion completing the electrical circuit to conduct electricity. The unit of measure for recording the electrical stability is volts.

Some of the important factors that influence the electrical stability are:

1. Water content. Higher water content decreases electrical stability because the distance between water droplets decreases. This allows for easier electrical circuit completion through coalescence and a reduction in the electrical stability.
2. Water-wet solids. Water-wet solids reduce the emulsion stability because they act like water droplets with respect to electrical conductivity. Clusters of water-wet solids conduct electricity, producing the effect of larger water droplets and a reduction in electrical stability.
3. Emulsion strength. Water droplets are larger and weakly emulsified in new and unstable mud systems, causing these fluids to have significantly lower emulsion stability values. Increased shear, temperature and the incorporation of some drill solids significantly increase emulsion stability in freshly prepared fluids. In existing fluids, increasing the emulsifier and wetting agent concentration improves emulsion strength. Water droplet size and emulsifier concentration governs this aspect of emulsion stability.
4. Temperature. The temperature at which the electrical stability measurement is made, changes the emulsion stability and electrical conductance. The electrical stability should be measured at the same temperature (120°F or 49°C) to provide relevant data.
5. Salt concentration. The electrical conductivity of water increases with increasing salt concentration. Increasing the salt concentration will usually cause a slight decrease in electrical stability.

In addition, adding powdered salt can cause temporary instability until the salt becomes solubilized into the water phase.

6. Saturation. When the water phase becomes salt-saturated, especially with Calcium chloride, salt will precipitate out from the solution as a water wet solid, which reduces the emulsion stability.
7. Weight material. High-density muds weighted with hematite or other special minerals often have lower electrical stabilities than muds weighted with Barite. This is essentially due to insolubility of Barium sulfate making the ionic atmosphere unavailable for the conductance of electricity

Freshly mixed, invert-emulsion muds usually have low electrical stabilities when shipped from the liquid mud plant, even though they are adequately treated with emulsifiers. The emulsions of these systems will tighten as they are exposed to down-hole temperatures and sheared through the bit. Sufficient shear will form a "mechanical" emulsion with little emulsifier or wetting agent. Once drilling has commenced, Barite, organophilic clay and drill solids will aid in further stabilizing the emulsion as these oil-wet fine solids are adsorbed at the water interface. The electrical stability of an oil- or synthetic-base fluid should be used to indicate trends. An electrical stability of > 300 in one mud may indicate a stable emulsion if the electrical stability has remained constant for several days. But an electrical stability of 500 in another mud might indicate major problems if previous values had been significantly higher. A well-defined downward trend or a rapid drop in electrical stability indicates the emulsion is weakening. Electrical stability measurements should be made and recorded routinely. These values should be plotted so that trends can be easily seen. Trends upward or downward indicate changes in the system. An analysis of sequential mud checks will indicate possible causes of the change. Equally important or may be an even more important feature of an oil base mud is the absence of free water in HTHP than the absolute value of ES. It has been observed in the field on several occasions that the mud of 300 ES shows more stability than a mud of 500 with traces of water appearing in the filtrate. The concept of the high value of ES being synonymous with a stable emulsion is a myth which has developed and prevailing in the industry for ages. What is more relevant is the absence of free water and a constant value of ES rather than a water contacting HTHP filtrate and a fluctuating high value of ES of a mud. Another feature which helps predicting the emulsion stability, is the quantity of solids in the mud, which can otherwise be sometimes also termed as an impurity in oil base mud. A stable emulsion has more to it than a strong emulsifier and finely dispersed internal brine with continuous oil phase clinging to it with the help of surfactant polar and non-polar parts. Although not completely scientifically understood and proved, it is a very common trait that higher quantities of solids always aid in the stabilization of emulsion. This essentially means that a stable emulsion is one which resists separation of phases and extrusion of water from it under bottom hole conditions and one indicator (not the only indicator) of such trait is the Electrical stability, which is not dependent upon the other promoters of stable emulsion- like solids.

Controlled Activity

Water activity (AW) is a measure of the chemical potential for water to be transferred between mud and shale. Activity is measured using the vapor pressure (relative humidity) of shale or mud.

The activity can also be estimated based on the chemical composition of the brine (salinity). Pure water has an AW of 1.0. Calcium chloride brines used in most non-aqueous emulsion muds have an AW between 0.8 (22% wt.) and 0.55 (34% wt.). Lower values for activity are generally more inhibiting.

Clay-containing formations and shale swell, and are weakened by the adsorption of water. The possibility exists that the water from an emulsion mud can alter shale if the AW of the shale is lower than the activity of the mud. The key to inhibition is to "balance" the activity of the mud with that of the shale, so that the adsorption of water onto the shale is theoretically reduced to zero.

It has become customary to maintain Activity of brine phase close to 0.75 on most occasions irrespective of the fact whether it is really required that low, nor even considering that such imaginary number can have negative impact also on the stability of the well-bore. After carefully measuring the activity of the rock and with the aim of comparing the same with the activity of the mud, it was logical to match both the activities so that there is no movement of solvent on either side. Thus, the concept of Balanced activity SOBM took place. In most of the cases it was found that the activity of the rock is close to 0.9 at shallow depth, which keeps gradually decreasing with an increase in depth. Therefore, scientifically speaking, if the salinity of the mud (directly responsible for the activity) and the rock are same, there will not be movement of solvent and no osmotic pressure is exerted on either side. This is ideal, as the adverse effect of solvent movement across the semi-permeable membrane is eliminated. The focus is then to gradually decrease the activity of mud with a corresponding incremental increase in salinity of the rock and continue to have balanced activity environment. This type of formulation not only creates chemical equilibrium, but also helps retain the native composition of the drilled cuttings. On the contrary the higher salinity of mud triggers the flow of water from the rock to mud, which makes the cuttings dry and crispy. Such extortion of solvent from the drilled cuttings can impart instability to the rock in the immediate vicinity by interaction of rock particles with-in themselves. The key to maintaining such instability at bay is ensuring that the activity of mud as well as rock always match during drilling. The rule is simple. Mud need not interact with cuttings and open bore hole in any form except for the functions expected of it. Like a dedicated servant to the hole, mud should obey the command of its master, who happens to be a mud engineer, and should not establish any relationship of chemical, physical and physico-chemical nature, that can have an adverse impact on the hole, be it short or long-term.

Factors that affect the oil-to-water or synthetic-to-water ratios are:

1. Base fluid cost

Oils and synthetic fluids are expensive. Lower oil- and synthetic-to-water ratios reduce the price of the final formulation.

2. Residual oil or synthetic fluid on cuttings

In some offshore areas, the permissible residual oil or synthetic fluid on the discharged cuttings is regulated. A low oil- and synthetic to water ratios reduce the amount of oil or synthetic residue on the cuttings, making it easier to meet specified limits. Invert emulsions with O/W or S/W ratios of 50:50 have been run in the North Sea to meet these specifications.

3. Residual brine on cuttings

In some locations, the amount of residual brine on cuttings may be restricted due to environmental considerations. All-oil or high-O/W or -S/W formulations can be used to reduce the amount of brine discharged on cuttings. Alternative internal-phase chemicals can be used to replace chloride salts and still obtain a low water-phase activity.

4. Rheology

The emulsified water in an invert emulsion acts like a solid and increases, the plastic viscosity as the O/W or S/W ratio decreases. The emulsified water also helps organophilic clays and fatty acid rheological modifiers yield. Special clays and alternative viscosifiers may be required for all-oil or high-ratio fluids (>90:10).

5. Fluid loss

The emulsified water in invert emulsions acts like a solid and lowers the fluid loss. Muds with low O/W or S/W ratios tend to have lower fluid losses than muds with high ratios.

6. Mud weight

The weight material in an oil or synthetic mud must be wet with the oil or synthetic fluid.

As the density of the mud is increased, the liquid phase volume decreases and the O/W or S/W ratio must be increased to provide additional base fluid to oil-wet the weighing material. The higher the mud weight, the more difficult it is to run low-ratio (high water) fluids.

7. Temperature

Emulsion stability decreases as well as increases with temperature. The initial emulsion strength increases as the oil base mud heats up (increase in temperature) due to mechanical agitation and once the emulsion is strong enough to impart its characteristics, then with subsequent increase in temperature the strength of emulsion weakens.

The O/W ratio remains constant when the mud is weighted up or solids are incorporated into the mud, even though the volume percent liquid is decreased significantly. A rapid decrease in the O/W or S/W ratio indicates an influx of salt water from the formation, and a pit volume increase should have been observed. When using oil or synthetic muds, all water hoses on the pits should be disconnected or plugged to prevent accidental contamination with water. Heavy rains may also contaminate the system with water. Contamination with water from any source will decrease the oil- or synthetic- to-water ratio and require that chemicals and oil or synthetic fluid be added to maintain the properties.

An increase in the O/W or S/W ratio is unusual unless base oil or synthetic is added to the system. If this is observed, carefully monitor pit volumes to be sure that formation liquids, such as crude, are not intruding and contaminating the mud system. Evaporation of the water phase has been reported with high flow line temperatures. This will increase O/W or S/W ratios. After the trend of increasing ratio is confirmed with more than one retort analysis, water should be added to the system to maintain a stable oil- or synthetic-to water ratio. If the oil- or synthetic-to water ratio is left on its own to increase too high from evaporation of water or any other mean, the salt in the water phase may reach saturation, precipitate and destabilize the emulsion.

The viscosity and HTHP filtrate will change with changes in the oil- or synthetic-to-water ratio. Changing the ratio is not used to alter either of these properties. What would appear to be a relatively minor change can involve significant increases in volume and cost.

For example, if the O/W ratio of a 14.4 lb/gal mud containing 30% solids is increased from 70:30 to 75:25, the oil, plus the Barite required to maintain the weight at 14.4 will increase the volume by 19%.

Flat Rheology Mud

Losses of drilling fluids during wellbore construction occur at various stages of operations including drilling, running casing or drilling assembly to bottom (tripping), and cementing. Drilling losses are usually related to the equivalent circulating density (ECD) exceeding the fracture gradient. Tripping and cementing losses are related to a combination of ECD, the pressure necessary to break the drilling fluid gel structure, and the equivalent static density at the time of the operation.

A creative technique to manipulate the drilling fluid system's degree of viscosity fluctuation under down-hole conditions has been used to reduce drilling fluid losses. This approach is reported to minimize the "unknowns" associated with drilling fluid behavior and is often referred to as "Flat Rheology" or an "apparent flat yield point" drilling fluid system.

Owing to its balanced yield point, which translates into efficient hole-cleaning across a broad spectrum of down-hole temperatures, the new system claims to deliver average rates of penetration (ROP) around 30 ft/HR higher than the conventional synthetic-base drilling fluids employed in offset wells. In its initial trial, the system drilled a 17 ½-inch section at ROPs as high as 150 ft/HR with zero losses. The system also exhibited excellent displacement, low dilution rates, and reduced requirements for chemical treatment. In addition, equivalent circulating densities (ECD) values were less than offset wells.

The major technical benefit of the new system is that while surface rheology numbers may at first appear higher than usual, the flat rheology profile allows the fluid to deliver consistent down-hole rheology parameters.

Engineered with lower clay content and employing newly developed emulsifiers, rheology modifiers and viscosifiers, the system was designed to address issues specific with drilling deep-water wells using synthetic-base drilling fluids. In deep-water, the cold temperatures in the riser elevate the viscosities and gel strengths of most standard invert emulsion drilling fluids. Further, operations such as tripping and breaking circulation result in greater surge pressures, which all-too-often fracture the formation leading to whole mud losses. While both performance and environmental characteristics make synthetic-base fluids ideal for the deep-water, whole losses of these premium fluids can be prohibitively expensive.

With a flat rheology drilling fluid, the mud properties maintain constant values at a mud-line temperature of 40° F; flow-line temperatures that typically are between 55-70° F in deep-water; bottom-hole circulating temperatures, and the standard API 150° F mud check temperature. Conceptually, a system behaving in this fashion can be expected to drill a well at lower equivalent circulating densities (ECD), encounter fewer problems drilling, and avoid/minimize lost circulation problems. With a flat rheology system, the on-site drilling fluid engineer can design a fluid that can be effective under all down-hole temperatures and pressure conditions. The cold temperature rheology does not become excessive and the high-temperature bottom-hole rheology parameters do not decrease to undesirable levels.

Since the properties do not have to be compromised at one temperature to satisfy requirements of a different temperature elsewhere in the well, the flat rheology profile greatly simplifies the design of the drilling fluid. Often in deep-water drilling operations, the surface rheology must be reduced to decrease the rheology and ECD when encountering cold temperatures. Consequently, what often results is, poor hole-cleaning, Barite sag or other down-hole drilling problems related to low rheology at bottom-hole static and circulating temperatures. Conversely, adjusting the rheology upwards to compensate for higher bottom-hole temperatures can lead to excessive ECD values, excessive gel strengths and pressure spikes that can break down the formation and lead to lost circulation.

Managing the rheology of conventional synthetic-base systems poses a very difficult challenge. With this new generation system, the unique flat rheology profile eliminates those difficulties and opens the door for more efficient drilling. From a hydraulics modelling standpoint, the new system allows the user to identify the optimal properties and design a mud system incorporating those properties over a broad range of temperatures.

- Flat rheology profile
  - ➢ Improved hole-cleaning
  - ➢ Reduced ECD values
  - ➢ Higher drilling rates
- Fragile gel structures
  - ➢ Lower surge pressures on pump startup
  - ➢ Reduced risk of mud losses running pipe and cementing
- Low clay content
  - ➢ Well-behaved rheology properties
  - ➢ Low ECD values
  - ➢ Fragile gels
- Environmental compliance
  - ➢ Acceptable for use like any other synthetic oil base mud.

Flat Rheology Typical Formulation

Purpose	Concentration
Emulsifier	6 – 8 lb/bbl
Oil Wetting Material	1 – 3 lb/bbl
Calcium Source	3 – 5 lb/bbl
Rheology Modifier	0.5 – 2 lb/bbl
Rheology Modifier	0.25 – 1 lb/bbl
Organophilic Clay	1.5 – 3.0 lb/bbl
Premium Organophilic Clay	0.5 – 1 lb/bbl
Viscosifier – Low Temperature Application	As needed
Thinner	As needed

How I look at the whole scenario of SOBM??

My first reaction to use of SOBM/OBM is, do we really need it or it is prevailing because of its drilling friendly behavior as well as a marketing gimmick of big companies to add to their revenue? If we should simply drill a hole, then SOBM certainly scores over WBM for its inert nature towards rock matrix which keeps the bore-hole stable for as long as we need. In another way, SOBM provides comfort of messing around the quality of borehole and yet get away unpunished. The single feature of oil being non-reactive to hydratable clay has been over-exaggerated to a level, where most of us tend to forget the other adverse effects of the use of SOBM! One can formulate SOBM and sleep peacefully without bothering about drag/over-pull, sticking tendency and fluctuations in mud properties over a reasonably long period.

If we are interested in drilling a well which will retain its integrity as well as the capacity to produce the hydrocarbon to its optimum potential, then one seriously needs to consider whether to use SOBM or not!! The very beneficial aspect of inertness of SOBM becomes a real limitation when it comes to pumping and placing cement behind the casing. Although all companies claim about changing the wettability of the open hole prior to pumping slurry, yet, it really is a big challenge to obtain a satisfactory cement job. The culture of recording CBL/VDL is gradually diminishing from industry and is almost extinct except on occasion where the casing needs to be perforated!

The poor cement job with presence of channels and soft set cement at irregular intervals due to the unchanged wetting characteristic as well as SOBM not getting completely removed almost invariably calls for remedial job when the well is put on flow. That time is too late and we end up spending up a huge amount of money either in repairing or in extreme cases abandoning the well and loosing the production.

The ease with which the SOBM gets lost in the rock, by its not reacting with the exposed rock is another area which eats away millions of dollars of industry in the form of the NPT, what to talk about the cost of the mud as well as the time and resources re-engaged in preparation of additional mud! In associating with drilling of > 130 Deep-water wells with many in water depth beyond 2000 m, the single largest reason for NPT observed was loss of circulation and poor cement job. It is not being claimed that with WBM there are no losses or there is no poor cement job, but the shear magnitude is astoundingly high with SOBM. With WBM even if the mud is not completely removed from the hole while displacing the slurry, the residual mud invariably assists the cement in setting to some extent, by its chemical interaction with slurry. Similarly, the same reason of reactivity of the continuous phase of WBM to impart instability to the hole becomes a desirable feature for curing losses when compared to SOBM. The reactive nature of WBM helps in the development of a semi-rigid fibrous structure in the thief zone and thus, resist tendency of loss circulation.

Who in the industry is not aware of the problems faced during kick detection in presence of SOBM? Instances are not uncommon when the intensity of influx (gas, water, oil from rocks) can be so severe that the entire annular mud loses its emulsion, leading into separation of both phases, leading to the generation of a huge chunk of solid mass, that can be unforgiving to the extent that the entire circulatory passage is blocked!

As if these features are not enough, we have contraction and expansion of oil with temperature and pressure further adding to our woes! The plug bumping in a cement job with SOBM is always tricky and we often end up with under-displaced slurry above the top plug, to be drilled later as well as having less than the planned height of the cement column behind the casing.

Everything else remaining same, SOBM does not flush the drilled cuttings out of hole as efficiently as WBM due to the slippery nature of mud. To add insult to injury, it offers very high ROP! One can easily imagine the result: Our greed will prompt us drill faster, but the un-flushed cuttings (accumulating over repeated cycles) can push the ECD beyond fracture gradient and the net result? Loss of mud in the formation!! Initiating a fracture can be independent of the type of mud, but SOBM helps in keeping the fracture open by its inert nature, which then gets wider by pressure exerted on it by upwardly moving column of circulating mud – leading to the accelerated rate of loss of mud in the rock matrix. The last nail in the coffin comes in the form of environment hazards of SOBM, strict norms of disposal of oil coated cuttings.

The use of SOBM is in fact a perfect example where everyone is comfortable during drilling, but eventually the well suffers. Alas, the well could speak! Or, the well is speaking and we don't understand its language? Or we are purposely turning a blind eye? Let's think….

HIGH PERFORMANCE WATER BASED MUD

This term has caught the attention of late as almost all big mud companies have resorted to promoting an alternative to SOBM and eliminate environmental related hazards of the use of SOBM. Concept wise, an attempt is being made by spreading the word around industry that it performs much better than conventional water based mud and on par with SOBM, if not better than them. The focus is on achieving shale inhibition by adding various specialized Polymeric additives with the claim that they delay the swelling of clays by various mechanisms and thus keep the borehole stable and intact for a period long enough to drill and case any perceived length of bore hole of any diameter.

Accretion, swelling and encapsulation are the fancy words always associated with such systems. Without going into finer details of the system (that is invariably available abundantly on sites of all major mud companies), the subsequent focus will be to dissect each component of the system and account each additive purely on grounds other than commercial.

The typical key components of a high-performance water base mud are

A Anti-accretion agent
B Encapsulating agent
C Anti-swelling agent

D Lubricant
E Primary shale inhibition agent

For a moment, if we refrain from the language of an expert which has many technical jargons and fancy words, that it is apt all the components together are expected to perform the role of keeping the drilled cuttings completely isolated from the mud, irrespective of its mineralogical composition.

Since all these forces are ionic in nature, charge-charge interaction of particles is bound to happen. Therefore, all the important components either alter the surface characteristics of the cuttings by clinging themselves on the exposed charged surface of cuttings to make them immune to electrostatic interaction with mud or intrude in between the layers of the clay lattice, bind the two separated lattices. This binding is often strong enough to keep the water molecules away, which otherwise occupy the space between clay lattice and cause separation of lattice, commonly termed as swelling or expansion of clay. The capillary forces which lead to invasion of filtrate through open pores of exposed rock are negated by providing an atmosphere in high performance mud which exerts equal and opposite force to eliminate osmotic inflow of water and negate solvent/solute interaction between rock and mud.

In a regular and conventional mud, all these functions are either left upon the Potassium ion (the most commonly used shale inhibitor) to take care, or are left unattended. When left unattended they still give a little window of time to start playing havoc with stability of well-bore. This time depends upon the clay type and content in mud. At times, we can be lucky to get away and case the hole before instability starts stopping us to do so without having NPT. With the development in technology and ever-increasing need to explore tougher and tougher regions for oil/gas production, there were occasions when the lack of adequate inhibition started hitting hard to industry, giving birth to SOBM. The environmental barriers getting stricter and stricter every day, then gave birth to a new thought process. Instead of using SOBM, let us mimic a water based system which can perform as well as SOBM, if not better. As expected, the focus was to tighten the grip of mud over the tendency of cuttings to dissolve in it. With a multi-pronged approach and systematic study in detail was conducted. Thus, a system evolved which was termed as "High-performance water based mud". Obviously, the name derives from the enhanced performance in terms of shale inhibition.

While it is true that High performance water based mud is superior to conventional systems, but the way it has been projected for capturing the market is something not completely professional. There are various commercial interests which over-rule the technically genuine requirement of the use of high performance water base mud. The story was same for introducing SOBM where SOBM also, was used in cases not calling for stringent shale inhibition!

By and large the age old conventional K, Cl / polymer system can provide enough inhibition to enable drilling a moderately reactive rock. Very few really do the hard work of exactly assessing the selection of right mud system, and by default - select a system whose drums is beaten loudly by mud companies. Thus, there is always a situation where the rock may not have reactive clay to the extent that it may warrant the use of high performance mud, yet- it is being used. It is at times an example of over-engineering a system. Another interesting feature observed in many wells is - despite the use of SOBM or high-performance water based mud, there has been well-bore instability related issues! All such cases conveniently get buried under the disguise of the mechanical nature of rock instability. If a conventional system is used and there is a genuine case of rock instability due to mechanical instability, the blame is conveniently put on conventional mud!

Strictly technically speaking, yes – high performance water base mud inhibits shale better and longer than conventional water base mud. However, is HPWBM a candidate for use on moderately reactive clay contained mud, is a matter of debate. Often in the absence of correct quantification of the nature of reactivity of the clay and efficiency (or the lack of it) of conventional water base mud to inhibit the clay, many operators fall a prey to revenue focused recommendations of mud companies and yet keep suffering from instability of well-bore being ascribed to reasons other than chemical instability / ability of the mud. Both high performance water base mud and SOBM can buy additional time and provide operational comfort as well as insurance against unplanned and unscheduled activity commencement with increased probability of a stable well-bore.

There are several cases of conventional mud doing a great job and high-performance water base mud, not living to the expectations of performance. The objective of this description is not to find fault with anyone, but provoking the reader to think and act what is technically most suitable without getting unduly impressed by tall claims, which disappear later. Looking beyond the horizon and not getting too much bogged down with finer intricacies, a corporate professional sitting at the top often plays /prefers to play safe and the decision, then tilts in favor of use of SOBM/High performance water based mud and there is nothing wrong in that. Yet, those top-notch professionals also need to accept the fact that their choice is not an absolute insulation from the NPT, which is what of paramount importance for them!

To sum up, the best of the available mud system is NO SUBSTITUTE to a sensible, knowledgeable, experienced and capable person at the right place to handle mud engineering activities of the company. It always more than re-pays to keep such professionals, irrespective of any compensation, because what they deliver back is way more than what they get.

Thankfully, the situation is gradually but certainly changing in favor of the use of WBM, a better option for providing a stable, long-lasting and healthy hole, which shall flow to its optimum period of its entire life without the need of any intermittent remedial job. The emergence of High performance WBM is a live example of this.

CHAPTER 11

DEEP-WATER DRILLING FLUIDS

INTRODUCTION

Deep-water drilling operations started as early as the 1960's. Activity steadily grew throughout the 70's and by the late 80's the activity, primarily in the Gulf of Mexico and Brazil. Interest in deep-water continued to grow spurring worldwide activity in such areas as Trinidad, Australia, Indonesia, Malaysia, Brunei, the entire west coast of Africa, West of Shetlands, the far North Sea, Norway, Atlantic coast of Canada and Egypt. The area of greatest activity has traditionally been the Gulf of Mexico, though other areas have shown much promise for future growth.

Deep-water drilling fluid is different than other conventional type due to following reasons:

- The temperature of the sea bed is low
- The column of sea water exerts additional hydrostatic head on the hole.

- The sediments immediately below the sea bed are very soft, un-consolidated, young, porous, Geo-pressured and permeable at times.
- The trapped reservoir fluid results in pore-pressure often reaching close to the fracture pressure of the rock.
- The possibility of the formation of gas hydrates is increased due to low temperature and high pressure.
- Due to the weak and fragile rock underneath sea bed, the ballooning feature is very common on Deep-water locations.
- The mud is more prone to down-hole losses due to the narrow margin between ECD and fracture gradient. Much required higher pump rate, which is needed to bring cuttings to surface, further compounds the problem to drill through.
- The volume of mud required to be handled is exceedingly high.

To sum up, the role to be performed by mud remains same as in any other location / environment, but the properties and composition needs to be altered to address / resolve the special conditions of the deep-water drilling location

Deep-water drilling operations face many additional challenges, some of them fluids related. Identification of the challenges and integrating the applicable deep-water fluid system must be thought out carefully. Deep-water wells present unique technical and operational challenges for drilling fluid systems, including but not restricted to:

- High daily operational costs
- Lost circulation
- Fluid selection and design
- Solids transport
- Wellbore stability, including low fracture gradients
- Shallow hazards
- Formation Ballooning
- Gas hydrates
- Salt
- Reservoir productivity
- Environmental compliance
- Large fluid volumes

- Low seabed temperatures
- Elevated bottom hole temperatures
- Annular pressure build-up
- Logistics

Shallow water flows, or pressurized gas zones, can present difficult technical challenges, deserving special focus in the planning and the execution phases of deep-water projects. These shallow hazards can cause lost circulation, hole-erosion, poor cement jobs, foundation instability, gas hydrates, casing buckling and in the worst cases, loss of hole!

Shallow water flows can occur in zones 500 to 3,500 feet below the seabed. Shallow gas sands are to be found at similar depths and require a sealing mechanism. Typically, wells are drilled riser-less through these zones, using seawater and Bentonite sweeps. Once a shallow hazard is detected there are two realistic alternatives: 1) spot a kill fluid and run casing or 2) drill ahead with a weighted fluid.

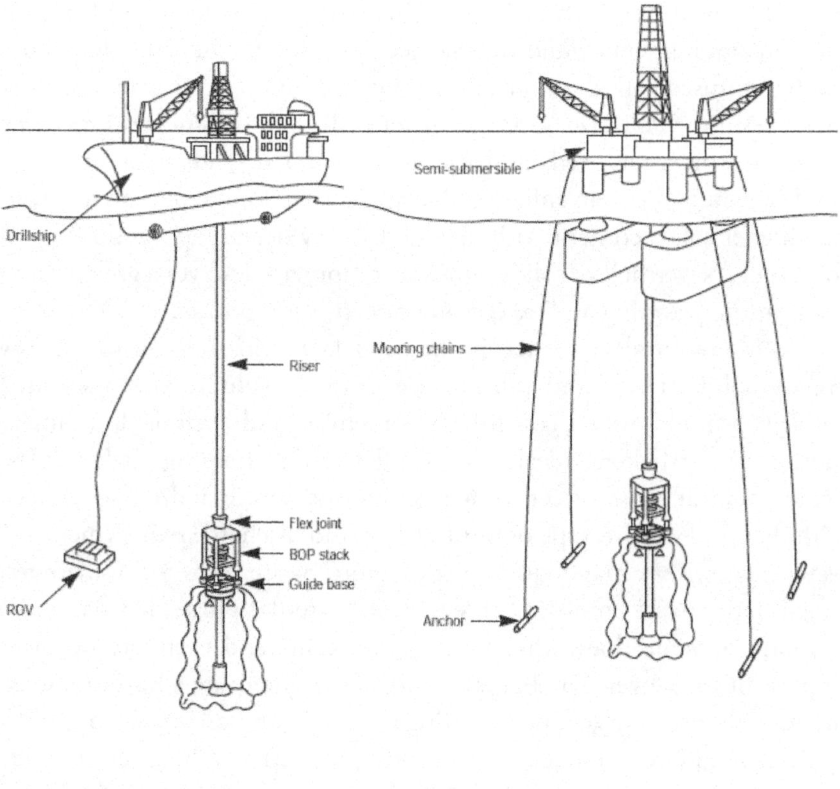

SELECTING DRILLING FLUIDS FOR DEEP-WATER

Drilling fluid selection for deep-water drilling operations can be an extremely complex matter due to many different considerations, including fluid compressibility/expansion as a function of wellbore temperatures, hydrate suppression, possible evaporate sequences, well path, environmental restrictions and perhaps the most importantly, well costs. Deep-water technical challenges also include shale reactivity, formation damage and drilling & completing in unconsolidated sands. Environmental and safety issues are, as always, of paramount importance. Often, the most practical result of the fluid selection process is the use of a fluid that delivers all the well-specific performance issues are essential. Inhibitive fluids with low dilution and consumption rates are of considerable benefit in minimizing rig logistics, cost and maximizing fluid performance. Typical riser volumes may range from 1,000 to 3,000 bbl (160 to 500m3) and circulating volumes from 2,500 to 5,000 bbl (400 to 800 m3), depending on water depth, hole size and total depth of the well. As demonstrated by this example, large fluid volumes are normally associated with deep-water wells. This is the result of multiple casing string requirements and/or production requirements. Fluid volume needed to fill the riser and pits should not be confused with dilution volume. Fluid dilution requirements are the direct result of hole-size (cuttings generated), Solids Removal Efficiency (SRE - cuttings removed) and the desired drilled-solids content of the fluid. The fluid type can enhance SRE and certain fluid types can incorporate higher solids content without significantly degrading the fluid's performance. Nevertheless, solids removal equipment deserves a great deal of focus in the planning and execution of deep-water projects.

Solids removal efficiency is the vital factor in the control of low gravity solids (LGS) and minimizing dilution volume and associated costs. Optimum solids control equipment would include four linear motion shale shakers, ideally preceded by four scalping shakers. The deployment and use of centrifuges should be considered in association with the mud weight, type of fluid system and discharge requirements. A water base system may require one or more Barite recovery centrifuges. A synthetic invert emulsion may have an external phase that has more economical value than other fluid components and therefore be most important to recover. Discharge of cuttings may require a host of equipment, including a dryer and centrifuge for LGS and oil separation. Each project should be evaluated on its economic merits. Whenever possible, it is desirable to utilize a drilling fluid which is environmentally acceptable for cuttings discharge and exhibits the required fluid characteristics.

AQUEOUS VS NON-AQUEOUS DRILLING FLUIDS

The first fluids decision to make for a deep-water operation is whether the fluid is to be aqueous or non-aqueous. Both types of fluids have been used with a great deal of success in deep-water operations.

The table below compares some of the weaknesses and strengths of aqueous and non-aqueous fluids. Each well must be looked at individually and the above components weighed to decide whether to use an aqueous fluid, e.g. if kick detection is of vital importance then an aqueous fluid is best as the gas is not soluble in water and so gas kicks are more easily detected. During the decision phase, one should also look at the weaknesses and their possible solutions.

Investigative Solutions to the Weakness of Aqueous and Non-Aqueous Systems

Strengths and Weakness of Aqueous and Non-Aqueous Fluids

Non-Aqueous Strength	Aqueous Strength
Faster ROP	Resistance to Lost circulation
Improved Wellbore stability	Better Environmental characteristics
Better handle on Gas hydrate suppression	Decreased probability of well-bore Ballooning
Improved Cuttings integrity	Ease of Reservoir characterization
BHA – bit balling, minimal	Gas solubility / kick detection is easy
Trip frequency decreased	The cost / bill is lower
Temperature stability over wider range	Infrastructure requirements are simpler
Excellent lubricity	Better quality of cement job

An effort is made here to objectively look at finding solution to weaknesses of each type of system and then re-assessing to arrive at a decision, which is free from any personal prejudice.

Aqueous		Non-Aqueous	
Weakness	**Solution**	**Weakness**	**Solution**
ROP BHA / bit balling	Bit hydraulics Bit selection Additives Proven deep-water systems	Lost circulation	Permanent seal High Fluid Loss LCM's Cross-link polymer
Wellbore stability	Inhibitive systems Proven deep-water systems	Environmental	Synthetic products Cuttings drying Skip & ship
Cuttings integrity => hole-cleaning	Maximized hydraulics Riser booster	Ballooning	Hydraulics software Pressure tool
Gas hydrate suppression	Ethylene glycol Spotting Procedures Salt	Reservoir characterization	Use of advanced technology, with base-line correction.
Trip frequency	Additives to enhance PDC bit performance Proven deep-water systems	Gas solubility	Kick detection procedures Knowledge of area
Temperature stability	Geothermal technology	Cost/bbl	Economic modelling, Overall well cost
		Infrastructure requirements	Economic modeling Risk Analysis
		Poor cement job	Good oil field practices and effective chemical treatment, to have water-wet rock

Mud Systems

Many different mud systems can and have been used in deep-water applications. They range from systems as simple as sea water-base Lignosulfonate muds to environmentally approved high-performance synthetic muds. By far, Partially Hydrolyzed Polyacrylamide (PHPA) /20% salt systems have been the most widely used water-base systems. Low viscosity synthetic fluids, have gained increased use in deep-water drilling. These synthetics have become popular due to the increased drilling performance and improved wellbore stability that can be achieved as compared to water base fluids.

Natural gas, CO2 and H2S are highly soluble in diesel, mineral oil and synthetic fluids. Gas solubility and its effect on kick detection and well control should be considered for all deep-water wells. Synthetic systems are expensive and are not recommended for areas where there is a high probability of lost circulation. Again, the proper drilling fluid selection, and correct design will result from considering the most important deep-water issues for a given project. As technology grows, new mud systems will surely be developed to surpass today's systems in addressing deep-water issues. Specific issues concerning mud systems will be attended to as they relate to the deep-water drilling issues discussed below.

Gas Hydrates

Gas hydrates are a key concern for operators drilling in Deep-water. Gas hydrates are "ice-like" mixture of Methane gas and water. At atmospheric pressure, fresh water freezes at 32°F (0°C). At high pressures, the water starts freezing at temperatures higher than its freezing point. When there is a presence of Methane at shallow depth, it results in the formation of a cage like structure where methane molecules are trapped between molecules of water. This structure very much is like a tetrahedron structure with the difference that there are two different moieties forming it. Both the molecules have similar spatial arrangements of their elements which resembles a tetrahedron. Such structures are termed as Gas hydrates. They can form at moderate temperatures also, although normally they naturally occur in the cold environment of arctic permafrost and deep-water, seabed deposits - usually at depths greater than

800 ft. They also occur naturally in the Gulf of Mexico at depths as shallow as 1,750 ft and at a temperature of 45°F (7.2°C). One cubic foot of gas hydrates can contain 170 ft3 of natural gas. Naturally occurring gas hydrates can pose a well-control problem when drilled, but gas hydrate formation in the drilling fluid is a more significant well-control problem in deep-water situations.

Gas hydrates can form in low-salinity drilling muds under pressure/temperature conditions as mild as 480 psi and 45°F (7.2°C), conditions which are common when controlling kicks in Deep-water. During well-control situations, hydrates can plug risers, BOP lines and choke-and-kill lines, interfering with effective well control.

Although reported cases of gas hydrates are few, the risk of losing the ability to operate BOP equipment adequately has been always present. For this reason, all deep-water mud systems must be formulated to suppress the formation of gas hydrates. Increasing the salinity of water-base muds is the common method used to suppress hydrates. The standard deep-water, water-base mud system uses 20% by weight salt to inhibit gas hydrates. Increasing the salinity of a water-base mud system will reduce the temperature at which gas hydrates can form at a given pressure. The amount of salt required depends on both hydrostatic and shut-in pressures and the seafloor temperature. Twenty percent salt muds are not sufficient for ultra-deep or arctic deep-water drilling. At higher pressures and colder temperatures, a combination of salt and either glycerol or water-soluble glycol is recommended for greater inhibition.

Diesel oil, mineral oil and synthetic systems provide excellent hydrate suppression. This inhibition is a result of the limited amount of water contained in them and the fact that the water phase generally has a high concentration (>25% wt) of Calcium chloride. Natural gas, CO_2 and H_2S are all soluble in diesel, mineral oil and synthetic fluids. This solubility increases the contact area with the emulsified brine phase and would accelerate the formation of gas hydrates, if it were not for the high salinity of the aqueous phase! Gas solubility and its effect on kick detection as well as well-control, should be considered for all deep-water wells. These systems are attractive in deep-water drilling because they also provide excellent shale inhibition and lubricity. They are expensive systems and are not recommended for areas where there is a high probability of lost circulation.

Geology/Reactive Formations

The geology of deep-water drilling is different from that on land and in shallow water. The formations, for example, are relatively young and hence more reactive. The clays and silts have not been altered by extreme heat or pressure and are not significantly de-watered. Sands are often unconsolidated and have not been compacted. Shallow clay formations are very soft and sticky. Cuttings from these formations can cause packed-off hole, plugged flow lines, balling of the bit, bottom-hole assembly and reduced Rate of Penetrations (ROPs). The mechanism that causes these problems is both mechanical (gouging from stabilizers) and chemical (hydration). Also, the young clays contain high volumes of water and can be extremely sticky and problematic regardless of the degree of inhibition. Swelling and dispersion of the reactive shale must be addressed when drilling in Deep-water. Synthetic, diesel, mineral, PHPA, enhanced chloride and Lignosulfonate systems all have been used in deep-water applications.

Synthetic and oil-base muds provide excellent inhibition, virtually eliminate problems with gumbo (due to the oil-wetting of sticky surfaces), and provide good lubricity and gas hydrate suppression. Water-base systems will require additives to increase performance and to minimize trouble with soft, sticky gumbo and for hydrate inhibition. Even so, water-base mud systems cost less and may be the system of choice when low fracture gradients and lost circulation are concerns.

Pore Pressure and Low Fracture Gradients

During deep-water drilling, challenges exist that are related to formation pore pressures and fracture gradients being very close at shallow depths. For deep-water applications, the fracture gradient and equivalent pore-pressure decrease as water depth increases. At extreme water depths (±10, 000 ft), this low fracture gradient (due to the lack of overburden) and low equivalent pore pressure make drilling impractical, even with un-weighted mud, due to annular pressure losses increasing the Equivalent Circulating Density (ECD).

Many schemes have been proposed to lighten the density of the mud column in the riser for these extreme water depth situations. In routine deep-water depths (1,500 to 5,000 ft), the low fracture gradients complicate well-control situations, since the casing seat will not hold a high shut-in casing pressure. The typical deep-water well uses frequent shallow casing strings to seal off low fracture- gradient formations. The low fracture gradients also present lost circulation problems from both surge and swab pressures. This is especially true with synthetic, mineral oil and diesel systems, which are compressible and tend to reduce the allowable fracture gradients. Surge, swab and ECD pressures are a significant concern for all deep-water drilling operations, particularly while running and cementing casing. Understanding the effects of temperature and pressure on hydraulics and drilling fluid rheology is very important in deep-water applications. Low water temperatures and the resulting low riser temperatures can result in elevated fluid rheology and high surge and swab pressures.

Riser Volumes/Large Casing/Logistics

As stated earlier, wells drilled in Deep-water require long risers and large diameter casing strings. The riser, large-diameter casing and larger hole sizes call for large mud system volumes. A 20-in. ID riser in 2,500 ft of water has a volume of 972 bbl. It is not uncommon for a deep-water drilling operation to have a circulating system of 4,000 bbl or more. These large systems require proportionately larger quantities of mud additives for maintenance and treating. The quantities of additives required and the limitations imposed by space and deck loads make inventory management critical. The distance of these wells from shore-base facilities and the logistics of operating offshore mean it may take a significant amount of time to deliver mud materials to the well. Other conditions, such as work boat availability and sea conditions can extend the delivery time even more.

Finally, minimum inventory requirements cannot be estimated solely from previous usage. Changing conditions and treatment requirements must be anticipated. Alternative product packaging should be considered. In many cases, bulk or liquid premixed delivery may be advantageous. Salt, for example, is required for the water-base systems typically used in Deep-water. Bulk delivery or 1-ton big-bag sacks provide more flexibility for product additions, storage and transportation. Salt also is available as brine, rather than in sacks or bulk bags. Many floating rigs have liquid storage capacity that can be used for brine, which will leave deck space for other products. A bulk-handling system for other products should be considered as an option for streamlining logistics, reducing product waste and cutting down on container disposal. Although these systems may require rig modifications, their potential for savings is significant in high-cost, deep-water operations. The use of computerized monitoring and product additions can make these systems very effective in providing a uniformly treated mud system. Proper estimates of materials usage and planning for product supply are necessary for a successful project. As with all the issues discussed in this chapter, they are a part of the total deep-water drilling system.

Low Flow Line Temperatures

As explained earlier, water temperature decreases with depth. Long risers surrounded by cold sea water will result in much colder mud temperatures and higher viscosity in the riser as well as in the flow line. The increased viscosity from temperature, particularly in oil base and synthetic muds, may limit the shale shaker screens to relatively large mesh sizes. Often, there is a temptation to treat the mud system to reduce the viscosity at the flow line, but this should be avoided, since it will reduce hole-cleaning in the riser. Circulating a third "boost" mud pump on the riser will limit the amount of cooling that occurs in it. A special HTHP rotational viscometer can be used to provide a more accurate profile of the effects of cold and hot temperatures and pressures on mud.

Formation Ballooning/ Down hole Mud-loss

To arrest / minimize the ballooning / down-hole losses tendency of the rock the formulation of mud is often not only tailored with the right amount of solids, but also on its ability to quickly form an impermeable barrier between exposed rock and mud with the help of filter cake.

Minimizing pore pressure transmission by keeping the lowest possible value of Fluid filtrate loss also helps to arrest the ballooning tendency of mud.

In addition to it, inclusion of formation-strengthening materials in the form of sealing and bridging agents helps to avoid ballooning feature.

As and when possible, it is also desired to engineer a fracture in the rock in a controlled manner and strengthen the rock matrix by filling the tip of the fracture with the correct amount and right size of sealing and bridging agents.

Stress Cage Concept:

Deep-water drilling throws a challenge of drilling through unconsolidated, soft and permeable rock matrix which is susceptible to accepting mud more readily than other types of environment. The situation is further compounded by the fact that SOBM, which is otherwise also more prone to down hole losses by its inherent nature, is widely used for drilling through such matrix. Thus, the probability of encountering downhole losses in Deep-water with SOBM is many folds higher than in conventional environments. Several innovative techniques have evolved to eliminate/minimize the adverse impact of this situation. Stress caging, first conceptualized and implemented by BP, is one such recently developed technique, being used to drill through such environment with least down hole losses and reduced NPT arising out of unstable bore hole. A brief description of the process is as follows.

THEORY

Stress Cages, is the name given to a method for strengthening the wellbore wall and effectively increasing the fracture resistance of rock formations. The technique can be used to drill above fracture gradient without inducing mud losses and has the potential to eliminate/minimize the number of casing strings. A major prize is accessing reserves in depleted reservoirs where the reduction in pore pressure has weakened the formation. Another application is in Deep-water drilling where the drilling window between pore pressure and fracture gradient is often narrow.

The range of application of stress caging may include following:

- Access to additional reserves (depleted zones)
- Reduced mud losses in deep-water drilling
- Loss avoidance when running casing or cementing
- Improved well control
- Elimination of casing strings
- Extending casing shoe setting depths
- Tackling over-pressured formations (avoiding losses in adjacent weak zones)

- An alternative option to expandable casing

Stress Cages achieve wellbore strengthening by changing the "stress state" around the wellbore rather than by altering the strength of the rock itself. This is done by allowing fractures to form in the wellbore wall, so that they act as wedges to compress the rock within a zone around the wellbore.

The fractures are sealed using bridging particles added to the mud system. The particles must hold the fracture open near the fracture mouth and they must seal sufficiently well to provide pressure isolation to prevent fracture propagation. Muds specifically designed to do this have been referred to as 'designer muds' or 'magic muds'.

Some general observations are:

- Fracture widths as small as 1 mm can produce an increase in effective wellbore strength of the order of 1000psi.
- A short fracture is best. A long fracture needs to be wider to achieve the same strength increase, and may be too easy to re-open.
- Softer rocks (low Young's modulus) will require larger fracture widths

Types of Application

There are many situations which could benefit from the application of Stress Cages. The following is a brief description of each type of application. There are two basic ways to engineer the treatment:

a) Continuous drilling – requires that bridging solids are maintained in the circulating system. This is an engineering challenge, in terms of solids control, logistics, and maintaining mud weight and rheology. There are ways to optimize this, which are described later. The advantage is that this method will protect the well bore continuously as we drill.

b) Pill applications – simpler and used to strengthen sections that have already been drilled (but have not already suffered losses).

Stress caging will not work if we wait for a loss event to occur and then apply the system – the fracture that has been created is likely to be large and hard to seal, and too easy to re-open.

Mostly stress-cage applications have been applied to the continuous drilling method, but each stress cage application will need to be tailored to the wellbore and drilling plan. Following are different situations where stress cage has possible application.

a) Casing Shoe Set in Sand with a Low Fracture Pressure

The casing shoe has been set in a weak formation and an FIT or LOT yields lower than required values. To drill the well to the next casing point will require raising the mud weight to a level which, when combined with ECD, will result in mud losses at the casing shoe. A sufficient kick margin will not be available to drill to the next casing point. To solve the problem, the sand interval below the shoe should be drilled and a stress cage pill can be spotted and squeezed to increase the fracture pressure. Subsequent drilling operations may require constant additions of bridging agents to build Stress Cages in the entire well bore.

b) Drilling Depleted Formations

The well is being drilled into a depleted reservoir with higher pore pressure formations above and perhaps also below the reservoir (usually shale). The overbalance required to control these adjacent intervals may exceed the reservoir fracture pressure (fracture pressure of depleted reservoirs is reduced due to depletion). In this situation, drilling into the depleted reservoir can result in massive lost circulation. A stress cage can be used in conjunction with prudent drilling practices to drill the depleted interval without severe lost circulation.

c) Drilling Pore Pressure Regressions

The wellbore drills from a high-pressure interval into a lower pore pressure interval. This is often caused by crossing into a fault, a fold, or new depositional environment. In this situation, a Stress Cage can be used to increase the fracture pressure of the subnormal pressure interval, and allow the well to reach total depth or to increase the casing setting depth.

d) Extending Casing Shoe Depth in Deep-water (narrow fracture window)

The extensive length of riser filled with weighted mud causes a higher than hydrostatic pressure on the formation at the last casing shoe. Since seawater (water depth) replaces much of the normal rock overburden pressure, lower than the fracture pressure exists, which causes lost circulation when drilling. The top-hole drilling procedure (pump and dump) can solve this problem by eliminating the riser and taking mud/cuttings returns at the seabed. As drilling progresses, a riser and BOP must be installed and the full hydrostatic pressure of the riser mud is then exerted on the exposed formation. In this situation, a Stress Cage can be built and carried for the entire section of the hole, which can strengthen the exposed formations and allow subsequent casing points to be set deeper in the well. The Stress Cage can also increase kick tolerance in the well bore.

e) Preventing Losses when Running Casing/Cementing

High pressures can be generated during casing or liner running, or when cementing. Often, these pressures will exceed the pressure generated during the drilling phase, leading to formation breakdown and unexpected losses. Stress Cages can prevent these types of loss situations.

g) Losses when Swabbing/Surging

The stress Cage technology could be applied where there is a history of induced losses caused by surge effects; e.g. where there are particularly narrow clearances (narrow annulus).

JOB PLANNING

Data Requirements

To plan a well for a Stress Cage application, the following data should be collected for the hole section(s) where Stress Cage will be used:

- Core Data (permeability, porosity, pore throat size, lithology, fracture permeability)
- Reservoir Data (pore pressure, permeability, porosity, horizontal stresses)
- Log Data (pore pressures, porosity, permeability, formation dip, and formation temperature)
- Offset drilling data (mud weights, hole problems, lost circulation, well deviation, mud systems, LOT/FIT data, flow-line temperature, hole size, ECD, solids control data, and lessons learned)
- LWD/MWD data (ECD, porosity, permeability, formation temperature, and hole deviation
- Offset Mud Recaps (LCM treatments, mud system requirements (SOBM, WBM), product concentrations, ECD, hydraulics, mud mixing capabilities, mud product logistics, mud properties requirements).

At time, the timely availability and/ or authenticity of the above-mentioned data can be a challenge. It is better not to have data than having a set of spurious data! In exploratory areas, the validity of the designated off-set well itself can be tricky. The information available for pre-drill pore pressure curve itself is often subjected to online correction, based on information revealed during drilling – leaving no scope for foolproof plan and invariably continuous course correction is the call of the day. Recording of FIT v/s LOT has several times been a matter of intense debate with followers of both schools of thoughts considering themselves right.

Therefore, one should rely more upon data available while drilling. Any previous information if correctly and timely available can - be a bonus!

Not to undermine the relevance of information being sought, the coordinated efforts of the team and continuous monitoring of every information gathered during drilling of the previous section, help the most to plan for drilling of the next section.

Pore Pressure vs Fracture Pressure Curve

It will be necessary to prepare a detailed pore pressure vs. fracture pressure vs overburden pressure graph to analyze the potential benefits when using Stress Cages in a well. Obtaining accurate data will be critical to success – very often, the fracture gradient line is only established after losses have occurred, when it is too late to apply the stress cage technique.

The main consideration when designing a Stress Cage application is the difference in the Sand/Silt Fracture Gradient. This is the area that Stress Cage is more reliable, since it refers to permeable formations. If the mud weight required to drill the well falls within the yellow shaded area of the pre-drill curve, then stress cages can be built and used successfully. Drilling above the shale fracture gradient line is a bigger challenge and there is limited experience in doing this to date, other than evidence from leak-off tests.

Mud Design

Most experience to date has been with oil based mud. Oil mud can be designed with a very low fluid loss compared to water based mud, which helps in the sealing action at the mouth of the fracture. But there are situations where it is preferred to use inexpensive water based mud and live with the risk of losses - e.g. if there are natural fractures or vugs which could be hard to seal and will almost certainly lead to large losses. Water based mud may also be needed for environmental reasons or technical reasons such as formation damage or for open hole completions.

Once the mud type has been selected, the following factors need to be considered:

a) Shale Inhibition: If reactive shale is present, a high inhibition level is required to ensure a good quality wellbore. This can be a challenge with water based muds and so should be carefully reviewed – Stress Caging will be difficult in a wellbore that is in poor condition. With oil mud, shale inhibition is usually very good, but note it is important to have the water phase salinity matching (or slightly higher than) the shale pore fluid salinity to prevent unwanted osmotic effects.

b) Mud weight: The mud design for stress caging is not particularly dependent on the mud weight. High concentrations of Barite in dense muds can assist in fracture bridging. At the other end of the scale, there is a minimum density limit because the bridging particles will contribute to mud density. Thus, Stress Caging in muds below about 1.1SG (9.17PPG) will be a challenge. In the continuous drilling, the mud weight will tend to increase with time as cuttings build in the system and additions of bridging solids are made. A schedule for mud maintenance will therefore be required.

c) **Rheology:** The rheology target is determined by hole-cleaning and ECD requirements, as in normal mud designs. Any bridging solids or fluid loss additives used for Stress Caging will need to be checked for their impact on mud rheology. With the continuous drilling method, bridging particles grind down over time, and drilled-solids build in the system. The rheology will climb unless there are proper mud maintenance procedures in place.

d) **Formation damage:** Consideration must be given to the types of additives used if drilling in the pay-zone. For cased and perforated completions, there is no requirement to run formation damage tests in the laboratory, provided filtrate invasion is kept minimal. For open-hole completions it is important to check filtrate damage as well as damage from the mud cake. Additives used to produce an ultra-low fluid loss, for example, could affect mud cake lift-off pressures or could be entrained in the mud filtrate and affect wetting. Marble used for bridging is commonly applied in pay-zones, is acidizable and so can be used with confidence. Graphite type bridging solids are not acidizable and so their impact on formation damage should be assessed by appropriate laboratory testing.

e) **Fluid Loss Control:** An ultra-low fluid loss mud is preferable for stress caging. Ultra-low fluid loss is defined as an HTHP fluid loss (at down-hole temperature) of <2mls in 30 minutes, preferably < 1ml. This is critical for stress caging in very low permeability rocks (shale). In permeable formations, it is less critical but still desirable. There are also side-benefits due to the lower fluid invasion (e.g. Reduced chance of differential sticking). By permeable rock is meant formations with a permeability > 1mD. Typically, shale has permeability much lower than this, in the Nano-Darcy range (10^{-3} milli-Darcy).

HTHP fluid loss values below 2 mls can be achieved realistically in oil muds. This is done using fluid loss additives such as gilsonite, asphalt or synthetic polymer. The synthetic polymer is preferred. Reducing the oil/water ratio (adding more water phase) can also reduce the fluid loss value, but will tend to thicken the mud and so is less suitable as a method especially at high mud weights. The presence of fine clays can assist in achieving the low fluid loss – usually this is automatic because of the use of an organoclay, or because of drill fines picked up during drilling. The emulsifier package can also affect the fluid-loss value and is another variable open to the formulator. There is a need to run screening tests to formulate the system, and demonstrate that an active mud can be converted into an ultra-low fluid loss system. Care is required to ensure that the additives do not produce excessive rheology or cause formation damage. 'Black powder' additives such as gilsonites and asphalts can mask hydrocarbons and may interfere with core analysis.

BRIDGING SOLIDS TYPE, SIZE AND CONCENTRATION

A) Solids Type

Experience points to a blend of marble and graphite as the best bridging mix. This is superior to fiber blends for Stress Caging. The marble acts as a proppant to hold the mouth of the fracture open, and the graphite is believed to have a sealing action. The graphite is resilient, which means it is deformable in a reversible manner – it can be compressed and will re-expand when pressure is removed. This property enables it to maintain a seal if the fracture flexes due to pressure changes.

Marble: a hard form is required (not chalk), so that the material does not grind down quickly. The best type is dolomite which in terms of mineralogy is a Calcium/Magnesium Carbonate mix. This tends to be harder than pure marble. In terms of mho value, a hardness level of 4 or more is preferred.

Graphites: This must be granular, resilient material. It is available from the major service companies with their own unique name, but is almost more or less same type.

B) Marble/Graphite Ratio

The optimum ratio of marble-to-graphite is still debatable, but laboratory tests and field experience suggest it is not critical. The recommended range is between 70:30 and 50:50 marble: graphite. Often, the ratio is determined by the sizes of marble that are available and how these should be blended to obtain a smooth distribution.

Stress caging has been achieved using marble alone and using graphite alone, but this is not generally recommended. If graphite should be avoided for formation damage reasons, marble alone will have to be used.

C) Size distribution

The solids must be large enough to bridge at or near the fracture mouth, and small enough to bridge any pores in the rock matrix to ensure permeable rocks are properly sealed by filter cake. In most situations, the fracture width will exceed the pore throat size so that the bridging mix can usually be designed around the fracture width.

As a general guide, rocks which are unconsolidated or weak will produce wide fractures (millimeters), and rocks which are strong will produce narrow fractures (<1mm). This is very much dependent on the pressure enhancement that is required. The bigger the increase in fracture pressure required, the wider the fracture will have to be. Modeling work is therefore very important to design the bridging mix.

Once the fracture width (and formation pore size) is known, the size distribution of the particle mix can be designed. In the absence of a model, the following general guidelines should be used. Let's assume a fracture opening width of 800 microns. The bridging solids' design would then proceed as follows:

i) Check the particle size distribution of the base fluid. For a Barite-weighted mud, the d90 size of the mud would typically be around 50 microns.
ii) Identify a source of Limestone that will extend the particle size of the mud from the base value of 50 microns to 800 microns (the opening width of the fracture). Thus, Limestone with a d90 of about 800 microns is required. This calls for blending 2 or 3 grades of Limestone to obtain a smooth distribution.
iii) Identify a source for the graphite material and check the particle size and quality of supply. Typically, the graphite will have a size range of 200-800 microns. The graphite size should

duplicate the Limestone, if possible. Avoid a fine material such as powdered graphite which is not really very effective.

iv) Decide on the blend: There is no ideal size distribution curve for stress caging. The size distribution is usually dominated by the Barite content which makes an ideal packing curves impractical, and the size distribution is, in any case, affected by the cuttings build-up in the system. Therefore, approximations should be made and a reasonable size distribution should be selected based on particle size information. This should then be confirmed by performing some laboratory test. Following would be an acceptable bridging mix for an 800-micron fracture opening using a Barite weighted mud:

Bridging solids blended equally	D 10 (microns)	D 50 microns	D 90 microns
Limestone	50	145	320
Limestone	300	560	840
Resilient graphite	150	350	550

D) Concentration

i) Pill Treatments

For pill treatments, high concentrations of bridging solids should be used. Laboratory data suggest that higher concentrations are more effective and it is usually practical to run with high concentrations in pill form. The sealing will be more rapid at high concentration, preventing the fracture from becoming too long/wide.

Typical mix for pill treatment:
Marble 1 (size as above): 20 ppb
Marble 2 (size as above): 20 ppb
Graphite (size as above): 20 ppb

ii) Continuous Drilling – coarse shakers

For continuous drilling applications, it will be difficult to maintain high concentrations. The bridging solids will usually be taken out by the shakers, and will also grind down with time, so that maintaining concentration can be a logistics battle. It is generally recommended to use fairly-high concentrations of solids (e.g. 30ppb) wherever possible, and to maintain these in the system by running coarse shaker screens. For example, to keep solids in the order of 500 microns in the system will require 38 mesh shakers. Sections as long as 500m in 8 ½" hole have been drilled in this way, by monitoring mud properties and making dilutions/running centrifuges to condition the mud when necessary.

Typical formulation for continuous drilling, coarse shaker screens
Marble 1 (size as above): 8 ppb
Marble 2 (size as above): 8 ppb
Graphite (size as above): 15 ppb

4 -6 sacks per hour of the coarse marble and 4-6 sacks per hour of the graphite should be added continuously whilst drilling to compensate for crushing of bridging solids. I.e. 100-150kg/hour of each.

iii) Continuous Drilling – fine shakers

If fine shaker screens are used, pre-loading the mud with the coarse bridging material will be ineffective and will simply blind the screens on the first circulation. In this case, treatment should be made by continuous addition into the active pit. The maximum practical addition rate should be used, unless modeling work dictates otherwise. The concentration can be calculated from the flow rate.

Laboratory Testing

Once the basic system has been agreed, it is necessary to check the properties of the mud, and check the sealing ability of the system. In addition to standard mud checks, the following should be carried out:

a) HTHP Fluid Loss

Perform fluid loss tests at equivalent down-hole temperature to measure the HTHP fluid loss with differential pressure equal to the difference between ECD and pore pressure. This should be < 2mls for the full designer mud, preferably < 1ml. Exceptions would be when drilling permeable formations (>1mDarcy) where the fluid loss value is less critical, or when using water based muds.

b) PPT Testing

The pore plugging test (PPT) can give an indication of the sealing performance of the mud. Aloxide discs are used as a filter medium and are available in a range of micron sizes. The disc should be selected to suit the anticipated fracture size (or pore size of the formation, whichever is greater). Fluid loss through the aloxide disc gives a measure of sealing performance. It can be a useful rig-site test for mud maintenance purposes. There are some limitations/drawbacks with the method. Firstly, the equipment operates upside down so bridging particles in the mud will sediment away from the disc. This means a high spurt loss is often seen and can be misleading. The test is a static (not stirred) so the particles are not kept suspended as the mud heats up in the test vessel. The test can be run at high differential pressures (2000 or 5000psi, depending on the equipment design) but should be run if possible at pressures which represent the field conditions to avoid erroneous data.

Implementation

The following lists some of the key operational points. Further aspects are discussed below.

1. Review the Stress Cage Plans with the rig team, mud company specialists for potential problems.
2. Make sure the method of application is understood – i.e. pill treatment, continuous drilling method using coarse shakers, or continuous drilling using fine shakers.
3. Ensure the rig can handle the logistics
4. Make sure sufficient bridging material is available and that the particle size is known.
5. Check quality of supply of bridging solids – avoid soft material that would grind down quickly.

6. Check the particles are not so big as to block nozzles or downhole equipment.
7. Ensure a limit is set on ROP when drilling and when underreaming. 50 fph is the upper limit for ROP for stress caging.
8. Take regular mud samples and check the fluid loss and sealing ability (e.g. using a PPT device).
9. Organize sieves at the rig that can be used to check the presence of bridging solids
10. Apply a clear plan for maintaining mud properties. Closely monitor mud rheology and density. Decide if regular mud dilutions will be needed, if centrifuges will be used, and if shaker screen sizes will need to be adjusted.
11. Be prepared to stop drilling to allow the mud to be reconditioned if necessary
12. Spot check the active mud system for particle size and concentration.
13. Avoid swabbing and avoid back-reaming if possible
14. If the hole opening is needed, run the device as close to the bit as possible
15. Increase mud weight slowly - do not permit surges to occur.
16. Minimize the presence of cuttings near fractures when increasing hydrostatic pressure (i.e. Circulate the hole clean)
17. Maintenance additions of bridging solids are required, even when using coarse shakers.
18. Develop lost circulation contingency plan.

Mud Pumps

Most mud pumps should be able to handle the concentrations of Calcium Carbonate and other bridging materials that are commonly used to create Stress Cages. Pump liner erosion or abnormal wear is not seen as a possibility as the graded Calcium Carbonate and other materials are less abrasive than Barite.

Solids Control

Ensure the sizes of the bridging solids are known, and that it is understood how they will be used and what sizes need to be retained in the circulating mud system.

NOTE THE FOLLOWING POINTS:

Pill Treatments:

- If possible, run coarse screens when circulating out the pill and incorporate the pill into the active system. Continue to run the coarse screens for a time to keep the bridging solids present. When drilling ahead, it is recommended to add 4-6 sacks/ HR of the bridging mix as a precautionary measure to reduce the likelihood of breakdown after the pill treatment.

Continuous Drilling

- Try to run coarse shakers that will keep the bridging solids in the mud. For example, this could mean 38 mesh to retain 500-micron particles in the mud. It may be feasible to allow this to happen, especially for short sections and small hole sizes. Surprisingly, even if all the drilled rock is incorporated into the mud, it may only amount to a few % solids at the end of the section.
- If fine shakers should be run and bridging solids are stripped out, ensure that continuous additions of bridging solid are made. But this normally results in low concentrations of essential solids, as depletion of desired particles' concentration not only happens due to stripping from fine shakers, but the retained solids also keep depleting due to churning/grinding action of bit upon them while being circulated.
- In high weight muds (>1.6SG/13.3PPG) rheology can climb quickly as solids build up. If coarse screens are being run, be prepared to stop drilling and condition the mud to bring it back into specification. This can be especially time-consuming due to high Barite levels.
- It is best to have the flexibility of using different screen sizes on multiple shakers and the ability of bypassing shakers.
- Centrifuges can be useful to help condition the mud when drill solids build to an unacceptable level. This may be more cost effective than dumping and diluting.
- If mud properties become unmanageable, then mud cleaners and de-silters/De-sanders should be available to treat the mud system

- A small circulating pit volume will minimize treatment time and cost when loading the system with bridging solids, but will make the system less tolerant to cuttings build-up.

Mud Mixing and Maintenance

The following are some guidelines for mixing and handling large volumes of Calcium Carbonate and graphite on the rig.

- Calculate the total volumes required for the job, including maintenance additions and contingency. Examine deck space and identify the best means of storage, transportation, and addition.
- If mud is to be supplied from a mud plant, consider having the bridging solid added to ease addition requirements at the rig. Check that the solids will not sediment out in a supply vessel (may require continuous agitation).
- When mixing bridging solids, avoid shear devices which will grind the particles down. Simple paddle-type agitators are best. To effectively extract the utility of shearing devices, it is preferred that first the mud be built by emulsifying the immiscible phases and emulsion strengthened by adequate shearing and emulsifier doses as required. The shearing can continue for other additives and can be stopped just before beginning to mix sized solids and thus get the best value for money, invested in the form of resources.
- If large volumes are required, use a deck hopper and bulk bags for adding the solids to the system, and consider other bulk addition methods.
- Ensure mud hopper(s) can handle the hourly simultaneous blending and pumping operations required for the planned program. Upgrade hopper systems as required. Typical mud hoppers and sack cutting can handle 40 sacks an hour per hopper.
- Ensure a 50 or 100 bbl slugging tank with agitation is available for Stress Cage applications requiring spotting and squeezing a pill.

Monitoring Bridging Material

When a bridging formulation has been agreed it should be possible to perform some checks to ensure the size distribution, concentration, and sealing performance of the system is maintained.

Measuring the particle size and concentration offshore can be time-consuming and difficult. It is not usual to have a laser particle size of the rig, and in any case, these devices are not straightforward to use with muds (especially oil muds) and can give misleading results. The simplest test offshore is therefore to use sieves. The following procedure is very simple and can be adapted as appropriate:

a) Identify the maximum particle size that needs to be in the mud (e.g. 600 microns)
b) Select a sieve size slightly smaller than this (e.g. 500 microns)
c) Pour a known volume of mud over the sieve
d) Observe the particles retained on the sieve (usually both graphite and marble)
e) If possible, wash with solvent (e.g. Water for water based mud, toluene/acetone mix for oil mud)
f) Allow to dry, and determine the amount of bridging solid retained by weighing

For a more thorough analysis, samples should be sent to a main laboratory for analysis. As a minimum, samples should be taken from the suction (i.e. Representing the mud being pumped down) at regular intervals. In investigative studies, samples should also be taken from the return line prior to passing over the shakers and from the return feed to the mud pits.

PPT should be run at the rig-site whenever possible. Ensure that the Aloxide disc has the appropriate pore size for the bridging required. For example, if the aim is to bridge 500-micron fractures, it is not appropriate to use a 90-micron Aloxide disc. Typically, for a 90-micron disc, the PPT should yield a spurt loss less than 1 ml and a fluid loss of less than 3 ml. Values will be higher for larger pore sizes.

Once drilling begins, the PPT test should be run every 12-24 hours to check if the mud is effective in bridging. The test is not a definitive test and has limitations as described earlier - but it will provide indications. If a PPT fluid-loss increase is seen on a test, then additional bridging material should be added to the system until the fluid loss decreases to the target range. Avoid overdosing with fine grade material – it is probably the larger material that has been lost or ground down.

If the strategy is to use solids control equipment, then constant additions of bridging material will be required. When the strategy is to strip out bridging material at the shakers, then addition rates can be calculated for the required concentrations by using the mud flow rate and hopper mix rate.

An innovative technique has been developed where a **stress caging pill** is used to strengthen the freshly drilled formation soon after carrying out a LOT. The intention here is to effectively seal the leaked rock because of the application of pressure beyond the FIT.

Following description covers the actual implementation plan

- After drilling through the shore-track, rat hole and 3 to 5 m fresh formation circulate and flush the cuttings out of the hole.
- Pump and displace a stress caging pill of 70 ppb Sized Ca Co_3 and 35 ppb Graphite to a depth just above the shoe area.
- After the recording of LOT is completed pump and allow the entire stress caging pill to be circulated across the zone which was subjected to pressure causing it to leak. The fracture so opened, although - not exactly engineered with input of reservoir parameters, do get filled up with the stress cage material thereby causing a strengthening of rock.
- This technique has been successfully deployed in many deepwater wells with a proven track record of success in the field.

Contingency Lost Circulation Material

Contingency materials should be in place to cure losses if the stress cage application fails, or if large natural fractures or vugs are encountered. For more details pl. Refer chapter on loss circulation control

Post Analysis

Planned versus actual results should be reviewed and the implementation procedure modified for the next well in the field.

Following guidelines for Stress Cage optimization during the post analysis process shall help and improve upon:

- Look for supporting evidence that the normal fracture pressure was exceeded and that stress caging took place. Sometimes actual pressures are quite different to the predicted/planned values.

- Review ECD and mud weight versus pore pressures and fracture pressures of the sands and shale intervals to assist with the above. Don't ignore cementing ECDs.
- Review open hole log data over the Stress Cage interval(s). Look at the shallow, medium, and deep resistivity logs data to see if invasion occurred across permeable intervals.
- Check caliper log data for indications of thick filter cake across permeable intervals which could indicate invasion.
- Review PPT daily checks and mud recaps to look for trends which indicate the effectiveness or lack of bridging material
- Review all loss zones versus offset data and determine if the stress cage was effective in stopping or slowing losses. Adjust bridging material for next well based on the results.
- Review rig operations to determine if the Stress Cage application caused operational and/or logistics problems.
- Use data and lessons learned to optimize Stress Cage application for the next well(s) in the field.

Conclusion about stress caging

No one can deny the validity of the concept in theory and is certainly an engineering marvel coming from a reputed international company. However, the number of variables and uncertainty associated with each of them at each step makes it almost impossible to implement stress caging strictly as per the recommendations. It is a dream come true sort of expectation level, both from planning team as well as implementing team to tick all the boxes right from the checklist!

Having said that, it still is a useful technique and provides fruitful results (may not be 100 % satisfactory) when compared to any contemporary technique proposed with the same intention.

Following are some of the field traits / tips to run the system effectively and further improve its performance / accountability-

- As far as possible the mud weight should range between an excess of pore pressure and below the breakout pressure to have effective ECD always above breakout pressure during drilling. Prior to POOH if hole conditions permit, the mud wt. to be raised close to effective ESD which is available from online data with the help of APWD tools.

- The 6 RPM reading should be roughly equal to hole size in inches or 9 whichever is more.
- Every time the mud weight is raised by 1 ppg or more consider raising the Activity by 0.01 value higher than before to off-set the impact of increased filtrate invasion and further increase hole-stability.
- In a typical case, the ECD values can be 0.3 ppg above the mud weight. If the difference between ECD and mud weight is more than 0.4 in the hole-size of 12 ¼" or more, consider either raising cleaning efficiency of mud by higher rheology or adjust ROP to keep ECD max 0.4, no matter even if the hole does not show any indication of instability.
- In the deep-water environment since the volume of circulating mud is larger, any occurrence of abnormal hole-condition takes a considerably longer time to surface up and once it surfaces up, then it is very difficult to overcome. Therefore, do not wait for symptoms to appear and strictly follow the practice of managing ECD below 0.4 ppg over the MW.
- A general correlation between hole size and admissible ECD is as follows (valid in vertical holes with max. Mud weight of < 11 ppg)

Hole size 12 ¼" & more	ECD max 0.4 ppg above MW
Hole size 8 ½" or more	ECD max 0.6 ppg above MW
Hole size < 8 ½"	ECD max 0.8 ppg above MW

- For every corresponding increase in mud weight of 1 ppg, consider additional 0.1 ppg ECD rise over and above the suggested values. For example, if the mud weight is 10 ppg and hole size is 17 ½" in a vertical hole, the max permissible ECD is 10.4 and optimum number is 10.3, whereas if the mud weight is 11.5 ppg then the max. Permissible ECD under identical conditions can be 12.0 ppg and optimum value is 11.9 ppg.
- In a deviated hole, these numbers are higher than the suggested values by a margin of 0.2 ppg maximum.
- These numbers are suggestive only and have been arrived at after analyzing data of over 100 wells in Deep-water environment. The bottom line is – lower the difference, better it is.

- However, if the difference in ECD and mud weight is > suggested values, it is a strong indicator of the likelihood of occurrence of problems related to hole stability and act soon by selecting following in best combination or isolation, whichever is feasible under the then existing conditions.
- Consider circulating the hole to decrease ECD, raise rheology, slow down on ROP, increase pump discharge or take any other feasible measure to bring down ECD to below suggested values even if the hole is not showing any sign of abnormality.
- Bring down inert solids concentration (only fine drilled solids) by running centrifuge and bleeding base oil slowly.
- Even if the fracture gradient and pore pressure values permit, it is not advisable to not to control ECD with-in suggested numbers as ECD beyond suggested numbers are indicative of accumulation of unwanted solids in the system and the same must be brought down to values to match with ECD as suggested.
- Consider the monitoring solids concentration and size of particles by simple sieve analysis of mud to help plan treatment accordingly.
- Whenever the centrifuge is run to decrease drilled solids, consider replenishing lost essential solids by constant adding of sized Calcium Carbonate and Graphite.
- Often it is difficult to manage this, as the rate of discarding of solids is invariably higher than the rate of replenishing. Under such circumstances the availability of LCM recovery mode of the shakers is very helpful.
- Changing over to SOBM from WBM in the hole should be personally monitored by an on-board engineer to ensure that the inter-phase contamination is kept to a minimum.
- If the hole contains Sea water and is to be put on SOBM, there is usually no need to have any isolation spacer between the two fluids and SOBM is invariable more viscous than sea water and hence, does not cause any fingering, thereby inter-phase is very sharp at shakers and can easily be spotted.
- While changing over from SOBM to sea water due to lighter fluid displacing the heavier and viscous fluid, there is always a requirement to pump sacrificial thick and weighted WBM spacer which will prevent the contamination of SOBM with sea water.

- At times, the cleanup pill train is required to either change wettability of rock or tubular and surfactants/spacers should be pumped with intermittent sandwiching of viscous sweep to improve the quality of the cleanup, as may be operationally required.
- While running completion string the level of cleanup required is very high and the final brine left in the hole should be as clean as 40 NTU turbidity and / or < 0.05 5 solids w/v. A dedicated cleanup program/pill train should be prepared for every such case.

Riser less drilling application (typical deep-water)

Riser-less drilling is the practice of drilling without lowering any riser on deep-water locations in dual mode to facilitate push the first casing as deep as realistically possible even if there are shallow water gas shows. This technique often is aimed at either penetrating deep with the same number of casing strings or to reduce the number of casings and yet complete the hole with desirable size diameter. The return is continuously discharged along with cuttings into the ocean, and yet keeping the drilled cuttings away from the pit area to enable un-interrupted drilling by pumping seawater to begin with at a discharge rate which is adequate to keep the bit ar4ea free of cuttings. Intermittent pumping of viscous sweeps of Bentonite /Guar gum are pumped to ensure that cuttings get flushed away farther from the drill - bit area and do not adversely affect the on-going excavation.

A stage comes where the matrix below has / can have hydrostatic head which cannot be controlled by plain sweater. At this stage, plain sea water is replaced by a heavier fluid which is cheap enough to discard and yet abundantly available to continue to drill ahead in pump and dump mode. A dual gradient mode of drilling is ideally suited under such circumstances, where the gradient above seabed is tackled by sea water alone, whereas, a weighted mud of just enough weight is used below the sea bed, which makes the gradient uniformly more than the expected hydrostatic head, but below the fracture gradient. This way, one can precisely manage the increasing hydrostatic head with increasing depth.

Process

Execution of such drilling mode requires adequate planning, effective coordination and open communication between various units of rig including but not restricted to drilling, Mud logging and shore-base professionals. The expected Pore pressure needs to be known as accurately as possible to decide on the fluid weight below the seabed. Logistics and pit space management are two big challenges as the volume of heavy mud can run into tens of thousands of bbls to continue to drill as well as to displace the open hole with heavy mud which will keep the hole intact during tripping prior to lowering casing. To save rig time, often drilling proceeds at a rapid rate. This has an impact on the volume of heavy mud required to be circulated through at a higher GPM, which is further compounded by large diameter of the hole at shallow depth.

One should always keep buffer stock of heavy mud to account for any unplanned event, or break-down. At times, the volume runs beyond the capacity of the rig to store, including brine storage tanks and the solution to such problem lies in keeping the supply boat handy with adequate volume of heavy mud, soon to be taken on-board as soon as pit space is available. Therefore, boat trips to shore should be planned very meticulously. It does not hurt to always top-up the boat storage tanks with heavy mud every time it makes a trip to shore-base.

Some useful tips are as below

- Ship large quantities of weighted WBM (16 lb/gal or 1.92 g/cc)
- Use CaCl2 brine - store in ballast tanks
- Blend heavy fluid on-the-fly composed of:
- Sea-water + weighted WBM slurry + heavy brine + polymers for fluid loss & viscosity control
- Continuous circulation while drilling with returns to the seabed

Data Needed for Planning

- Pore Pressure vs. depth
- Hole Geometry

- Estimated ROP
- Required Flow Rates
- Rig capacity
- Boat capacity
- Weather

Benefits

Riser less drilling addresses

- Shallow hazards with a dual gradient approach
- Provide improved wellbore stability
- Push surface casing deeper than with Seawater + Sweeps
  - Higher frac gradient
  - Potential to save casing string
- Batch drills, surface holes

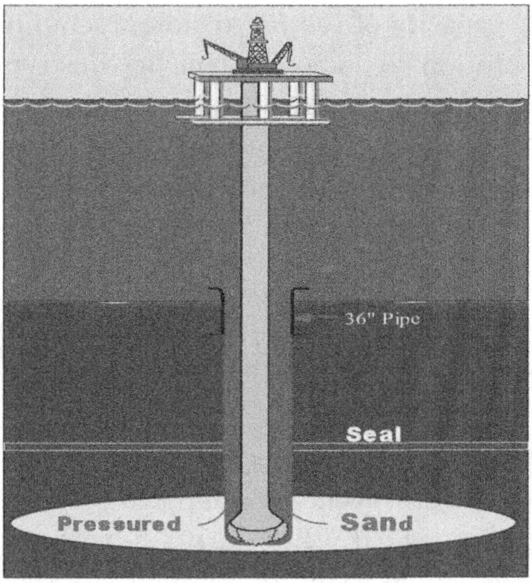

Typical benefits of using this application:

- Mitigating shallow hazards while continuing to drill (dynamic kill)
- Improved wellbore stability by utilizing mud rather than seawater, pushing the surface casing deeper (due to being able to increase the mud weight)

- Batch drill the surface holes (cutting weighted mud back with seawater decreases logistics problems saving rig time)

This system works via the dual gradient theory. Since there is a column of seawater on top rather than a column of mud, EMW is less than expected. Also, as we go deeper down the hole, the EMW slowly increases because we are increasing our column of heavy mud.

Different mud companies facilitate the same type of arrangement in principle to facilitate uninterrupted supply of homogeneous blend of a pre-determined weight of fluid by enabling blending of two different densities of fluid converging in a pit from two paths. This pit is often connected to a hopper for online mixing of additives to control other parameters of fluid (viscosity and fluid loss etc.). When a heavy mud of 17 / 18 ppg loaded with Bentonite is cut back to a lower number with the help of sea water, it can often result in separation of Barite which eventually settles at the bottom of the pit. This not only necessitates the cleaning of the pit, but the weight entering the well-bore could be erroneous, causing instability in the well-bore by the flow of matrix fluid due to dropped hydrostatic head.

To overcome such situation, one should continuously and gradually add a viscosifying polymer at the junction / at the vortex of agitator and thus stop the separation of barite from heavy mud.

Liquid type of polymers are very useful, as one can plan to straight away administer a slow stream of such liquid even without hopper from top of grating of pit or any source which facilitates this mixing. If the first shallow casing is pushed to a reasonably deeper depth, then it makes sense to add fluid loss control agent also through hopper at a manageable rate for homogeneous mixing of additives. At times, the ideal situation may also fall short of accurate requirement as far as F/L control is concerned. The alternative option is to properly treat the displacing fluid for water loss control and let that stay in the hole during tripping. As far as possible the viscous pills to be made up from PHB only as they are a better choice over polymer sweeps. As a compromise, one can also pump polymeric viscous sweeps. This happens when PHB consumption cannot match with replenishment because PHB requires time to yield and polymers provide a quick rise in viscosity. If polymer pill is used during drilling, it is preferable to once sweep the hole with large volume of PHB sweep before displacing the open hole with heavy kill mud prior to lowering casing / tripping. The figure below helps unserstand, how on a typical deep-water rig, the equipment can be set up to facilitate continuous blending and pumping.

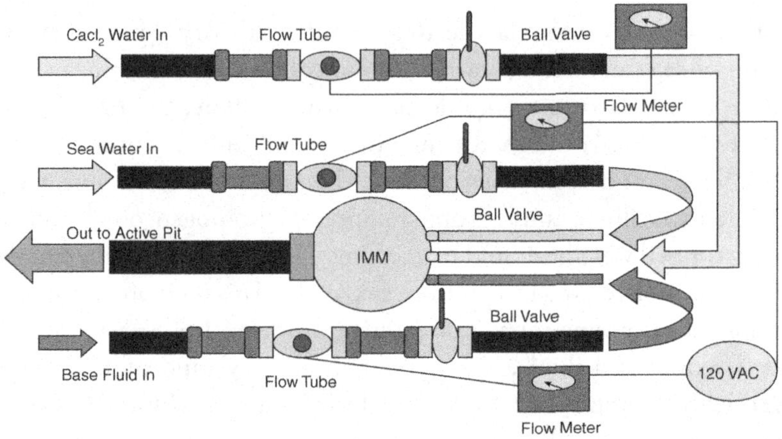

This type of arrangement allows for "continuous blending and pumping" adjustments to mud weights, flow rates, and polymer concentrations. The16 ppg mud enters the system through the bottom leg and seawater through the second leg. The streams are controlled using a flow meter and valve combination, and are mixed together using mixing device. The equipment provides excellent mixing for consistent products as well as the flexibility to change and control flow rates, mud weights, and polymer additions. Because all the rig sites are different, it is important to design the blending equipment to be as adaptable as possible.

Summary

- A reliable & proven process is available to mitigate shallow hazards
- Comprehensive preplanning is paramount to success
- Seawater and Calcium Chloride are viable cutback fluids according to case specifics

Chapter Summary

Deep-water drilling is a complex and costly undertaking. The issues and problems are unique and require careful planning and consideration.

Due to the high cost structure, contingency planning is suggested. High-performance drilling fluids are required to optimize hole-cleaning, borehole stability and the inhibition of gas hydrates. When combined with good drilling practices, they have the potential to deliver the greatest likelihood of success at minimum well costs. While short-listing the appropriate drilling fluid for Deep-water drilling/development/completion no consideration should be given to the cost of the fluid, if there are clear tangible benefits seen in the use of that fluid. The savings far out-weigh the cost of drilling fluid, especially with rig operating cost touching almost a million dollar per day. A properly selected and executed mud if saves even 15 minutes per day, it more than re-pays the amount spent - no matter what is the cost of handling / running the system. Caution also needs to be exercised in clearly defining the benefits. The rig time saved can be a cumulative effect of many variables, of which mud system could be just one and the benefits should be wisely, proportionately and justifiably accrued to all those variables. This exercise is not simple as it may appear, but a thoughtfully selected mud system with equally thoughtful execution can save great amount of money by eliminating down hole complications as well as minimizing NPT. One should never forget the ultimate bigger aim of bringing hydrocarbons to surface without adversely impairing the potential of the rock, of which drilling is just one step!

Therefore, any mud system which has a probability of causing ir-repairable and permanent formation damage to the extent that the rock cannot flow to its full potential should not be used, However convenient it sounds for drilling. One should always give due importance to the use of non-damaging nature of the fluid when drilling through reservoir part of the well which is intended for exploitation by completing the same. Theoretically speaking, it pays to undertake thorough formation damage core flood tests before selecting the fluid. If the extent of damage is not strikingly different between two fluids, then due consideration MUST be also given to the ease with which the well can be completed in a mud. Thus, the final fluid to be used should fulfill all or most of the following essential requirements.

1 Minimum / Zero damage to the rock
2 Convenience for completion operations

3. Providing adequate comfort and stability to the hole during drilling phase.
4. Should be robust and simple along with the fact that the system ingredients should be easily available and completely understood by professionals handling/managing the same in the field.

The market today is flooded with fancy systems and one need not get carried away with tall claims, which often fail to get transformed into reality. Any system which is robust, time proven and is technically feasible only should be considered for detailed investigation before using in the field.

CHAPTER 12

LOSS OF CIRCULATION

This chapter is specially written in detail to assist field engineers in minimizing lost circulation problems. Lost circulation is a complex subject, and experience of drilling in the rock is the most useful tool to prevent/cure losses in that area. Loss prevention/curing mud loss is more a scientific art than just engineering science. The chapter is intended and being projected as a 'user-friendly' tool, which can be used at the planning stage of a well as efficiently as during execution. If there is single largest and most challenging situations which tests the skill-set of mud engineers and patience of the rest of the rig crew, it is definitely curing and dealing with loss circulation!

INTRODUCTION

Loss of circulation refers to a situation where the drilling fluid pumped down-hole does not return to surface, partly or completely, and instead, enters the rock matrix.

Lost circulation can be broadly classified into two major categories based on the cause of the loss.

- *Natural losses:* These occur when the rock matrix has very large pores, or contain natural fractures or voids.
- *Induced losses:* These occur when a fracture is created due to the hydraulic force exerted in the wellbore, exceeding its strength to withstand, thereby yielding and accepting the fluid.

We normally drill with an overbalance, and hence losses of whole mud can occur into naturally occurring fractures or voids. The situation of induced losses is more complex, but is not uncommon in today's challenging wells. In deep-water environments, tectonically stressed areas, or in depleted reservoirs, the ECD is often very close to the fracture gradient. This means that small increases in effective mud weight (e.g. from ECD increase) can cause formation breakdown and hence trigger losses.

1.2 Natural Losses

Natural losses can occur in three formation types:

- High matrix permeability formations (gravels, coal pleats & coarse sands)
- Cavernous and/ or vuggy formations.
- Formations with conductive natural fractures or faults

Matrix Permeability

For the whole fluid to be lost to the permeable formations, the pore openings must be large. In theory, the pore size would need to be at least three times larger than the median particle size of the solids in the fluid. Since the average particle size is around 50 mm in most types of mud the pore size would need to be of the order of 150 mm for matrix losses to occur.

The types of formation where this applies include coarse sands and gravels. The losses in these formations are commonly called *seepage* losses and can normally be treated using conventional LCM materials as bridging agents.

These structures form when portions of the formation are dissolved or decomposed over geologic time. The voids generally form in dolomite or limestone formations and can range in size from small "worm-hole" networks, to *extremely* large caverns. Lost circulation, which occurs in a cavernous formation can be the most difficult type to overcome. The size of the void and the total volume can make it difficult to bridge the openings or fill the void with any lost circulation material pill or cement plug. In addition, these voids often exist filled with water or hydrocarbons, which does interfere with remedial techniques.

Losses to caverns are generally very large and are frequently preceded by "drilling breaks" as the bit drops through the void. To cure such losses, it either requires filling the entire void or plugging the cavern near the wellbore, a task that can range from very difficult to nearly impossible. In many cases, drilling through a cavern will require either tolerating/ managing to live with the loss or using an alternative technique - such as air drilling or mud-cap drilling.

Natural Fractures

Natural fractures and faults can occur in any type of formation. They are commonly found in tectonically disturbed areas surrounding salt domes. In such areas, it is generally accepted that lost circulation is a matter of fact and preparations to handle such problems are made a standard part of drilling plan. Natural fractures can sometimes be widened by excessive hydraulic pressure. They then thus also behave like induced fractures. Loss rates can be sudden and severe, and the treatment by non-conventional methods such as gunk squeezes, cross-linked polymers or special cements may be required.

1.3 Induced Losses

Induced losses result from the creation and extension of fractures by the drilling operation. Induced fractures are formed when the fluid pressure in the annulus exceeds the fracture gradient of the rock. This causes the formation to part, opening a fracture. Losses frequently occur during pressure surges and are characterized by their intermittent nature and flow back, as the fracture closes when the pressure is reduced.

Unlike natural losses, which first occur at the bit, induced loss often occurs in the *weakest* formation. For a given formation, rock strength generally increases with depth so that the location of the fracture is often just below the previous casing shoe depth. This attribute of induced fractures complicates the identification of the loss zone and the placement of materials designed to combat the problem. Anticipating induced losses at the planning stage is clearly important.

Induced losses can happen if the mud weight selected is simply too high, but they can also occur when the equivalent circulating density (ECD) is high, or when tripping through a tight hole, drilling too fast, or any other situation which causes a temporary pressure surge.

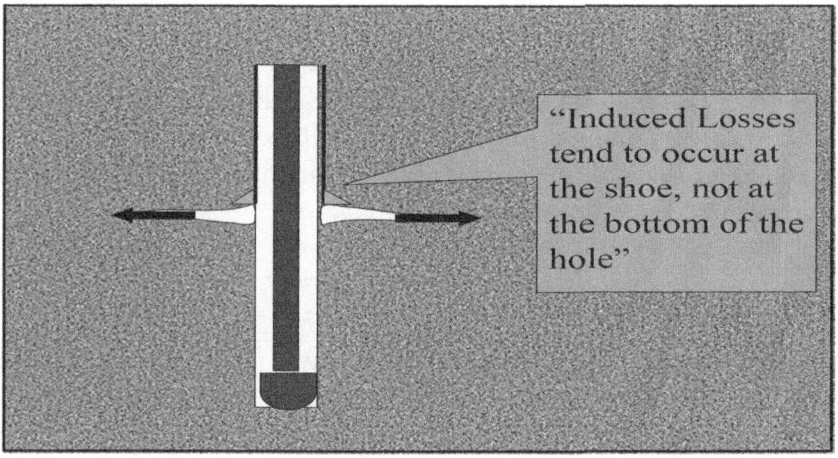

Curing losses in induced fractures can be troublesome, unless the ECD can be reduced. They are often characterized by the fact that loss rates are high when circulating, but reduce or cease when the mud is static (ECD effect). Thus, the fractures close and re-open again, making sealing by conventional LCMs more difficult. This is often referred to as ballooning or breathing and can lead to a back-flow of fluid and treatments into the well bore as the fracture closes. Another feature is that the propagation pressure needed to grow the fracture is generally much lower than the pressure required to initiate it. Consequently, losses can occur suddenly and accelerate away rapidly to give high loss rates.

Raising Fracture Pressures

If it were possible to raise the fracture gradient (effectively increase rock strength), this could greatly reduce the likelihood probability of induced losses. Many recently developed hypotheses/ theories suggest that the fracture gradient can be further enhanced by sealing the fracture and propping it open. In this method, the hoop stress is increased and the rock around the well bore is put into compression. This should have a strengthening effect (see diagram). Another idea is that it may be possible to chemically strengthen rocks, for example, using various specialty products available in the market. In theory, these methods should allow higher mud pressures to be tolerated without losses, but such treatments are not 100% established in the industry so far.

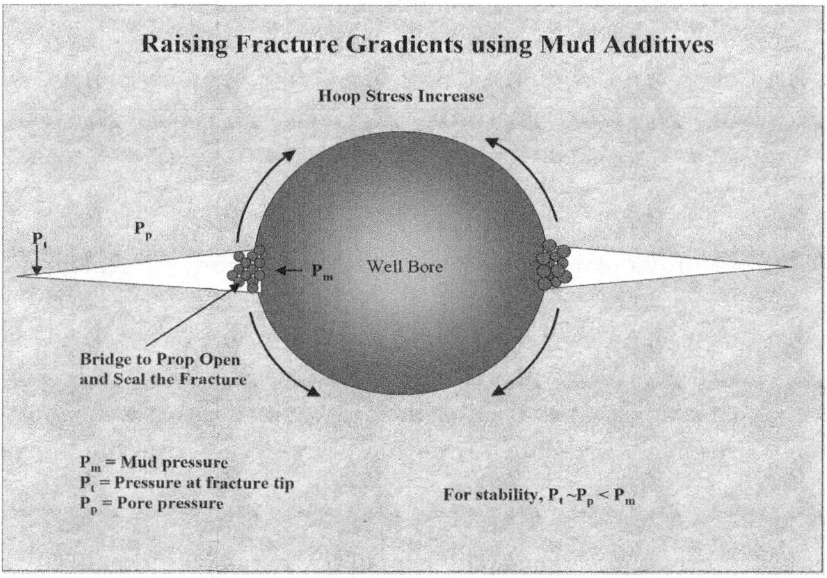

LOSS PREVENTION AND PLANNING

Introduction

Lost circulation is generally not 100% preventable. There are occasions when the down-hole conditions cannot be reliably predicted and fluid will be lost to the wellbore. In much of cases, losses are preventable and the first step in prevention is planning. The geological sequence and pressure profile must be studied to develop the casing plan and for the selection of the appropriate fluid densities as well as selecting the right type of fluid. At this point the potential for losses should be examined and a strategy developed. Every operation from spud to abandonment must be assessed for potential losses, and appropriate materials as well as procedures be kept ready at the well site.

2.3 Understanding Loss Prevention

The following section discusses in detail the factors which should be considered as part of the planning and risk assessment process.

- **2.3.1 Well Profile**

Rock stratigraphy is determined by nature, but it may be possible to select a well trajectory which minimizes the exposure to potential loss zones (faults, weak formations etc). The flexibility in the design will depend upon the well objectives, rig location, economics and possibly environmental or political factors.

When selecting the well profile:

- Identify the natural loss zones (eg faults) from the geology and try to avoid them
- Consider the pore and fracture pressures and try to relocate the well if it appears 'undrillable' in first place.
- Check offset well data for the occurrence of losses. One should ensure the authenticity of the information as well as the circumstances under which losses occurred (or did not occur). Losses arising out of incorrect/sub-optimal operating procedure often does not get highlighted in an offset well information!!

- **2.3.2 Casing Setting Depths**

Optimizing casing setting depths requires the following factors to be considered:

- It is normal practice to minimize the number of casing strings, but in difficult wells this will increase the likelihood chance of losses. As far as possible, avoid planning too close to the fracture gradient.
- Allow for ECD effects in the design.
- Minimize the amount of hole occurring between the casing shoe and a known or potential loss zone – this will make the losses easier to cure.
- Set casing as close as possible to any producing horizon.
- Allow for contingency string(s) in the design

- **2.3.3 Mud Weight**

Mud weight is a critical factor with respect to loss prevention as it is the driving force for losses. The pressure acting on the formation at a given depth is directly related to the mud weight:

Hydrostatic pressure = $h\rho g$ = height x density x gravitational constant

In oilfield units:

Pressure (psi) = depth (ft) x mud wt. (lb/gal) x 0.052

The mud weight (density) required is usually determined by wellbore stability and well control requirements. Mud weight must be sufficient to satisfy these needs, but not excessive as this will promote losses. Plans should be in place to clearly indicate whether mud weight can be safely reduced in the case of losses.

Although the hydrostatic pressure is set by the mud weight, the sides of the hole will see higher (or lower) pressures due to pressure changes in the annulus. While circulating, there is an annular pressure loss which equates to an increased density referred to as the Equivalent Circulating Density (ECD). This increase can range from zero to several ppg depending upon conditions. The annular pressure can also change when tripping in and out of the hole (swab and surge effects), when changing the BHA or drill-string design, and when starting/stopping the mud pumps. Clearly, these pressure effects are critical in cases where the mud weight is close to the fracture gradient.

The main factors to consider when evaluating annular pressures are:

- Rheology
  - ⇨ Any increase in the Plastic Viscosity and Gel Strength will increase the ECD
- Flow Rate
  - ⇨ An increase in flow rate can increase the ECD. If the hole is poorly flushed, increase in flow-rate will first decrease ECD and then increase!
- Hole Size/Annular Clearance
  - ⇨ Smaller annular clearance increases the ECD
- Annular Restrictions
  - ⇨ Any restrictions in the annulus will increase the ECD

- Cuttings Loading
  ⇨ High cuttings loading in the annulus (poor hole-cleaning) will raise ECD
- Tripping Procedures
  ⇨ These affect SWAB and SURGE pressures
- PWD Tool
  ⇨ A pressure while drilling tool will record down-hole pressures and should be used.

Although these factors are important for consideration, but more important is to have a complete picture and not decide just based on the effect mentioned. Let us say: Increased flow rate increases ECD, which means that we should decrease flow rate for decreasing ECD and minimize the probability of occurrence of losses. This is not always true. At times increase in flow rate further promotes hole-cleaning because of which ECD drops due to cleaner annulus.

- **2.3.4 Hole size**

In general, the smaller the hole size the higher the ECD in the annulus. As far as possible, avoid abrupt changes in hole sizes. For example, it is not preferred to drill a 12 ¼" below 20" casing, since 26" rat hole left below the 20" shoe will see a constriction in hole size straight from 26" to 12 ¼". Such areas provide a site for cuttings to accumulate and can create conditions of down hole losses in the immediate vicinity area.

- **2.3.5 Drill-Pipe Size**

Replacing 5" drill-pipe with smaller pipe will reduce ECD (and can reduce torque), but checks are needed to ensure that hole-cleaning will still be adequate and pipe strength is sufficient.

- **2.3.6 Hole-cleaning**

Drill cuttings in suspension add to the density of the drilling fluid. Cuttings accumulation is a function of ROP, flow rate and rheology of the drilling fluid. The planned fluid viscosity and flow rates should have been selected for the anticipated ROP, to prevent excess solids loading of the annulus. If the ROP exceeds the fluid's capacity to transport cuttings this will result in increased solids loading and could lead to losses. Consider using hole-cleaning models to estimate a controlled instantaneous drilling rate. Even drilling for short periods of time at high ROP can overload the annulus and lead to high bottom hole pressures.

- **2.3.7 Tripping and Pumping Practices**

Tripping into the hole will impart an additional pressure. The factors that affect the degree of this force are:

- Tripping Speed
- BHA properties (flow by area, Stabilizer design and quantity)
- Drilling Fluid Viscosity

Tripping into the hole causes Surge pressures and there is an increase in the risk of losses. The pressure will increase if the 10 minutes and 30-minute gel strengths are excessive. The surge pressures should be calculated using the programmed and actual fluid properties to determine the maximum tripping speeds.

Starting circulation will cause additional pressure in the wellbore; the more viscous the fluid, the greater the pressure. Therefore, lower gel strengths and the gradual application of the mud pumps will reduce the risk of losses.

Minimize SURGE pressures by:

- Breaking circulation at the shoe and roughly every 300m / 1000' while tripping in the hole. Circulate for at least 5 minutes before continuing the run.
- Rotating the drill string before initiating circulation.
- Bringing the pump up to speed slowly after connections.
- Pumping-out the first few stands off-bottom when tripping out.
- Controlling trip speed in open hole. Surge pressure calculations should be performed and the maximum pipe running speed established before trips.
- In Deep-water wells keep booster pump continuously running, whenever possible, to circulate the riser volume of mud. This breaks gel and negate the loss tendency of fluid.

Note that **SWAB** pressures (POOH) should also be considered. Tripping out of the hole should aim to minimize swab effects as the piston effect could suck filter cake or bridging material from the wellbore wall. On re-entry, losses can occur.

- **2.3.8 Fluid Design**

The following are the fluid parameters to consider when designing the mud system:

- Mud Weight
- Rheology
- Filtration Control
- Mud type/cost
- Mud pretreatments for loss prevention

Mud weight: Should be optimized keeping a proper balance between hole stability and tendency to lose.

Mud rheology: If ECD needs to be minimized, this may mean using low solids systems such as clear brines or minimizing low gravity solids in conventional muds. It may mean using special additives in WBM/SOBM to improve temperature stability and/or control rheology.

Filtration control: High filtrate loss produces poor quality thick filter cake, which can restrict the annulus and thereby increase the ECD and hinder hole-cleaning. A high filtrate loss can also lead to fracture propagation and hinder remedial actions. Good filtration control is required.

Mud type/cost: If severe losses are likely it is important to consider water based mud rather than oil based. With a barrel cost of up to 10 times for synthetic oil muds compared with water based, losses of the order of 1000 bbls or more can amount to huge mud bills for oil muds.

Pre- treating the Circulating System

Pre-treatment consists of adding certain types of lost circulation materials to the drilling fluid system in preparation for anticipated losses.

This is a powerful technique, which is most effective and easy to manage when losses are classified as seepage. In this case small-to-medium sized bridging agents are usually effective in controlling the losses. The most common pretreatment is to add granular/assorted Calcium Carbonate and Graphite. The pore or fracture size needs to be estimated and then particles with an appropriate size can be used. For an effective seal, particles should be at least as large as the maximum pore or fracture width; i.e. the d_{90} of the particle mix should be at least this size. The system should also contain a range of smaller sized particles to fill the gaps between the larger ones. The dose of Graphite normally ranges from 8 to 10 ppb. It is necessary to use the best quality graphite which is characterized by high resiliency to ensure that the pores which were closed by the intrusion of graphite during circulation do not open-up again when the pump is switched off. Another equally important factor is to use Calcium Carbonate of high hardness to ensure minimum attrition of particles, thereby retaining their effectiveness throughout the course of drilling. ***Around 40 ppb of bridging and 6 to 8 ppb of sealing particles are generally needed as a minimum to ensure efficient blockage.***

The following guidelines procedure should be followed to finalize a pre-treatment plan:

- Estimate the maximum pore or fracture width.
- Select hard-grade Calcium Carbonate (dolomite or marble) with a narrow size distribution with a d90 size at least as large as the max pore/fracture width.
- Check this will not cause excessive shaker blinding or block down-hole equipment. If required install coarse screens on shakers to retain the essential solids in mud
- Add 40 ppb of this Carbonate to the mud.
- If the mud lacks smaller particles at the start, add some smaller sized particles as well.
- Use a particle size analysis equipment if possible to check the mud size distribution; in some cases, the mud may already have the right size and concentration of solids so that no Carbonate additions are required at all.
- Whilst drilling, add 1 jumbo bag per 5 hours of the large Carbonate to maintain particle size

- Use a PPT (permeability plugging test) device if available, to check and monitor the sealing performance of the mud. This equipment is a modified HTHP cell which uses permeable discs of known pore size to simulate the formation being drilled. It provides a way of monitoring the bridging performance of the mud. PPT total filtrate values of 10mls or less are good.

The Graphitic has a large particle size (200-800 microns) and so needs to be continually added to maintain levels. Use of Graphite must be carefully planned with active input from the reservoir and geophysics group to ensure that there are no issues related to formation damage or interpretation of logs due to closed pore spaces.

The addition of asphalt and gilsonite or other fluid loss products will reduce fluid loss through the mud cake and so can be beneficial. Care needs to be taken if considering these types of products in the pay-zone, due to the possibility of irremovable formation damage.

As a rule, preventative measures should be considered as a first option, especially for seepage losses (low loss rates). If properly implemented, this method avoids lost drilling time.

- **2.3.9 Being Prepared for Losses**

If loss prevention measures should fail, procedures and materials need to be in place to deal with the losses. The outcome of the risk assessment will give a view on the likelihood of losses. Even if the likelihood is very low, contingency plans need to be in place and a minimum of materials ready on the rig to deal with unplanned events.

The key points are as follows:

a) Use the risk assessment to predict the types of losses and where they could occur
b) Identify suitable products and procedures
c) Ensure sufficient LCM materials are available
d) Check inventory to build or replace lost fluid
e) Ensure appropriate expertise (personnel) will be on the rig
f) Discuss the plans at the pre-spud meeting
g) Prepare at least 200 bbl pumpable LCM pill
h) Keep one small pit clean for cross-link polymer pill pumping, whenever possible

Selecting the best products and procedures for curing losses is not easy. The mud contractor usually plays the key role in this, but there are many potential pitfalls. The following discussion helps.

CURING LOSSES – FIRST STEPS

Step 1 "Establish the Loss Rate"

As a matter of routine, the monitoring of pit levels and returns should give early indications of losses. Once losses are observed, the loss rate should be carefully monitored as it could change rapidly.

Loss rates are expressed in barrels per hour (bbl/hr). The general definitions for loss rates are as follows:

Loss Rate Definitions

Loss severity	**Loss rate in bbl/hr**
Seepage	< 12.5
Partial to moderate	13 to 45
Severe	45 to 130
Total	<130

This classification is very loosely bound where often boundaries do trespass.

Step 2 "Check on Surface"

It is important to confirm that the losses are occurring down the hole and not at the surface.

The following procedure is recommended

1. Check the solids control equipment to ensure that no new equipment has been placed online and that the discharge rates are normal.
2. Check to ensure that no mud has been dumped, transferred or otherwise removed from the system.
3. Check all pipe joints; connections, and valves for leaks.
4. Find if any of the lines/valves of the riser joints is also leaking into the sea bed. This is important as otherwise, even though we

are not actually losing in the rock we will assume so and spend time on curing losses without any fruitful results.

Step 3 "Is the Loss Rate Sustainable?"

If it is established that the losses are occurring down-hole, decide if the loss rate is sustainable. If inexpensive mud is being used or the losses occur near TD of the section, it may be deemed acceptable to 'live with the losses' even if loss rates are fairly high. There are special cases where drilling with high losses is done such as losses in massive Carbonates. The option to drill ahead with losses should always be considered, but it is a decision which should be carefully made. Consider well control and implications for cementing once the liner/casing has been run.

Step 4 "Characterize the Loss"

The following checks can be carried out to characterize the type of loss.

1. Check the drilling program and check with personnel - is this a potential zone for *natural losses* (e.g. natural fractures, vuggy formations)? If so, LCM treatments will be needed to cure the loss.
2. Check for *induced losses* by considering the following options:
   - Back off drill rate (ROP)
   - Reduce pump rate
   - Stop mud pumps for a short time
   - Thin down the mud

Discuss the options in the context of the type of well, being drilled. Don't experiment and jeopardize other well requirements (e.g. Hole-stability, well control, hole-cleaning and Barite sag).

If one of these methods, cures the losses, the losses were induced by ECD effects and may not require LCM treatment. As another indicator, check PWD data (if available) and note if there was an increase in ECD just before the losses occurred. If so, this is indicative of induced losses. If losses are induced, consider reducing the mud weight to get a permanent cure. If these methods fail to control the situation, LCM treatments will be required.

Consult with mud logging service providers and discuss the trends seen in the last few hours; consider pit level indicators to establish the type of loss zone:

Pit Level Indicators:

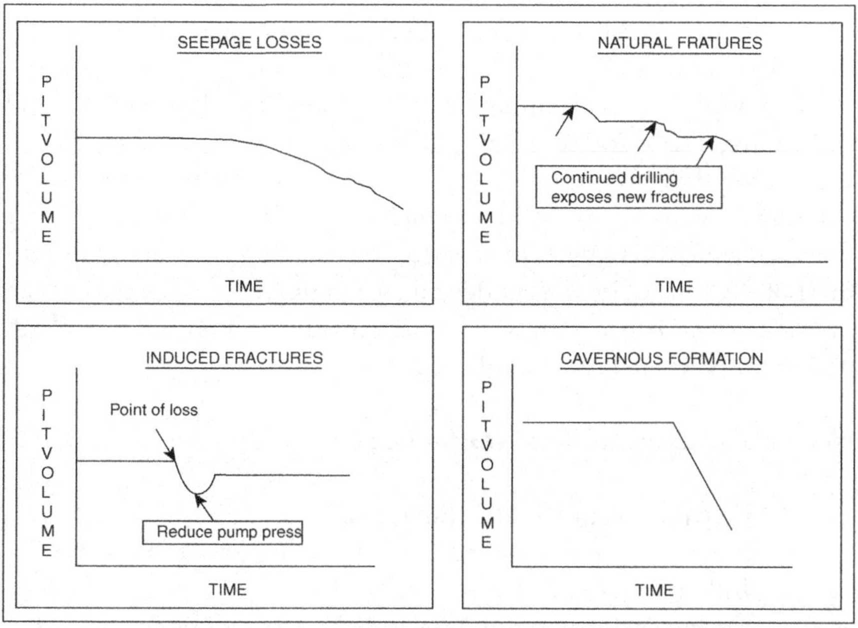

Pores/matrix: In this case the pit volume decreases gradually with time. The loss rate may increase during drilling as more formation is exposed. The loss rate can be low or high depending on the formation permeability.

Natural Fractures: A sudden decrease in pit volume occurs. The loss rate depends on the width and extent of the fracture and can be low or catastrophic. If there are small multiple fractures, a stepwise decrease may be new fracture surfaces are exposed.

Induced: Losses caused by increased annular pressure. Reducing pumping rate, stopping pumps, reducing ROP or optimizing rheology in case of excess rheology/gel strength can cure the problem. Dynamic losses high, but static losses low.

Cavernous/Vuggy: A sudden decrease in pit volume and high loss rate. It is difficult to distinguish this from a large natural fracture, but geological information should provide clues.

Step 5 "Locate the Loss Zone"

The accurate location of the loss zone is critical to enable the effective deployment of LCM treatments. Loss circulation pills will fail if they don't apply to the right zone. Locating the loss zone is not always easy.

It is important to examine the type of formations being drilled and the drilling circumstances leading to initiation of loss. Drilling breaks, such as the drill bit, dropping into a cavern, are obvious indicators of entering a matrix which will more likely accept fluid. Drilling a highly fractured zone where bit torque varies abnormally can be another indicator of the zone of loss. A sudden change in the rate of penetration can indicate the bit is penetrating a formation with high porosity and high permeability where losses could occur.

Note that losses are usually ***"On Bottom"*** **if....**

- They first occur while drilling ahead.
- The loss is accompanied by a notable change in torque or drilling roughness.
- The loss is due to natural fractures, caverns, or pores.

Losses are usually ***"Off Bottom"*** **if...**

- They first occur while tripping, wiping or increasing mud weight.
- They are the result of an induced fracture

If losses are off the bottom it can be difficult to locate the exact zone. The most likely place is at the base of the previous shoe where the formation is normally the weakest. It helps to double-check if the mud weight has been raised close to, or above, the fracture gradient at depth, just below the last casing shoe

If the loss zone cannot be located by the above steps it is probably best to attempt a cure anyway by pumping LCM pills. Try spotting pills near the bottom, and near the top of the hole, or in other suspected zones and cover areas, where even a slightest hint is available for occurrence of loss.

If this fails, there is the option, then to to run surveys:

Surveys

There are many survey methods available to assist in the identification of loss zones. While of great utility, all the survey methods have the limitations of the same nature, as follows:

- A lot of time is required to run the survey and accurately locate the loss zone
- The tools are not always readily available
- A considerable amount of drilling fluid is lost doing the survey.
- The results are often difficult to interpret.
- The tool could be lost in the hole.

The following paragraph describes survey types:

PWD/LWD

If a PWD or LWD tool is being run consider using it to help to locate the loss zone. With PWD, a discontinuity in pressure would be expected when comparing above and below the loss zone. If a LWD tool is being run, trip data from LWD-resistivity can sometimes identify loss zones and show healing on subsequent trips. Invasion of oil mud is easily picked up by the resistivity.

Other popular survey methods, tools are

Temperature Survey

Spinner Survey

Tracer Survey

Pressure Transducer

Acoustic Log

Hot Wire Survey

The description below, followed by different procedures to cure the losses, is meant to give a proper understanding and background of mud loss. The nature, volume and concentration of LCM pill to control mud loss will have to be decided based on the prevailing situation at the time of planning for the job. Almost all / most of the jobs described here are highly specialized and invariably an experienced specialist is highly desirable on the rig while execution of the job, who should be associated with the team from inception stage. Every job has its own limitations and strength and author neither does preferentially recommend any specific package, nor endorse for any specific mud service company. The names of the package are only for illustration purpose and are retained only to familiarize the readers with popular techniques. There are no claims being made about the superiority of one package above the other. Nevertheless, beyond the efficacy of the package, single most important aspect which can differentiate between success and failure is the competence of the team. It will certainly more than repay to have the best available manpower to ensure the success of the job, which is also dependent upon the selection of the right package for the right situation. Selecting the correct treatment method, having the right team and timely availability of resources, cures any type of loss. Of late, software modelling has come to our rescue in terms of execution details. Nevertheless, one cannot shirk away from the fact that software modules work based on the data input as well as the correct understanding of the questions being asked by the model.

4.1 Introduction

There are many conventional lost circulation products in the market, ranging from simple granular materials such as ground marble, to nut shells, cellulosic fibers, and plate-like systems like micas, flaked Calcium Carbonate, or shredded plastics. There has not been a definitive study to prove that one product is more effective than another. It is not too surprising, therefore, that in most applications, a mixture of different products is used in the hope that at least one of the components will do the job! To some extent, the selection is based on local experiences, cost, availability and personal preferences. This does not matter, provided some important general rules are applied. By simply applying the basic principles correctly, it is possible to bring more certainty to the curing of losses.

The following rules should be followed:

- **Use an LCM mix with the sufficient particle size of the loss rate encountered**
- **Consider formation damage issues when selecting product type**
- **Use a good engineering approach**
  - **Sufficient pill size**
  - **Sufficient concentration**
  - **Check the LCM won't block nozzles**
  - **Use correct placement techniques**

4.2 LCM Size

Traditionally, LCM treatments are formulated based on particle size. The terms 'fine', 'medium', 'coarse' and 'extra coarse' are usually used to describe the components of an LCM pill, e.g. '15ppb medium Carbonate' or '10ppb graphite'. These terms are misleading because the size ranges have not been defined and can be very different for different types of product. For example, an extra-coarse grade Carbonate could be of similar size to a fine grade fiber. This makes it difficult to standardize on pill formulations, and could explain why performance is so variable from job-to-job.

Size definitions for LCM. Pl. refer to LCM listing table ahead

Too often, the numbers also overlap, coalesce in the absence of stricter guidelines. However, the classification is like splitting the continuous spectrum of light energy in various types based on wavelength. We have UV rays, light rays IR rays with further sub-division of near IR and far IR, etc. In reality, it is a continuous spectrum of energy which has only been classified for convenience of studies. Similarly, the sizes have been split into Fine, Medium and coarse category since inception, without any scientific reasoning and hence they tend to overlap, merge at two ends where one type ends and other type starts. Not worrying or getting bogged down by numbers, the objective is to use judgement and combinations which will best fit in the case under consideration. This comes from experience and that is the main cause to refer to loss cure more as an intuitive art than a perfect science.

* **Note** that particle sizes are best measured using sieve analysis, especially for the larger particles. This will better reflect their plugging ability and allow to some extent for different particle shapes (aspect ratio). Laser sizes tend to give an average spherical diameter even for needle-shaped particles, which may not accurately reflect their sealing ability.

The d90 is a measure of the large particles present. It is defined as follows:

d90 value = 90% of the particles are less than this size

For example, a material with a d90 of 100 microns would be classified as 'medium grade' in the above table. The d90 is a good parameter to use when referring to pore and fracture sealing as it tells us that there is a reasonable concentration (10% of the particles) which are large enough to form an initial bridge. The d90 should approximate to the maximum pore or the fracture width which needs to be sealed.

In most cases the fracture/pore size is not known. When this happens, the LCM size selection must be based on the loss rate. The following table is a useful guide in selecting the size to match the loss rate:

Selecting LCM size is simple. Higher the loss rate, coarser the particle.

LCM formulations should contain a blend of products such as mixtures of coarse and medium grades. This gives a range of smaller particles to fill the gaps between the larger ones. Most importantly though the largest size must be sufficient for the pore/fracture width and loss rate encountered.

4.3: Formation Damage Considerations

When the loss zone is in or close to a potential production horizon, the use and subsequent removal of LCM could result in formation damage. Therefore, the selection of suitable LCM is critical, particularly if open hole completions are planned, perhaps involving OHGP placements or sand control screens. In these cases, it is best to incorporate LCMs that have a proven history and can be removed by down-hole treatments such as acid and hypochlorite treatments.

- **Calcium Carbonate** can be subsequently removed using acid (typically 15% HCl).
- **Cellulose fibers** can also be used, with caution. While only about 40% soluble in acid, the fibers can usually be oxidized and removed by a 2% to 5% hypochlorite solution at high pH.
- **Sized salt particles** (e.g. Na Cl) can be used in saturated salt fluids. The salt can be dissolved later using water or dilute brine.
- **Oil-soluble resins.** Crude oil or condensate can, in theory, dissolve the resin when the well is brought on production, or treatments with diesel can be applied. This sort of treatment requires some careful pre-planning and laboratory testing to judge the suitability.

Note: Productive formations or injection zones usually use a drilling fluid specifically designed to prevent formation damage, often referred to as a "Drill-In Fluid". These fluids may already incorporate one or more of the materials to minimize fluid invasion and hence formation damage.

4.4 LCM Listings

The table below lists size data and formation damage information for various LCMs. The products are categorized in terms of the above size definitions, which should help in selecting equivalent materials and finding substitutes when needed. The data are based on sieve analyses, wherever available. Quality control can be an issue, especially from region-to-region where products are locally manufactured. Therefore, there is no guarantee that the sizes will be exactly as given in the table. Particle size checks are recommended before using a product.

Different Ca Co3 sizes

LCM Grade	*Microns (d50)*	
Ca Co3 (coarse)	*220*	*Used in all exploratory wells*
Ca Co3 (medium)	*120*	*Used in all exploratory wells*
Ca Co3 (fine)	*36*	*Used in all exploratory wells*
Ca Co3 (sized)	*200*	*Designed for specific field but can be blended with other size Ca Co3*
Ca Co3 (117)	*117*	*Designed for specific field, but can be blended with another sized Ca Co3*

Generic name of the product	Size in microns d 10	D 50	D 90	F,M,C	Whether can be used in reservoir rock
$CaC0_3$	1	10	25	F	Y
CaC03	45	53	63	F	Y
CaC03	3	13	120	M	
CaC03 Flakes	38	75	125	M	Y
CaCo3	22	50	200	M	Y
CaC03	75	180	355	C	Y
Flakes CaC03	75	250	355	C	Y
CaC03	60	200	455	C	Y
Graphite	80	255	524	C	N
Mica	100	345	700	C	N
Mica flakes	260	650	850	C	N
Walnut shell	435	890	1125	C	N
Almond shell	350	765	985	C	N
Cellulosic fibres	400	900	1165	C	N
Polymeric pill	560	1120	1545	C	N
Blended LCM	200	435	715	C	N
Shell/fibre/ cellulose	315	780	1235	C	N

1. Gel Flakes

Particle size distribution

CALCIUM CARBONATE FLAKES (COARSE)	
	% CUMMULATIVE RETAINED
4 MESH	1.0max
8 MESH	40-50
12 MESH	80-90

CALCIUM CARBONATE FLAKES (MEDIUM)	
	% CUMMULATIVE RETAINED
16 MESH	1.0max
20 MESH	40-50
25 MESH	80-90
60 MESH	90 min

CALCIUM CARBONATE FLAKES (FINE)	
	% CUMMULATIVE RETAINED
30 MESH	5.0max
60 MESH	45-55
100 MESH	70-80
200 MESH	85-95

2. Cross Polymer pills

CROSS LINKING pill	*Comes with a complete package of accelerator/ retarder.*
Formaplug	*Comes with a complete package of accelerator/ retarder.*

3. Special Acid Soluble LCM pills

HTHP COMPLIANT LCM PILL	*Used for excessive loss and when acid soluble LCM is desired*

4. Assorted LCM used in the mud/LCM pill

Gel K seal	*Blend of mica, walnut, fiber*
G Seal	*Synthetic graphite used to plug pore throats*

4.5 Treatment Procedures

Procedures are given below for treating loss rates up to 120 bbls/hr. The procedures show generic pill formulations for fine, medium, coarse and extra coarse LCM. The user would need to refer to the table above to select chemicals. In practice, the chosen mud company normally suggests the pill formulation.

Remember that treatments for very high loss rates (greater than 100 bbls/hr) will call for specialized treatments which are discussed in subsequent discussion.

Loss rate bbls/hr	Non-productive zone	Productive zone
< 12.5	Procedure 1	Procedure 2
13 to 45	Procedure 3	Procedure 4
46 to 135	Procedure 5	Procedure 6

Procedure #1

Seepage Losses (< 12.5 bbl/hr)

1. Firstly, decide if the loss rate is acceptable/ sustainable. At low loss rates and with inexpensive mud it may be alright to drill ahead without treatment.
2. If the losses are thought to be induced, reduce ECD (flow rate/ viscosity, ROP) and consider a mud weight reduction. Ensure hole-cleaning is maintained. If these approaches are successful this confirms the losses were induced. Drill ahead - LCM treatment may not be required.
3. If losses continue, drill ahead and add 5 to 10 sacks per hour MEDIUM LCM to suction pit. Calcium Carbonate is recommended, or it may be substituted with another MEDIUM grade LCM. If the mud does not contain fines, add 2 sacks/ hr Fine LCM in addition to the Medium grade, until the fines build in the system.

4. If seepage continues, increase LCM particle size and quantity, and consider different blends of materials. E.g. mix 10 sacks per hour of a combination of MEDIUM fiber and MEDIUM Calcium Carbonate.
5. Check that mud weight and viscosity restrictions are not being exceeded. Monitor shaker screens, for blinding.
6. If the seepage continues and is unacceptable, locate the loss zone and spot minimum 50bbl pill and pull up and wait 2 to 4 hours.
7. The pill may contain

- 135 ppb LCM blend up to mud weight 9.3ppg – 9.7ppg
- 200 ppb LCM blend up to mud weight 9.7ppg –10ppg
- 230 ppb LCM blend for mud weight > 10ppg

E.g.

50ppb Calcium Carbonate (C)
50ppb Calcium Carbonate (S)
10ppb Calcium Carbonate flakes
15ppb gel flakes

If curing losses to SOBM environment always remember that LCM should be treated with 7ppb wetting agent. - made up using mud from the active system.

If the loss zone cannot be located change shaker screens to larger mesh size and treat the entire system with 15 to 25 lb/bbl MEDIUM/FINE LCM blend in the ratio 2:1 MEDIUM to FINE.

The viscosity of the LCM pill should be very high. A thumb rule is: the pill should not flow freely through marsh funnel

8. If seepage continues and is unacceptable refer to **Procedure #3**

Procedure #2

Seepage Losses to Productive Formations (< 12.5 bbls/hr)

1. If the losses are thought to be induced, reduce ECD (flow rate/viscosity, ROP) and consider a mud weight reduction. Ensure hole-cleaning/well bore stability is maintained otherwise pack-off will occur. If these approaches are successful this confirms the losses were induced. Drill ahead - LCM treatment may not be required.

2. If losses continue, drill ahead and add 5 to 10 sacks per hour MEDIUM LCM to suction pit. Calcium Carbonate must be used unless other materials have been approved for use in the pay zone (such as certain cellulosics). If the mud does not contain fines, add 2 sacks/hr FINE Calcium Carbonate in addition to the MEDIUM grade until the fines build in the system.
3. Check that mud weight and viscosity restrictions are not being exceeded. Monitor shaker screens for blinding.
4. If the seepage continues, locate the loss zone and spot minimum 50bbl pill and pull up and wait 2 to 4 hours. The pill should contain 70 - 100ppb Calcium Carbonate or another approved LCM blend, e.g.

   30ppb MEDIUM grade Calcium Carbonate
   30ppb MEDIUM grade cellulosic (if approved)
   15ppb FINE LCM (Calcium Carbonate)
   - Made up using mud from the active system.
   Alternatively, if loss zone has not been located change shaker screens to larger mesh size and treat the entire system with 35 to 40lb/bbl MEDIUM/FINE Calcium Carbonate blend in the ratio 2:1 MEDIUM to FINE.
5. If seepage continues and is unacceptable refer to **Procedure #4**

Procedure #3

Partial Losses (13 to 45 bbl/hr)

1. Consider reducing mud weight and ECD, if possible. Drill slower, reduce circulation rate and rheology (be sure to still provide adequate hole-cleaning and maintain well bore stability), trip in hole slowly, break circulation while rotating and working pipe, etc. Evaluate and eliminate any annular restrictions which may be causing an imposed annular pressure, such as balling of bit / BHA, mud rings, cuttings beds, etc. These actions may cure the problem if the losses are induced.
2. Drill ahead, but consider treating the entire system with 35 to 40 lb/bbl MEDIUM LCM. If the mud contains very few fine solids, add 5ppb FINE grade LCM initially as well, until the

larger LCM starts to grind down and produce the smaller particles. Check shaker screens for blinding.

3. If losses continue and loss zone identified, spot 50-100 bbl COARSE LCM pill and pull up and wait 2 to 4 hours:

The pill should be made using active mud containing:

- 80 lb/bbl COARSE LCM (e.g. Carbonate, fiber, or nut shells)
- 60 lb/bbl MEDIUM LCM
- 10 lbs/bbl FINE LCM
- 25 ppb Graphite

Note: Be sure bit nozzles and any MWD equipment open spaces are large enough to allow larger LCM to pass without plugging.

4. If partial losses continue, repeat pill procedure, spotting larger volume pill. Drill ahead and add 5 – 10 sacks/HR MEDIUM LCM to maintain the sealing capability of the mud. Check bit nozzles/MWD can take the LCM and shaker screens are not blinded.
5. If losses are not controlled, refer to **Procedure #5.**

Procedure #4

Partial Losses to Productive Formations (13 to 45 bbls/hr)

1. Consider reducing mud weight and ECD, if possible. Drill slower, reduce circulation rate and rheology (be sure to still provide adequate hole-cleaning and maintain well bore stability), trip in hole slower, break circulation while rotating and working pipe, etc. Evaluate and eliminate any annular restrictions which may be causing an imposed annular pressure, such as balling of bit / BHA, mud rings, cuttings beds, etc. These actions may cure the problem if the losses are induced.
2. Drill ahead, but consider treating the entire system with 35 to 40 lb/bbl MEDIUM LCM. Use Calcium Carbonate unless other LCM has been approved for the pay zone. If the mud contains very few fine solids, add 5ppb FINE Calcium Carbonate

initially as well, until the larger LCM starts to grind down and produce the smaller particles. Check shaker screens for blinding.

3. If losses continue and loss zone identified, spot 50-100 bbl COARSE LCM pill and pull up to the top of the pill and wait 2 to 4 hours:

The pill should be made using active mud containing:

80 lb/bbl COARSE Carbonate (or blend 20 lb/bbl with 10 lb/bbl cellulosic if approved)

50lb/bbl MEDIUM Carbonate (or blend with cellulosic if approved)

10lb/bbl FINE Carbonate

Note: Be sure bit nozzles and any MWD equipment is large enough to allow larger LCM to pass without plugging.

4. If partial losses continue, repeat pill procedure, spotting larger volume pill. Drill ahead and add 5 – 10 sacks/HR approved MEDIUM LCM to maintain the sealing capability of the mud. Check bit nozzles/MWD can take the LCM and shaker screens are not blinded.
5. If losses are not controlled refer to, **Procedure #6**

Procedure #5

Severe Losses (46 to 135 bbls/hr or more)

1. Consider reducing mud weight and ECD, if possible. Drill slower, reduce circulation rate and rheology (be sure to still provide adequate hole-cleaning and maintain well bore stability), trip in hole slower, break circulation while rotating and working pipe, etc. Evaluate and eliminate any annular restrictions which may be causing an imposed annular pressure, such as balling of bit / BHA, mud rings, cuttings beds, etc. These actions may cure the problem if the losses are induced.
2. Identify loss zone, spot 50 –100 bbl cross linking polymer pill, pull up and wait 2- 4 hrs: The pill should be made using active mud.
3. Following details are mentioned here, giving but the generic description still holds good for any other cross-linking polymer which is in use and has similar mechanism of action.

The following guidelines shall help proper preparation and placement of CROSS LINKING Polymer pill to maximize the probability of curing losses.

Field Mixing procedure for CROSS LINKING Plug

Make sure the mixing pit is cleaned properly and mixing / transfer lines are flushed with fresh water. Confirm this by checking salinity and pH of the fresh water

Mix in the following order: -

The following should provide 100 plus 12.5 ppg pill

- Take 75 bbls fresh water
- 9 sx 50 lbs Retarder, No mixing limit
- 180 sx CROSS LINKING Plug (3 to 4 min per sack, check for fish eyes)
- Make sure the fluid is homogeneous
- Mix 40 ppb Sized Calcium Carbonate
- Add enough Barite to get the desired weight

Keep the fluid agitated, but no mixing / shearing. The fluid can be kept in this state for any length of time while keeping agitator "ON "and switching the mixing pump "OFF "

Ensure no leak in the pit

Make sure mixing line is clear prior to the last step

Just before pumping add 90 sx activator @ 1 min / sack and place the pill at desired depth in the quickest possible time. Do not hold the pill in the pit after mixing activator, as it shall otherwise set wherever it is! Thus, it is essential to ensure that rig is geared up for pumping the pill before mixing the activator. If due to any reasons the pill is not pumped down the hole after mixing the activator, it should be dumped overboard soon to avoid setting of pill in the pit.

Clean out the pit lines soon after pumping.

Note:

In deep-water applications, wherever bottom hole temperature is below 80 Deg C there is no need to add retarder. However, this is mentioned in the formulation because of anticipated long pumping time (consider the riser length). If the pumping time is expected to be < 60 minutes, retarder may as well be eliminated from the formulation. It does not hurt to have the retarder in the pill except the fact that the setting time may be high (4 hours or more, lost rig time!!!). Retarder normally doubles the setting time. In high weight pills (13 ppg or more) presence of retarder helps in addition of Barite as it has a secondary thinning effect.

It is strongly recommended to perform a pilot test in the lab to confirm the setting time of the plug once activator is mixed. The lab samples should be loaded with base oil (only few ml) to prevent evaporation. It will help decision making whether to have retarder or not, when two pilot tests are carried out – one with and another without retarder. Having known the setting time of both the type of pills, a decision can then be taken as to what should be the composition of the pill, which should be governed by the expected pumping time (keep margin for any unforeseen delay in pumping / placing).

Depending upon the type of mud in use, select the correct spacer design and include the same in the sequence of pill pumping at appropriate place as mentioned below

SPACERS BEFORE AND AFTER CROSS LINKING PLUG PILL

SOBM:

Prepare approximately 25 bbl surfactant pill with 3.0 % v/v Solvent and 2.5 % v/v Surfactant.

Pump 10 bbl pre-flush followed by CROSS LINKING plug and 3 bbl after flush followed by regular mud.

Although CROSS LINKING Plug pill has been found to resist minor SOBM contamination, it is preferable to keep the pill isolated from SOBM for improved effectiveness.

WBM:

In WBM environment the spacer is only for minimizing the impact of dilution of pill conc. Thus, a hi-vis pill made up of ~ 3 ppb Xanthum gum polymer in fresh water can be used before and after (20 bbls each) the pumping of CROSS LINKING Plug, followed by displacement with the normal drilling mud. The weight of the spacer pill should be same as the weight of the LCM pill and it should not contain any salt. The entire weight should be either made up with CaC03 or with Ca Co3 and Barite.

CROSS LINKING Plug density guide (per barrel)

Density	Drill Water	Retarder **	Cross-linker	Activator	Barite*
10.1	0.835	6.0 lbs	100 lbs	50 lbs	0 lb
10.5	0.827	6.0	100	50	21.78
11.0	0.809	6.0	100	50	51.21
11.5	0.793	6.0	100	50	80.64
12.0	0.777	6.0	100	50	110.64
12.5	0.759	6.0	100	50	139.49
13.0	0.745	6.0	90	45	177.27
13.5	0.728	6.0	90	45	206.51
14.0	0.711	6.0	90	45	235.76
14.5	0.695	6.0	90	45	265.00
15.0	0.695	6.0	80	40	298.96
15.5	0.678	6.0	80	40	328.09
16.0	0.661	6.0	70	30	371.12
16.5	0.647	6.0	70	30	399.89
17.0	0.637	6.0	70	30	436.1

** **Confirm by pilot test**

* **The Barite conc. is shown for the composition without CaC03, therefore the actual requirement in the pilot test and field mixing is going to be less than mentioned. This fact must be remembered during mixing the pill in the lab / field.**

The Laboratory mixing instructions are for CROSS LINKING Plug for Pilot testing

Refer the density guide table for knowing the amount of additive to be added in 1 lab barrel. 1 ppb field units are, 1 gm of additive in 1 lab barrel, which is 350 ml, Suppose the recommended dose is 100 ppb in field, it means in the lab 100 gms of that additive should be mixed with 350 ml water

The following should provide 350 ml fluid

- Add fresh water to the Hamilton beach mixer
- If required, add retarder at this stage.
- Add CROSS LINKING Plug to the breaker while stirring at "medium "setting. Add steadily and slowly, in one minute. It contains polymers that require hydrating and after it is all mixed, leave it for 20 minutes at 'Medium "setting.
- Mix 40 ppb Sized Ca Co3
- Add Barite (if and as required) or any other weighing agent as specified. This can be mixed as fast as possible and after mixing is completed, this set can be kept aside if not immediately required.
- Just before carrying out the lab pilot test, add "Activator "as fast as practicable.
- Use the mixed material as soon as all the activator is in as the chemical cross - linking starts as soon as the pH rises to 10
- To check the setting time, the best way is to preheat the HTHP cell, to expected BHT and then fill the same with pill to be tested.
- Check after two hours, it should have gelled up

PUMPING / PLACEMENT PROCEDURE FOR CROSS LINKING PLUG PILL

Before pumping the CROSS-LINKING Plug pill, refer to the spacers pumping details mentioned separately, as spacers differ depending upon the type of mud in use. As suggested the pill should be sandwiched between the correct type and volume of spacers to have its maximum impact in controlling the losses.

1. With the bit above loss zone pump CROSS LINKING Plug as fast as possible
2. Once in the loss zone, depending upon the nature of the losses, spot the pill preferably balanced, withdrawing the string as the pill exits the bit.
3. If losses are induced ENSURE THE PILL IS SQUEZED INTO LOSS ZONE leaving ~ 3 bbls in the well bore
4. Pull above the zone TWO STANDS, circulate to surface to clear any remaining plug in the string while maintaining the overbalance to hold the pill in the formation
5. If losses are extreme (no returns), spot the pill in loss zone, pull back two stands and circulate to surface as slowly as possible.

This should allow the pill adequate time to stay in position and allow the string to be cleared of any remaining plug material

6. Once the pill is in position and other static condition ALLOW THE PILL TO CURE UNDER RECOMMENDED SET TIME AND BHT.
7. When CROSS LINKING Plug has set, run to the bottom and drill out the plug in the well bore.
8. When the CROSS-LINKING Plug pill has been pumped out of the pit, ensure that the lines and pits are cleaned and washed with drill water.
9. Ensure one pit is kept pre-cleaned by default.
10. Consider the possibility of pumping surfactants from cementing unit.

General guidelines to help the successful preparation / placement of pill

1. Establish BHST at loss depth, as accurately as possible. This will improve the probability of success and the determination of the exact conc. of the components of the pill.
2. Estimate the average pumping time for the pill, add a margin for unforeseen delay.
3. Perform Pilot tests with and without retarder at BHST.
4. Plan for a setting time of +/- 3 hours, depending upon the expected pumping time.
5. Take all precautions to thoroughly clean the complete system, which is going to be the surface circulatory path of pill to ensure success.
6. Once the tanks are all lined up, first prepare spacer pills, both before and after pill – as the case may be, depending upon the type of mud in use.
7. Then prepare CROSS LINKING Plug pill and do not mix and activator without confirming with all the partied involved to ensure that the rig is ready to pump pill.
8. As far as possible, pump the spacers just before mixing the activator. This will ensure that there is the minimum gap between activator mixing and pill pumping and it is very essential to do so.
9. Coordinate with all the team members and pump pill as fast as possible, once it is ready after mixing the activator.

10. Keep all the sacks of activator near hopper for easy access and fast mixing in minimum possible time.
11. Ensure that enough chemical to prepare at least 75 bbl is available on the rig.

Such pills should be tried twice and more attempts need to be evaluated depending upon the progress if any, made in curing losses and the need to bring down losses. The minimum volume of such pills should be 40 bbls. The maximum volume should normally be > twice the open hole volume. The ability to divert the contents of the pill in the loss zone goes a long way in curing losses with just required volume of pill and hence great care should be exercised in deciding the depth at which string is to be kept prior to resuming pumping and displacement of pill.

Very often such pills are not prepared in advance and hence there is always a time gap between deciding to pump cross linking polymer pill and pumping the same. During that time, consider pumping an assorted pill of Ca Co3/gel flakes/marble chips to give another chance to healing losses with following composition

Coarse Ca Co3: 80 ppb (300 to 700 microns Ca Co3)
Medium Ca Co3: 60 ppb (< 300 microns Ca Co3)
Gel flakes: 40 ppb (flaky Ca Co3)
Marble chips 30 ppb (> 1000 microns)
Graphite: 30 ppb (regular or coarse, particles ranging from 200 to 800 microns)

Note: Be sure bit nozzles/down-hole equipment are large enough to allow such LCM to pass without plugging. Be prepared for shaker blinding.

If losses now reduce from heavy to moderate, repeat the pill treatment and allow 2 to 3 hours of soaking time for the pill to heal the loss zone. If losses remain unchanged pump Cross linking pill as mentioned above.

Procedure #6

Severe Losses to Productive Formations (41 to 120bbls/hour and more)

1. Consider reducing mud weight and ECD, if possible. Drill slower, reduce circulation rate and rheology (be sure to still provide adequate hole-cleaning and maintain well bore stability),

trip in hole slower, break circulation while rotating and working pipe, etc. Evaluate and eliminate any annular restrictions which may be causing an imposed annular pressure, such as balling of bit / BHA, mud rings, cuttings beds, etc. These actions may cure the problem if the losses are induced.

2. Identify loss zone, spot 50 –100 bbl LCM pill, pull up and wait 2- 4 hrs. The pill should be made using active mud.
3. Identify loss zone, spot 50 –100 bbl cross linking polymer pill, pull up and wait 2- 4 hrs. The pill should be made using active mud.

The following guidelines shall help proper preparation and placement of CROSS LINKING Polymer pill to maximize the probability of curing losses.

Field Mixing procedure for CROSS LINKING Plug

Make sure the mixing pit is cleaned properly and mixing / transfer lines are flushed with fresh water. Confirm this by checking salinity and pH of the fresh water

Mix in the following order: -

Following should provide 100 bbls 12.5 ppg pill

- Take 75 bbls fresh water
- 9 sx 50 lbs Retarder No mixing limit
- 180 sx CROSS LINKING Plug (3 to 4 min per sack, check for fish eyes)
- Make sure the fluid is homogeneous
- Mix 40 ppb Sized Calcium Carbonate
- Add enough Barite to get the desired weight

Keep the fluid agitated, but no mixing / shearing. The fluid can be kept in this state for any length of time while keeping agitator "ON "and switching the mixing pump "OFF "

Ensure no leak in the pit

Make sure mixing line is clear prior to the last step

Just before pumping add 45 sx activator @ 1 min / sack and place the pill at desired depth in the quickest possible time. Do not hold the pill in the pit after mixing activator, as it shall otherwise set wherever it is! Thus, it is essential to ensure that rig is geared up for pumping the pill before mixing the activator. If due to any reasons the pill is not pumped down the hole after mixing the activator, it should be dumped overboard soon to avoid setting of pill in the pit.

Clean out the pit lines soon after pumping.

Note:

In deep-water applications, where bottom hole temperature is normally below 80 Deg C there is no need to add retarder. However, this is mentioned in the formulation because of anticipated long pumping time (consider the riser length). If the pumping time is expected to be < 60 minutes, retarder may as well be eliminated from the formulation. It does not hurt to have the retarder in the pill except the fact that the setting time may be high (4 hours or more, lost rig time!!!). Retarder normally doubles the setting time. In high weight pills (13 ppg or more) presence of retarder helps in addition of Barite as it has a secondary thinning effect.

It is strongly recommended to perform a pilot test in the lab to confirm the setting time of the plug once activator is mixed. The lab samples should be loaded with base oil (only few ml) to prevent evaporation. It will help decision making whether to have retarder or not, when two pilot tests are carried out – one with and another without retarder. Having known the setting time of both the type of pills, a decision can then be taken as to what should be the composition of the pill, which should be governed by the expected pumping time (keep margin for any unforeseen delay in pumping / placing).

Depending upon the type of mud in use, select the correct spacer design and include the same in the sequence of pill pumping at appropriate place as mentioned below

SPACERS BEFORE AND AFTER CROSS LINKING PLUG PILL.

SOBM:

Prepare approximately 25 bbl pill with 3.0 % v/v Solvent and 2.5 % v/v Surfactant.

Pump 10 bbl pre-flush followed by CROSS LINKING plug and 3 bbl after flush followed by regular mud.

Although CROSS LINKING Plug pill has been found to resist minor SOBM contamination, it is preferable to keep the pill isolated from SOBM for improved effectiveness.

WBM:

In WBM environment the spacer is only for minimizing the impact of dilution of pill conc. Thus, a hi-vis pill made up of ~ 3 ppb Xanthum gum polymer in fresh water can be used before and after (20 bbls each) the pumping of CROSS LINKING Plug, followed by displacement with the normal drilling mud. The weight of the spacer pill should be same as the weight of the LCM pill and it should not contain any salt. The entire weight should be either made up with $Ca\ C0_3$ or with $Ca\ Co_3$ and Barite.

CROSS LINKING Plug density guide (per barrel)

Density	Drill Water	Retarder **	Cross linker	Activator	Barite*
10.1	0.835	6.0 lbs	100 lbs	50 lbs	0 lb
10.5	0.827	6.0	100	50	21.78
11.0	0.809	6.0	100	50	51.21
11.5	0.793	6.0	100	50	80.64
12.0	0.777	6.0	100	50	110.64
12.5	0.759	6.0	100	50	139.49
13.0	0.745	6.0	90	45	177.27
13.5	0.728	6.0	90	45	206.51
14.0	0.711	6.0	90	45	235.76
14.5	0.695	6.0	90	45	265.00
15.0	0.695	6.0	80	40	298.96
15.5	0.678	6.0	80	40	328.09
16.0	0.661	6.0	70	30	371.12
16.5	0.647	6.0	70	30	399.89
17.0	0.637	6.0	70	30	436.1

** **Confirm by pilot test**

***The Barite conc. is shown for the composition without CaC03, therefore the actual requirement in the pilot test and field mixing is going to be less than mentioned. This fact must be remembered during mixing the pill in the lab / field.**

The laboratory mixing instructions for CROSS LINKING Plug for Pilot testing

Refer to the density guide table for knowing the amount of additive to be added in 1 lab barrel. 1 ppb field unit is 1 gm of additive in 1 lab barrel, which is 350 ml, Suppose the recommended dose is 100 ppb in field, it means in lab, 100 gms of that additive should be mixed with 350 ml water

The following should provide 350 ml fluid

- Add fresh water to the Hamilton beach mixer
- If required, add retarder at this stage.
- Add CROSS LINKING Plug to the breaker while stirring at "medium "setting. Add steadily and slowly, in one minute. It contains polymers that require hydrating and after it is all mixed, leave it for 20 minutes at 'Medium "setting.
- Mix 40 ppb Sized Ca Co3
- Add Barite (if and as required) or any other weighing agent as specified. This can be mixed as fast as possible and after mixing is completed, this set can be kept aside if not immediately required.
- Just before carrying out the lab pilot test, add "Activator "as fast as practicable.
- Use the mixed material as soon as all the activator is in as the chemical cross - linking starts as soon as the pH rises to 10
- To check the setting time, the best way is, to preheat the HTHP cell to expected BHT and fill the same with pill to be tested.
- Check after two hours, it should have gelled up

PUMPING / PLACEMENT PROCEDURE FOR CROSS LINKING PLUG PILL

Before pumping the CROSS-LINKING Plug pill, refer to the spacers pumping details mentioned separately, as spacers differ depending upon the type of mud in use. As suggested the pill should be sandwiched between the correct type and volume of spacers to have its maximum impact in controlling the losses.

1. With the bit above loss zone pump CROSS LINKING Plug as fast as possible
2. Once in the loss zone, depending upon the nature of the losses, spot the pill preferably balanced, withdrawing the string as the pill exits the bit.
3. If losses are induced ENSURE THE PILL IS SQUEZED INTO LOSS ZONE leaving ~ 3 bbls in the well bore
4. Pull above the zone TWO STANDS, circulate to surface to clear any remaining plug in the string while maintaining the overbalance to hold the pill in the formation
5. If losses are extreme (no returns), spot the pill in loss zone, pull back two stands and circulate to surface as slowly as possible. This should allow the pill adequate time to stay in position and allow the string to be cleared of any remaining plug material
6. Once the pill is in position and other static condition ALLOW THE PILL TO CURE UNDER RECOMMENDED SET TIME AND BHT.
7. When CROSS LINKING Plug has set, run to the bottom and drill out the plug in the well bore.
8. When the CROSS-LINKING Plug pill has been pumped out of the pit, ensure that the lines and pits are cleaned and washed with drill water.

General guidelines to help the successful preparation / placement of pill

1. Establish BHST at loss depth as accurately as possible. This will improve the probability of success and the determination of the exact conc. of the components of the pill.
2. Estimate the average pumping time for the pill, add a margin for unforeseen delay.
3. Perform Pilot tests with and without retarder at BHST.
4. Plan for a setting time of +/- 3 hours, depending upon the expected pumping time.
5. Take all precautions to thoroughly clean the complete system, which is going to be the surface circulatory path of pill to ensure success.

6. Once the tanks are all lined up, first prepare spacer pills, both before and after pill – depending upon the type of mud in use.
7. Then prepare CROSS LINKING Plug pill and do not mix and activator without confirming with all the partied involved to ensure that the rig is ready to pump pill.
8. As far as possible, pump the spacers before mixing the activator. This will ensure that there is the minimum gap between activator mixing and pill pumping and it is very essential to do so.
9. Coordinate with all the team members and pump pill as fast as possible, once it is ready after mixing the activator.
10. Keep all the sacks of activator near hopper for easy access and fast mixing in minimum possible time.

Such pills should be tried twice and more attempts need to be evaluated depending upon the progress if any, made in curing losses and the need to bring down losses. The minimum volume of such pills should be 40 bbls. The max. volume should normally be > twice the open hole volume. The ability to divert the contents of the pill in the loss zone goes a long way in curing losses with just required volume of pill and hence great care should be exercised in deciding the depth at which string is to be kept prior to resuming pumping and displacement of pill.

Very often such pills are not prepared in advance and hence there is always a time gap between deciding to pump cross linking polymer pill and pumping the same. During that time, consider pumping an assorted pill of Ca Co3/gel flakes/marble chips to give another chance for healing losses with following composition

Coarse Ca Co3: 80 ppb (300 to 700 microns Ca Co3)
Medium Ca Co3: 60 ppb (< 300 microns Ca Co3)
Gel flakes: 40 ppb (flaky Ca Co3)
Marble chips 30 ppb (> 1000 microns)

Note: Be sure bit nozzles/down-hole equipment are large enough to allow such LCM to pass without plugging. Be prepared for shaker blinding.

- If losses now reduce from heavy to moderate, repeat the pill treatment and allow 2 to 3 hours of soaking time for the pill to heal the loss zone. If losses remain unchanged pump Cross linking pill as mentioned above.

Note: Be sure bit nozzles/down-hole equipment are large enough to allow larger LCM to pass without plugging. Be prepared for shaker blinding.

Curing Losses - Loss Rates > 120bbls/hr

There are many proprietary treatments for dealing with such high loss rates, ranging from ordinary cements to specialized cements, 'gunks' and polymer systems. With such a wide choice of options available it can be difficult to make a selection. The situation is also complicated by the fact that some of the systems are virtually identical chemically, and yet suppliers will claim unique benefits of their system. It is not surprising, therefore, that different operational centers tend to have their own preferred treatments based perhaps as much on effective marketing strategies as on technical evaluations. In practice, the exact chemical system chosen is not critical, provided it is the right type of system for the job and is engineered correctly. Most failures occur because of engineering errors, such as those listed below, rather than inappropriate chemistry:

- Insufficient pill volume is pumped
- Setting time is miscalculated
- Contamination has occurred
- The pill has not reached the right place
- Engineers are not familiar with how to run the treatment
- Appropriate equipment is not available

Clearly, it is most important to ensure the equipment and skills needed to perform the job are in place, and to have a set of detailed guidelines for carrying out the treatment. This really should be done at the planning stage of the well.

5.2. Engineering Principles

Many of these treatments will require specialized mixing and pumping procedures. The distributor should be contacted for detailed instructions on the mixing and placement of the material. Some general guidelines are as follows:

- Locate the loss zone and design sufficient pill volume
- Establish how the material will be delivered down-hole (i.e. by a pumping unit or with the rig equipment); make sure the equipment is set up.
- Count pump strokes and be sure to know where the treatment is at any one time
- Some treatments are polymer systems which are triggered by cross-linking agents. Setting time is usually very much dependent on temperature. Therefore, accurate down-hole temperature prediction is very important. It is recommended a pilot test is carried out to confirm cross-linking time at down-hole temperature and pressure. Note that the temperature will normally be less than the static BHT due to cooling, which has occurred due to the losses. So, setting times can be longer than expected. Remember that pills pumped down will take a considerable time to heat up. This should be the deciding factor for finalizing the composition of the pill with respect to adding retarder or not.
- Monitor drill-pipe pressure and casing pressures to reduce the possibility of induced losses and to verify the displacement of the LCM from the pipe.
- Squeeze the treatment into the formation if possible, but do not try too hard, as it can sometimes further deteriorate the situation.
- Pull the drill-pipe clear of the calculated top of the pill and circulate for 5 minutes to wash the bit. Pull out to the casing shoe - otherwise the string could become cemented in.
- Wait the required time for the treatment to react and take effect.
- Reciprocate string while BHA is in open hole.
- Use a pill with a density equal to the active mud to avoid migration.

5.3 Treatment Selection

As noted above, there are many available options so that selection is difficult. Conventional LCMs discussed earlier and are not effective for very high loss rate situations (>100bbls/hr). They work by bridging, and at >100bbls/hr losses the fractures/voids are generally too large for the materials to function. The alternative is to use chemical systems which set and seal the loss zone. There are several types of system available:

- Cements/Modified Cements
- Gunks and Reverse Gunks
- Dilatant Slurries
- Silicate systems
- Combination of cement and cross-linking pill treatment

The following points should be noted:

- Ordinary cement forms a hard set which tends to be brittle. If flexing of fractures is likely to occur, a more flexible plug (e.g. polymer plug or gunk type system) is more appropriate.
- The cement used for extreme losses should be preceded with a temperature triggered type of pill, which helps cement to flash-set, Otherwise, huge volumes would need to be pumped.
- Gunk and reverse gunk treatments are cheap and often effective cure in very high loss situations. Although they have the advantage of being insensitive to temperature (setting time), but they do not withstand high differential pressures. A gunk pill strengthened with added cement powder is generally a better option (see DOB2C), although this is only suitable for WBM applications.
- When considering x-linked polymer treatments for sealing, ensure there are bridging solids present, which will help to screen out and seal any fracture. Many polymer systems already contain such solids. Ensure the treatment can still be pumped through the BHA (if this is intended). Be aware of the importance of circulating temperature.
- Curing losses in the pay-zone requires special attention. It should normally be assumed that damage will occur in the region where the pill is pumped. The selection guide below indicates the few treatments which are claimed to be used in open hole completions. For example, acid-soluble cement can be used. Chemical systems used to remove the treatment would only really have a surface effect. It is most effective to try to engineer a treatment with bridging solids to minimize the invasion of material into the rock matrix. Of course, if the formation will be perforated or fractured later, the damage shall minimize.

5.4 Product Description

- ### 5.4.1 Cement Slurries

Description

Cement slurries can be used as an effective lost circulation remedy in some situations and can be used in combination with other techniques as modified cements (see later). It is not usually the first course of action as there are several risks and drawbacks.

Benefits:

- Well-established technology, procedures well-known
- High density slurries easy to formulate if required
- If correctly placed, forms very effective/permanent seal, especially for fractures

Limitations:

- Will tend to screen out on pores so may be difficult to inject and seal rock matrix (better for fractures).
- Will require very large volumes if losses are into large fractures. Thumb rule says that the slurry volume planned should be thrice the open hole volume where losses are experienced.
- Set time is very sensitive to temperature; it is crucial to check thickening times. Cooling of the wellbore by mud circulation and mud losses will mean the down-hole temperature in the loss zone is likely to be lower than the estimated BHST. This can mean that cement slurry may take more time to set than calculated/estimated/predicted.
- Premature setting can lead to cementing the BHA or pipe into the hole.
- Forms a hard set which can lead to accidental sidetracks when drilling out the cement following treatment. Re-entering the zone before the pill gains full strength can help prevent side-tracking, but at some risk of getting stuck or washing away too much of the pill. Consider softer materials to avoid this.
- Spacers are needed to avoid mud contaminations which can dramatically affect setting times and compressive strength. Oil

mud contamination can greatly-weaken the cement. Some contaminations (eg CaCl2) can act as a rapid accelerator.

- Not normally suitable for pumping through the BHA (high risk), although can be done but make sure the nozzles will not block. Normally, necessary to POOH and remove the bit/BHA to minimize risk.

- **5.4.2 Modified Cements**
- **Description**

There are modifications that can be made to the basic cement slurry to try to improve its suitability for loss control. The cement must stay in place for long enough time to solidify and seal off the loss path, so there is a benefit in trying to avoid excessive flow of cement into the formation. The penetration of the slurry must proceed inside, at least, to some minimum distance into the formation, otherwise losses will return when the well bore is drilled out. The tools available to control these processes are, bridging additives, initial gel strength, and the rate of compressive strength development of the cement. Loss of effective sealing against thief zone during further drilling. The slurry should set soon after the planned thickening time to increase the probability of curing losses. Compressive strength development after displacement of slurry should be very fast to not only save rig time, but also minimize the probability of falling-off, of set cement when operations are resumed.

Compared with muds, cement slurries will form solid bridges more quickly since they have a higher solids content and inherently less fluid loss control. But bridging of the loss zone will be enhanced by adding bridging additives. Suitable products are Gilsonites, crushed coal, Calcium Carbonates, or other LCMs like fibers or flakes. The materials must be non-reactive in the cement, suspend easily, and encompass a broad particle distribution range. The cement contractor can advise on suitable concentrations.

A separate cement program for loss control should be given to rig.

Limitation: Systems containing bridging solids will not be suitable for pumping through the BHA due to the risk of plugging. The setting time of cement treatments tends to be very temperature sensitive.

Special cement formulations like Magnesium based cements and thixotropic cements can also be effective. Magnesium based cements will harden even if partially contaminated with mud. Thixotropic cements make use of a gelling effect at low shear; i.e. they form rapid gels to prevent an excessive loss of the cement to fractures, and their gel characteristics resist the cement being washed away from the wellbore. ECD needs to be considered with these systems.

Another modification uses a silicate solution, perhaps containing conventional LCM, which is pumped ahead of the cement with a fresh-water spacer. The LCM helps to prevent excessive loss of the silicate. Then, when the cement meets the silicate it flash-sets resulting in either partially or fully sealing off the loss zone.

Benefits: Some of these modified systems form softer cements which can be more easily drilled out with less risk of sidetracking. Sidetracking is at high risk, if the loss zone is in a soft formation and the cement treatment forms a hard plug.

- **5.4.3 Gunks (WBM)**

Description

There are variations to these systems (e.g. See DOB2C below or Flexiplug). The basic system forms a fairly soft plug. The addition of cement produces a firmer set which can withstand differential pressure and has a longer lifetime – this is known as DOB2C. Essentially the gunk treatment involves mixing Bentonite, Bentonite/polymer, or Bentonite/cement powder blends into an oil such as diesel. This mixture is pumped down the drill-pipe whilst water based mud is pumped down the annulus. When the two streams meet, a thick gunk (gel) is formed as the Bentonite or other additives hydrate. This is then squeezed into the formation. The hardness of the gunk depends on the relative pump rates and the concentration of the pill (as well as the pill type). Typically, the mud and the pill would be pumped at similar rates and the pill would contain around 200-400ppb Bentonite. In some situations, environmental regulations may preclude the use of diesel oil. Low toxicity oils such as synthetic oils may be used in place of diesel, but should be tested in the laboratory first.

A cement unit is used for displacing the gunk squeeze. The pumping unit displaces the mixture down the drill string and the mud pump pumps fluid down the annulus. The slurry reacts with water, and hence, precautions to keep water from prematurely contacting the slurry are needed. An oil spacer is pumped in front of and behind the pill.

Benefits:

- Temperature is not very important
- Soft set avoids sidetracking problems (easy to wash/drill out)
- The slurry will pass through most BHAs
- Procedures are well established
- Suited to large fractures/voids

Limitations:

- Placement – need to generate gunk in the right place
- Flow rate and mix are important to achieve suitable consistency
- Suitable experience required
- Avoid premature mud contamination
- Possible environmental restrictions dealing with oily returns
- The effect can be short-lived due to the softness of the plug (simple gunk)
- May not work with high salt muds (>50,000 mg/l chlorides)

A generic description of a Gunk job, with the broad procedure is given below. It is recommended that the rig team plans their own job, while using this as a guideline, including the volume numbers mentioned here, which are only for illustration purpose. They are neither accurate nor fit for ay situation, unless re-engineered by the team solving the loss.

- **Gel-Polymer Gunk Squeeze Procedure**

Formulation (for 25 bbls)

- 18 bbls Diesel/Base oil
- 750 lbs PHPA
- 4,800 lbs Bentonite (density is approx. 10 lb/gal)

- **Mixing Procedure**

Mix in a cementing batch tank, not in the pits

◊ Ensure the cement unit, batch tank and all associated pipe work is completely drained and then flushed thoroughly with diesel.

◊ Add PHPA to the diesel first, then add the Bentonite. Preferably add the PHPA through a mixing jet (not directly into the tank) at a very gradual rate while ensuring that lumps are not formed.

◊ If the mixture becomes too thick to add the required quantity of Bentonite, stop mixing and use the slurry as it is

◊ Test the gunk by mixing a sample of it 50/50 with the mud. Check the quality - **it should have a consistency, like chewing gum within less than a minute.**

1. Follow the pumping procedure as devised by the rig team in consultation with the cement service provider and mud engineer

Comments

The gunk pill is inert until it is mixed with water or water based mud. **IT MUST ONLY CONTACT WATER BASED FLUIDS AS IT EXITS THE BIT.**

- **DOB2C (DIESEL OIL BENTONITE TO CEMENT)**
- **PRODUCT INFORMATION AND APPLICATION**

DOB2C is Gel - cement slurry that is mixed with diesel oil. It is typically used to cure severe loss circulation situations. DOB2C when contacted by water or a water base mud will react, releasing the diesel and hydrating the cement. This reaction begins ~ 1 min after mixing. The degree and rate of reaction depends on how much water mixes with the DOB2C.

- **SAFETY PRECAUTIONS**

Diesel Oil Slurries should be mixed with adequate ventilation. The equipment used to mix the slurry must meet requirements to handle combustible materials.

Variations on this system which use alternative oils to diesel are possible, but would require pilot testing.

- **MIXING PROCEDURE**

1. Mixing on location in a cement batch mixer.
2. Blend the required Cement and Bentonite at the bulk plant. Load this blend into a cement bulk truck.
3. To assure fluid integrity and avoid potential incompatibility situations all mixing and pumping equipment must be free of water. Circulate the batch mixer with warm diesel and isolate that material for use ahead and behind the DOB2C.
4. Measure the required volume of diesel fuel into the mixer.
5. Begin weighting up the diesel with the Bentonite cement blend.
6. One barrel of DOB2C is made up as follows:
   - 0.7 bbls Diesel Fuel
   - 110 lbs of Bentonite
   - 207 lbs of Class G cement
   - 1 gal per barrel of wetting surfactant can be used to disperse and enhance the water-wet state of the cement.
7. Confirm the final density with a pressurized mud balance.
8. Allow the slurry to mix for 15 minutes prior to pumping.

- **DOB2C (DIESEL OIL BENTONITE 2 CEMENT) PUMPING PROCEDURE**

It is important for the success of this operation to have a clear understanding of the loss rate. Both dynamic and static rates should be reviewed just prior to doing this job and incorporated into the final design. Various mixtures of drilling mud and DOB2C result in various degrees of a set product. Annular and drill string injection rates will be based on what the loss rate is at the time of the job. Treatment volume will be based on a review of all available wellbore/loss data with a typical minimum recommended volume to be 40 bbls, depending upon tubular and annular clearances alternative volumes may be utilized.

This job is highly specialized and cementing companies as well as cementers have a better grip on such jobs. Although a mud engineer can decide the recipe, but the success/failure often depends upon how precisely the slurry is pumped, and how accurately is it placed at loss zone.

- **5.4.4 Reverse Gunks (OBM)**
- **Description**

This is the reverse procedure compared with ordinary Gunks. Typically, organophilic clay is mixed in an aqueous suspension and pumped down the drill-pipe. The oil mud is pumped simultaneously down the annulus and gelation occurs when the two fluids meet. The gunk is squeezed into the formation. Note that it is not possible to add cement to strengthen the pill (unlike DOB2C used for WBM applications).

Benefits:

- Temperature is not very important
- Soft set avoids sidetracking problems (easy to wash/drill out)
- The slurry will pass through most BHAs
- Procedures are well established
- Suitable for large fractures/voids

Limitations:

- Placement – need to generate gunk in the right place
- Suitable experience required
- Avoid premature mud contamination
- Mixing time/placement often greater than 6 hrs
- The effect can be short-lived due to soft plug formed
- The type of clay used to make the treatment can be quite critical

Procedures:

Formulations (per 1bbl) for unweighted reverse gunk	
Water	0.7 + bbl
Caustic soda	1.5 lbs
Ligno sulphonate	3.5 lbs
Oil swelling clay	➢ 240 lbs

This formulation is provided as a guide only. The yield obtained will vary greatly dependent on the clay used. Pilot testing of the pill formulation is strongly recommended. The formulation given above will have a density of approximately 1.45 sg. The slurry density can be increased to that of the active mud weight by viscosifying the water/clay slurry with XC polymer and adding Barite.

Mixing / Displacement Procedure

The pill should preferably be prepared in the cement batch tank. Ensure that the batch tank and all associated lines are clean by flushing thoroughly with a water based rig wash followed by large volumes of water.

Add the ingredients in the order shown, utilising as much shear as possible to disperse the organophilic clay. A pill of approximately 20bbl is recommended.

The resultant slurry is preferably pumped through open ended pipe, but may be pumped through the bit if it is not practicable to POOH. The pumping / displacement procedure should be carefully planned and strictly followed to prevent gelation within the drill-pipe. Before the displacement the string should be pulled back to a height sufficient to leave the pill volume of open hole.

- **5.4.5 Dilatant Slurries**
- **Description:**

Dilatant slurries exhibit the characteristic that more force that is applied to move them, the harder they become to move or deform. Beach sand is a good example of a dilatant slurry or fluid. If a foot can sink into the sand by the action of water washing over it, and then if we try to remove it quickly, the sand appears to de-water and "sets up" around the foot such that it is almost impossible to remove. Contrarily, if one attempts to remove the foot slowly, it will come out of the sand with minimal effort.

This example demonstrates a dilatant fluid's characteristic of shear thickening. That is, the higher the amount of shear, the thicker the system becomes. In general, dilatant slurries are prepared by mixing very high concentrations of fine solids into a minimum amount of liquid. As the fluid flows into higher shear rate regimes (e.g. a fracture tip), the resistance to flow increases dramatically. This mechanism can be applied to high-solid slurries such as Barite plugs. A pumpable mixture of a solid and liquid can be prepared. This mixture generally consists of slightly less than 50% solids. It can be pumped into place and, as it de-waters itself by the liquid fraction leaking off into the formation; it causes the solids to increase in concentration. A dilatant condition occurs when the remaining slurry "sets up" as pressure tries to force it to move.

Benefits:

- No temperature limitations
- Firm but easy to drill
- Simple to mix
- Can be pumped through the BHA if additional LCM not used (not MWD in general)
- Tolerant to mud contamination
- Can be used with oil or water based muds

Limitations:

- Beware of pill settlement in pits that are poorly agitated
- Avoid cement contamination
- Mixing time and placement can be more than 6 hrs
- Cement pump and blender required at >12ppg (1.44sg)
- This dialatent treatment additive reacts with drilling mud and creates a barrier at the face of the lost circulation zone. It does not penetrate the matrix, but bridges at the mouth of the fracture - or at the mouth of the pores in high permeability zones.

An oil based spacer is needed (diesel): at least 1000 ft of spacer ahead of treatment and 500ft behind. Treatment volumes should be suited to the loss rate encountered and may vary from 10 – 100bbl. Experienced specialist only can advise on detailed procedures.

Dilatent sluury can be formulated with cement to provide compressive strength. Substitute 30% of the dry blend mix with cement. Conventional LCMs can be added, but the treatment is probably then not suitable for pumping through the bit.

Benefits:

- Can be pumped through most BHAs (provided coarse LCM not added)
- Sets quickly (fast reaction with mud)
- Wide temperature range 75 °F (24 °C) – 400 °F (204 °C) downhole; ambient T > O °C
- Success rate 75%

Limitations:

- Sensitivity to water/mud during mixing
- Danger of drill-pipe becoming stuck if left in the annulus
- Requires specialist help to ensure success
- Batch tank and blender required for mixing – mud pits must not be used
- Not recommended for severely washed out holes

General Treatment guidelines:

Follow these guidelines for a Dilatent squeeze treatment:

1. With the end of the work string **above** the thief zone, circulate or bullhead the treatment fluids to the end of the work string.
2. Resume displacement, while simultaneously pumping down the annulus with mud at a rate that will result in a 50% mixture of mud and Dilatent treatment going into the thief zone. The jetting action from pumping through a drill bit will provide more intermixing than open-ended pipe alone, thus reducing or eliminating any delay in the reaction.
3. Once the treatment volume has exited the work string, stop pumping down the annulus and begin squeezing through the drill-pipe only. The treatment is most effective when pumped as slowly as possible.
4. If the well goes on vacuum (psi goes to 0), continue squeezing down the annulus. Both running and hesitation methods of squeezing can be used.

- **5.4.7 Dilatent slurry OBM**
- **Description**

This treatment reacts with drilling mud to create a barrier at the face of the lost circulation zone. It does not penetrate the matrix, but bridges at the mouth of the fracture - or at the mouth of the pores in high permeability zones. The treatment is a variation on the standard reverse gunk system.

A typical recipe involves mixing 80lb of Dilatent slurry OBM dry blend with 16 gal of fresh water and 20-gal latex, to make I bbl of slurry. Approximate slurry weight is 9.2ppg, although multiple density ranges are available. The slurry must not be mixed until ready for use as it will gradually set. A water based spacer is needed: at least 1500 ft of spacer ahead of treatment and 1000ft behind. Treatment volume 10 – 100bbl.

The treatment is pumped down the work string and as it exits mud is pumped down the annulus to react with the treatment. To perform a squeeze treatment, stop pumping mud down the annulus once all the treatment has exited and begin squeezing through the drill-pipe only.

Benefits:

- Can be pumped through most BHAs
- Sets quickly (fast reaction with mud)
- Wide temperature range 75 °F (24 °C) – 400 °F (204 °C) down-hole; ambient T > O °C
- 75% success rate
- Can give minimal penetration into rock matrix and so avoid excessive formation damage (if perforating)

Limitations:

- Avoid cement and rust contaminations
- Must be made when required (cannot be stored for long)
- Danger of drill-pipe becoming stuck if left in the annulus
- Requires specialist help to ensure success
- Batch tank and blender required for mixing – mud pits must not be used
- Not recommended for washed-out wellbores

General Treatment Procedure:

Follow these steps for a Dilatent slurry squeeze treatment:

1. With the end of the work string **above** the thief zone, circulate or bullhead the treatment fluids to the end of the work string.
2. Once the back side is closed, resume displacement, while simultaneously pumping down the annulus with mud at a rate that will result in a 50% mixture of mud and Dilatent slurry treatment going into the thief zone. The jetting action from pumping through a drill bit will provide more intermixing than open-ended pipe alone, thus reducing or eliminating any delay in the reaction.
3. Once the treatment volume has exited the work string, stop pumping down the annulus and begin squeezing through the drill-pipe only. The treatment is most effective when pumped as slowly as possible.
4. If the well goes on vacuum (psi goes to 0), continue squeezing down the annulus. Both running and hesitation methods of squeezing can be used.

- **5.4.8 Silicate type system**
- **Description**

The Silicate fluid is injected and combines with divalent metal ions in brine (pumped ahead of the fluid or in the formation) to form a gel plug. Often a cement squeeze is performed directly behind the Silicate type system. The cement flash-sets when it contacts the silicate gel. A typical treatment would involve:

a) Pump freshwater spacer (150ft inside work string)
b) Pump a slug of Calcium chloride (10%) brine
c) Pump freshwater spacer (150ft inside work string)
d) Pump slug of Silicate fluid, as per requirement depending upon severity of loss, advised by on board expert
e) Pump freshwater spacer (150ft inside work string)
f) Pump cement

Volumes are selected to suit the well geometry, but typical pill volume is 20bbls of Silicate fluid

Density range: max 11.8ppg

The lowest temperature at which it is pumpable: 75 °F. Works at high temperatures?

Benefits:

Can be used with WBM or OBM

- Only about 2 hours to prepare job

Limitations:

- Silicate systems have high pH and need careful handling
- Contaminations can cause premature gelation
- Cannot be mixed in mud pits

- **5.4.9 One sack crossl-inker**
- **Description**

This is a one-sack blend type of cross-linked polymer system technology, suitable for high loss situations, particularly where large cavern or vugs are involved.

Typical volume of pill: 1.5 times estimated volume across thief zone
Down-hole temperature range 85-400°F
Density range 8.4 – 18.0 ppg
Preparation time: 1 hour for 20bbl weighted pill.
Can be premixed without adding the cross-linker and stored until required.
Setting time is about 4 hours
Total time for the job, including blending, pumping and squeezing is about 5 hrs
Mud engineers need special training to engineer the system correctly
Like all losses treatments, proper planning is required
The success rate is high at around 90%

Benefits:

- One sack blend (polymer, LCM, cross-linker)
- Suitable for all mud types
- Fairly quick treatment
- The pill can be easily drilled out (soft) decreasing sidetrack possibilities
- Can be mixed in mud pits if clean (but batch tank preferred)

Limitations:

- Need to know down-hole temperature for procedure to work
- Environmental restrictions could limit applications in some areas due to chrome (III) cross-linker
- Sensitive to divalent brine contaminations
- Not compatible with cement
- Must be pumped immediately after mixing (cross-linker already included)
- Due to larger particle sizes and quantity it must be pumped open ended

Another variant of the above product, is also a one sack blend of polymers, crosslinking agents, but has larger fibrous lost circulation materials designed to plug the matrix, and naturally fractured or vugular zones. When activated with time and temperature, it produces a rubbery, ductile, spongy, soft-set gel that effectively prevents loss of fluid to the formation. The principal component responsible for loss control here is a fibrous cellulose containing a wide variety of particle sizes. It has application in very high loss situations where there are caverns or vugs as it contains large particles. It requires spacers of about 20 to 30bbl in front of and behind the pill. Spacer formulations are dependent on mud system type water- or non-aqueous-base fluid.

Barite or hematite can be added to weight the pill. The setting time depends on temperature so pre-screening work is required to optimize the formulation. Two products are available to control the setting times, the retarder and an accelerator chemical. For example, 10ppb of the retarder is used if the BHT is 200°F. Below 100°F no retarder is required. The accelerator is used in situations where the make-up water is at low temperature.

Placement:

The open-ended string is placed across the loss zone and the pill is pumped as fast as possible until the whole pill has cleared the drill string. **It is important not to leave any pill in the pipe.** To begin squeezing, pull above the pill height, close the annular preventer and begin applying pressure. If pressure is noted, hold for at least 4 hours to obtain a firm set. Allow about 4 hours for the pill to obtain maximum strength.

HTHP COMPLIANT LCM PILL

All the products described so far suffer from the inherent disadvantage of losing their strength with time and exposure to increased temperature. This necessitated further research and the outcome of this work resulted into a new package of LCM named as "HTHP COMPLIANT LCM PILL", which can resist the degradation of a pill at elevated temperature. The utility of the pill is enhanced by the fact that it keeps acquiring higher compressive strength as it gets exposed to higher temperature with time. Therefore, there is no chance of pill getting degraded later. It has been demonstrated that the pill is acid degradable as a CROSS LINKING plug (> 95 % acid soluble in H Cl)

This work method provides for mixing **50 bbls** of HTHP COMPLIANT LCM PILL-HT Loss Circulation material to be able to pump **35 bbls HTHP COMPLIANT LCM PILL-HT down-hole** as well as the mixing of **50 bbls of viscous spacer** to enable lead and tail pills of 15 bbls each to be pumped with the HTHP COMPLIANT LCM PILL treatment.

Description

The work method below describes the mixing and pumping of 50 bbls of the HTHP COMPLIANT LCM PILL HT loss Circulation material (LCM) using the rig mud pits and 50 bbls of viscosified spacer using the rig slug pit. Two high pressure pumping units should be available always when pumping the treatment.

Procedures

Mixing and placing HTHP COMPLIANT LCM PILL-HT as well as LT, is a specialized job. Much depends upon the loss rate prevailing at the time of curing losses and hence no fixed procedure can be written. It involves mixing and viscosifying polymer solution in water first, then followed by various components of HTHP COMPLIANT PILL. Invariably on all such jobs, an expert is available from the service providing company, who has devised this technology.

C: -OPERATING PRINCIPLES

1. A mixing system with a re-circulation line/agitating paddle is required.
2. Ensure that appropriate weighing scales are available for weighing the additives (where appropriate).
3. A testing kit including sample bottles and pH indicator is advised for mixing of treatment fluid.

4. E: -ADDITIONAL INFORMATION

1. After adding the HTHP COMPLIANT LCM PILL powder, NEVER re-circulate or shear for any periods of time unless specified by the expert, under whose supervision the entire job should be carried out on site.
2. Never leave HTHP COMPLIANT LCM PILL in the lines with centrifugal pumps running.
3. Never mix HTHP COMPLIANT LCM PILL without confirmation of down-hole temperature.

The HTHP COMPLIANT LCM PILL does not have a generic description which suits all occasions. The treatment plan is case specific and expert / specialist has software to select the most suitable doses of various components. It requires extensive pre-planning and pilot testing with the available chemicals to select the best formulation and maximize the probability of curing losses.

It is reiterated here that "HTHP COMPLIANT LCM PILL" has two sets of products – HTHP COMPLIANT LCM PILL-LT and HTHP COMPLIANT LCM PILL-HT and based upon the expected BHST/BHCT the correct set up is to be used. LT and HT represent low and high temperatures respectively. By and large, if the BHCT at application depth is < 150 Deg C it can be catered by LT and temperature above 150 Deg C require HT package.

Merits:

The range of products can cater to various higher ranges of temperature, which other cross-linking polymers cannot.

The strength of the set fluid increases with more and more exposure to high temperature, which ensures that the set fluid will not deteriorate at any stage and the zone once sealed will be permanently sealed forever, unless acidized deliberately for flowing back the producing rock.

Demerits:

Very much dependent upon correct prediction of bottom hole temperature

Even the slightest variation in temp can either prematurely set the pill in string or delay the setting of pill in thief zone. There is not enough margin for any approximation.

It requires highly skilled and experienced group of personnel on-board. The timely availability of such experts can be a limiting factor for execution of the job at a short notice.

Acid soluble cement for loss control

This pill has been used by the author in conjunction with technical support from cementing service provider, who happens to be the original inventor of the technology. Sized Ca Co_3 is blended with API Class – G cement in the ratio 2.5: 1 and slurry is prepared and pumped down hole in thief zone. This slurry can then be reacted with H Cl which will dissolve the Ca Co_3 part of set pill and allow passage of hydrocarbon through a producing rock. This composition has been used on walls under conditions of complete loss of return and was partly successful in minimizing losses.

In an extreme situation, a CROSS LINKING plug slurry can be and was pumped down the hole and the same was displaced with an acid soluble cement slurry to maximize the probability of curing losses and that has potential to provide relief in total loss conditions to the extent that losses can be brought down to manageable levels to pull the string out of the hole from a hopeless situation of complete loss of return.

Another product which has been identified with a potential for curing massive losses is **Recron-3S.** It has polymeric fibers with excellent property of remaining in suspension and forming a network aggregate to help seal the pores of the rock. These fibers can be blended with conventional type of pill to maximize the chance of regaining circulation when the damage of the rock is not an issue

Being a very specialized job, much depends upon the way the job is executed in the field. If not properly dispersed these fibers can plug the surface lines and cause disruption of the flow path. It is therefore recommended that a pilot recipe must be prepared in the lab on the rig before executing the job. The mixing of fibers should be as per the recommendations of the expert specialist to ensure that the fibers can be exploited to their potential for curing losses.

- **5.4.20a Solids free cross-linker**
- **Description**

Solids free cross linker systems comprise of a viscous liquid polymer system designed to control moderate to severe losses in the production zone, thus preventing expensive completion fluid from leaking into the pay zone.

It has dual design, one for all non-zinc completion fluids, while another is for Zn-based completion fluids. Both the systems are pre-cross-linked, which eliminate the need for special blending equipment and an additional engineer. Supplied and palletized for a 10 bbl application to the rig, the system is prepared by adding 1 pallet of material to ~6 bbl of viscous liquid polymer completion fluid in the cementing unit (1 pallet of material = 4 bbls). The system is gently stirred with a paddle-like mixer to chop-up the cross-linked material. The pill should be 'clumpy' or 'stringy' in appearance. Once chopped-up, the pill is pumped. Small spacers (3-5 bbls) may be implemented to prevent the dilution of the system.

Once inside the wellbore, it forms a soft, cross-linked plug that does not penetrate the formation. The cross-link is broken with a mild acid treatment. An internal breaker can be used to destroy the viscous carrier fluid, thus minimizing the formation damage.

Typical application: 10 bbl / 70 ft interval
Temperature Range: 100 - 325°F
Density Range: 8.5 – 18.0 ppg
Preparation Time: Can be prepared simultaneously as we pump.
Suitable for completion fluid only

Does not require special equipment or personnel

Benefits:

- Solids-free
- No special training required
- No special mixing equipment required
- Can be easily cleaned-up with acid
- Environmentally Friendly – contains no heavy metals

Limitations:

- Can be over sheared and lose the effectivity.
- Should not be used for >1000 psi differential pressure without consulting an expert technical Representative

CONCLUSION: The industry has seen a big advancement regarding developing such products in last three decades and every major service provider has come up with his own set of products. Despite being useful, their description does not find place here, because, being specific to the company - and the book aims to be independent of any company. By and large, all major types of recently developed category of products have been covered by providing a generic description and every generic description has a specific package developed by major service providers. The intention here is to familiarize the readers with various types of products and technology available, and the basic chemistry behind them, as much as known in public domain. The story signifies the saga of never-ending efforts to keep pushing the limit of excellence to newer height with every innovation.

Subsequent discussion will cover more generic composition of loss control which were used in yesteryears, but still find sporadic applications, occasionally.

• 5.4.25 Barite Plug

Barite is an inert solid with a specific gravity of 4.2 to 4.15. It will settle in a fluid medium with insufficient gel strength. When agglomerated together, it can form a solid plug, causing the annulus to plug or seal. This fact is utilized in making a Barite plug. This pill can be used to shut off the gas zone when the wellbore takes a kick. Lower the viscosity of the medium, better the settling rate. Likewise, higher the weight of the Barite plug more effective is the bright plug.

The volume of the pill should be enough to cover > 100 m of fill. Longer pills can have issues in terms of pipe plugging since pipe needs to be pulled out through the plug.

The maximum recommended slurry weight is 19 ppg. The slurry should be more than 16ppg.

pH of the slurry should be maintained at 8.0 to 10 by using Caustic soda to assist settling.

6-8ppb chrome Ligno-sulfonate must be used as dispersant.

Table-1 Water and Barite Dose

Slurry density	Barite SX/BBL	GAL water/BBL
14.0	3.10	33.2
15.0	3.7	31.7
16.0	4.2	30.1
17.0	4.8	28.6
18.0	5.3	26.9
19.0	5.94	25.5

PROCEDURES

1. Clean and flush all the mixing lines. Clean the slug pit. It is preferred to mix this pill in cementing unit if Barite access to the cementing unit is possible.

2. Take the required amount of Drill water. Keep the agitator running and both the mixing pumps running.
3. Add caustic soda to adjust the pH from 8.0-10.0
4. Add 6-8ppb of Chrome Ligno sulphonate.
5. Add the required quantity of Barite.
6. Do not stop pumping in between. Displacement to be carried out through the mud and not water, to avoid any premature settling. Pump the entire amount of slurry from the drill-pipe and P/O 30 mts above the expected top of the pill. Circulate to check if the string and annulus are clear. If possible, rotate the pipe while pumping the pill.
7. Carry out a Pilot test before pumping the slurry in the wellbore. The Yp of the slurry should be zero or negative for effective settling.

Last, but not the least, in one of the many unforgettable experiences of handling severe losses, it was learnt that all solutions need not get recorded for future references. After struggling for almost 10 days to recover complete loss of return which swallowed 10 cement plugs, 24000 bbls mud, 2500 bbls LCM pills and unmeasured volume of sea water without a drop reaching to surface, the trick which then smoothened the process of curing losses wass to drop LCM balls through OEDP from rig floor. The balls were made up of bentonite, cement, mica, walnut and just enough water to be able to convert this mixture

into the shape of a ball. The entire rig crew and galley staff were together preparing such balls and ~ 25000 such balls were dropped in the hole. This herculean affort showed a visible impact when se water level could be seen inside the d/p to around a metre above the bottom, pior to which there were no signs of any droplet! This charged up the-otherwise frssutrated rig team and partial returns were observed after several hours of continued ea water pumping above the LCM balls.

Such a long chapter deserves closing remarks. Not getting spiritual or philosophical, curing losses does require patience and preparedness to try new stuff, even if it doesn't seem to be fitting in the umbrella of logic, simply because loss of mud is also naturally illogical! Logic says anything we pump in should return. If it's not returning then it is illogical and illogical situations can have illogical solutions! So, be prepared for surprises, trust yourself, trust your team, and most importantly, EARN THE CONFIDENCE OF YOUR TEAM!

CHAPTER 13

LOSSES WHILE CEMENTING AND OTHER SPECIAL SITUATIONS OF UNCONVENTIONAL DRILLING REQUIRING TAILORED DRILLING FLUID FORMULATIONS

6.1 Loss While Cementing

Loss of mud and / or cement slurry during the pumping and placement of slurry in annulus to case the open hole, is a very common feature in oil well drilling applications. Restricted annular passage due to lowering of larger side casing pipe, high specific gravity of the cement slurry and increased gel strength of drilling fluid, are the principle causes of such occurrence. Subsequent discussion aims at providing useful practical information to minimize/eliminate the occurrence of such event.

Guidelines

1. Calculate maximum ECD for circulating cement between the pipe and the formation. This value is particularly important when circulating cement in constricted casing annuli.
2. Condition the drilling fluid prior to running casing to minimize surge pressures. Reduce the plastic viscosity and the gel strength of the fluid to the lowest possible values without settling of solids or Barite.
3. Calculate the ECDs at various flow rates when circulating cement behind casing. Select the rate that will produce an effective displacement without breaking down the formation.
4. Always break circulation slowly to avoid undue stress on the formation.
5. Watch fluid pits for loss of fluid.
6. Gradually establish full circulation prior to landing casing.
7. Decide whether to reciprocate the pipe based on location of the loss zone and the top of the cement. Maintain optimum flow rate with the casing hung off in the wellhead.
8. Maintain sufficient fluid volume in the pits to displace the casing contents even when total losses have occurred during circulation prior to cementing. In some cases, seawater may be used to displace the cement.

Planning

The problem of losses, while cementing can be minimized by planning. The only effective method for handling losses during cementing is adequate pre-planning and impeccable execution of the plan.

Lost circulation problems while cementing had been a recurring problem due to a substantial reduction in hydraulic diameter after lowering of casing pipe. The mud in the hole remains without adequate circulation during casing lowering which further aggravates the already tough condition. Often it may require "foam cement", lightweight slurries, or stage cementing to limit ECD and final hydrostatic pressure for successful and safe completion of the cement job.

The situation of lost circulation, if prevailing, should be cured prior to running casing. Effective treatment of the loss zone prior to cementing increases the maximum permissible ECD during the cement job by increasing the effective leak-off pressure at the loss zone. A formation integrity test after the lost circulation treatment will provide an updated maximum ECD value. If an interval is suspected of significant seepage, an LCM pill should be placed across the interval before running casing or cementing.

Minimizing Pressure on the Formation

1. Reduce the ECD, while pumping the cement slurry, to the lowest level commensurate with producing a quality cement job. This should be adjusted with the planned thickening time of the slurry to mitigate last minute failure arising out of the settling of slurry at the undesirable interval / place.
2. If a reduction in the fluid density is possible prior to running casing, consider this an option for lower ECDs. Lowering the yield point will reduce swab and surge pressures. Software models can make surge-swab calculations to determine the maximum safe fluid properties for running casing.
3. Minimize casing and liner running speeds to avoid surge pressures that may fracture the formation.
4. Cement density can be reduced by the addition of common extenders, i.e. Sodium silicate extender, Bentonite extenders, micro spheres, gilsonite. It should be remembered that reducing cement density also reduces compressive strength and bond strength. Lower density slurries can also reduce mud displacement efficiency.
5. Another method to reduce ECD would be to use a light weight flush, ahead [or instead of] of a weighted spacer. Either water or oil may be used depending on the fluid system. As a precaution, be sure the minimum overbalance pressure is determined so that the well is not made unsafe at any time while pumping the flush.

6. Cement plugs used to seal a loss zone often fail because of their density. Lower density cements are available. Cement treated with gilsonite has found wide use in bridging "thief" zones that were accepting cement slurries and preventing the achievement of the desired full sheath of cement around the casing. The low specific gravity [1.05] and sealing characteristics make gilsonite effective against losses to permeable zones, natural and induced fractures, and caverns. Ceramic micro spheres can reduce the density as low as 9.0 lbs/gal. They have been added to reduce the slurry density down to 10 lbs/gal to cement production liners. Foam cements have also been used for reducing the cement slurry density below 10 lbs/gal. Typical densities have ranged from 8.0 to 9.0 lbs/gal.
7. Other LCM commonly mixed with cement to stop losses include mica, calcined shale, crushed coal, Calcium Carbonate, Graphite and cellophane flakes.
8. Reactive flushes are sometimes used to stop cement losses. In this case a chemical such as Sodium silicate is pumped ahead of the cement with a spacer of 5 to 20 barrels of fresh water between the cement and the flush. Typical treatment volumes are 20 to 50 bbl. The purpose of the reactive flush is to quickly react with the cement, thus stopping further losses. These treatments are not recommended for liners or stage cementing applications.
9. The design of any lost circulation treatment pumped ahead of the cement should consider the possibility of formation damage based on whether the loss zone is in or close to a potential production horizon. In this case, it is best to incorporate the use of acid soluble materials in any fluid treatment considered.
10. To select the correct rheology for the spacer to effectively and completely displace the stagnant mud out of the hole from behind the casing during attempts to circulate prior to start of cement job, if a rough idea can be obtained about the likelihood value of gel 10 for the mud inside the hole, the probability of complete displacement of mud is drastically increased. One simple way of doing this is to grab a sample of mud from the active pit just prior to start of casing lowering and keep the same static in the lab. At regular interval of 4 hours the Gel10 value of this sample be measured and the values recorded till the casing is lowered to the bottom. This will be close to the value

of gel 10 of the mud in the whole. Taking this as a benchmark number the weighted spacer is such designed that its Gel 10 value is at least 10 % more than that of mud in the hole. This weighted spacer of specific rheology can then be fed into the software model to select the correct pumping rate and negate induction of losses and yet ensuring that the displacement of gelled mud is completed more effectively.

6.2 Living with Losses

In some situations, the lost circulation problem is so severe that it cannot be economically cured. In these situations, the driller must "live with the losses". The following paragraphs describe methods that have been used to ameliorate lost circulation without the use of lost circulation materials. It must be remembered that utilizing these methods involves risk of the drilling operation.

If one of the methods is used, it must be carefully and thoroughly planned to ensure the safest possible outcome.

- ### 6.2.1 Drilling Blind

Often, losses are such that drilling blind (with no returns) is used to drill through the loss zone and into a competent formation prior to running casing.

This is a potentially dangerous situation where well control and hole problems can occur. To drill continuously without returns the first requirement is an adequate supply of water. Not only is it important to pump water down the drill-pipe, but a second pump must be used to pump water down the annulus. This keeps some hydrostatic pressure and flushes cuttings away from the well and into the loss zone.

Sweeps are required to occasionally move cuttings away from the bit and up the hole, so a reserve of several hundred barrels of viscous mud for sweeps is recommended. Closely monitor the pump pressure and drill string torque and drag for indications of cuttings pack off and the need for a sweep. In addition, it is recommended to control drill at one 30 ft joint per hour and to pick up every 15 feet to monitor the condition of the hole. The drill string should be kept moving always and surveys should not be run while drilling blind. Prior to connections, either run a sweep or spot viscous mud above the BHA to keep cuttings from packing off around the BHA or bit. For well control monitoring, it is important to assign a person to monitor the flow line. If circulation is regained or a change in the loss rate is observed, stop and check for flow. If the well is flowing, close the BOP and observe shut-in pressures to check for a kick. When drilling blind, always be prepared to pump cement to control formation fluids and pressures.

As a variation on this which arguably is a safer option – see mud cap drilling.

- **6.2.2 Flow Drilling**

Flow drilling is a special case of under balanced drilling where formation fluids can enter the annulus and be circulated from the well. This technique is used most often in fractured or vuggy Carbonate reservoirs where massive lost circulation occurs in the production zone, such as the Austin Chalk area. In this situation, most lost circulation remedies would cause formation damage.

The flow drilling method controls formation pressures with back pressure applied to the annulus by a rotating head (or RBOP) and the choke to limit the rate of flow from the formation. Oil is usually skimmed from the mud in special surface pits where the mud can be isolated and processed. Flow drilling uses heavier mud or brine to kill the well prior to and during trips. It is a tricky operation where deciding on what mud weight to use and learning how to manage the amount of back pressure takes some trial and error learning. Often in flow drilling applications, the formation pore pressure is lower than that of the hydrostatic pressure of water. In this instance, the under balanced condition does not occur, until lost-circulation reduces the annular mud level and hydrostatic pressure and allows flow. Flow drilling is often used to help identify potential production as additional fractures are encountered.

- **6.2.3 Aerated Mud**

Aerated mud uses the two-phase flow of air and mud to reduce the ECD to the 4 to 8 lb/gal range. This technique is used most often when drilling in sub-normally pressured intervals with massive lost circulation which cannot be readily resolved. Generally, the ECD of aerated mud is maintained only slightly higher than the pore pressure. This will prevent an influx of formation fluids or gas into the circulating mud system. Equipment on the surface separates the air and mud, so that the mud can be processed through the surface system and recirculated. Aerated mud drilling requires using an air drilling compressor and other equipment, such as a rotating head and mist and air/mud separator. High and possibly severe corrosion rates, unsteady flow in large-diameter holes, and erosion from turbulent flow are all potential problems when using aerated mud.

There are several different methods used to achieve aerated mud:

- "Down the drill-pipe" aeration – injecting both air and mud into the standpipe
- "Parasite" aeration – using a parasite tubing string run outside the last casing string to inject the air and pumping mud as normal. The mud and air mix at the casing shoe and aerated mud flows up the annulus
- "Dual casing micro-annulus" aeration – where a temporary casing is hung inside the last cemented casing so that air is injected down the two casings annulus and flows into the mud at the inner casing shoe then back up the drill-pipe to the inner casing annulus as aerated mud.

Aeration by injection of both air and mud into the standpipe is the most common. For parasite aeration, the air and mud flows are more independent and easier to control, plus the air pressure requirement is usually lower. Parasite injection requires the extra cost and time for the tubing plus extra equipment for pressure control. In addition, the ECD is not as low as can be obtained with standard aeration due to the restricted air volume capacity of the tubing and the depth of injection being higher. For dual casing micro-annulus aeration, the last cemented casing must be large enough to allow the temporary inner casing to be run. This may require a larger casing size and previous hole size.

- **6.2.4 Foam**

Foam drilling uses mainly air (or gas) with a water and foaming surfactant slurry to form and circulate foam with an ECD in the 2 to 4 lb/gal range as the drilling fluid. Foam is applicable in very low-pressure zones where massive lost circulation occurs and is difficult to cure. When foam drilling, the air and foam slurry must be metered within a narrow range of ratios and requires a rotating head and extra mechanical equipment. The foam slurry must be pumped with a mid-size pump at about 25-75 gpm.

Two types of foam are "stable" and "stiff" foam drilling. Stable foam uses mainly water and a surfactant (commonly called soap) to form a stable air-in-water foam mixture with the air. Stiff foam uses a mud-like formulation as the foam slurry with Bentonite and polymers to form a long-lasting foam. Stiff foam is good for drilling large holes where air volumes are insufficient to affect proper cleaning with a stable foam. Foam has excellent hole-cleaning characteristics and uses less air volume as compared to air or mist drilling. Since the foam slurry cannot be recovered and reused, water and foam chemical costs are high.

• 6.2.5 Air or Mist

Air or mist drilling provides the lowest possible ECD. Compressed air or gas is circulated at high flow rates to remove cuttings. An ECD of 0 to 2 lb/gal is possible with air and mist drilling. It is most applicable in dry formations, where the well produces little liquid and massive lost circulation and networks of vugs or caverns are encountered. Large capacity high-pressure air compressors, a rotating head, and other mechanical equipment are required.

Air drilling is commonly referred to as "dust" drilling as the discharge from the well is dust. Mist drilling uses water and a small amount of 1% foam solution injected to help remove small amounts of produced fluids from the well with discharge from the well, being mist or froth.

Many Coal bed methane (CBM) wells deploy foam drilling.

• 6.2.6 Floating Mud Cap Drilling

Theory

Floating mud cap drilling is a process used when total loss of circulation has occurred and resulted in the intrusion of reservoir fluids into the wellbore. The well is kept under control by a column of weighted mud that is continuously pumped into the annulus. Water is used as the drilling fluid. The water and drill cuttings are lost in the zone. The intruding fluids are kept under control by the hydrostatic pressure of the mud column in the annulus.

The density of the mud is selected to be as close as possible, but very slightly over (~250 psi) the formation pressure. Because the mud is constantly lost in the zone, the hydrostatic pressure must be maintained by the addition of fresh mud to the annulus. Generally, an inexpensive, relatively viscous mud is used. In theory, the hydrostatic pressure of the mud column at some height will be balanced by the formation pressure. This will cause the mud to "float" against the formation pressure, just above the lost circulation zone. The basic requirement is to keep the mud weight in a range that will force a small volume of mud into the loss zone. If the mud weight is too light, formation fluids will migrate into the well bore, decreasing the hydrostatic head and potentially resulting in a well control problem. If the mud density is too high, a stable annular pressure will be difficult and expensive to maintain, also potentially allowing the flow of fluids into the well and resulting in a well control situation.

Operation

Mud cap drilling cannot take place without a large dedicated water supply. It is essential that pump rates, which will clean the bit and annular velocities that will convey the cuttings into the loss zone be maintained. Generally, annular velocities above 120 ft./min will be required. That equates to a flow rate of slightly less than 250 gallons per minute in an 8.5" hole with 5" drill-pipe. At this rate over an 8-hour drilling period, about 2,860 barrels of water will be required.

The procedure for implementing the floating mud cap will vary depending upon the drilling conditions. For example, the conditions for controlling the well will be slightly different for the deep-water environment, where the concern for hydrate formation exists, than for an onshore operation. Nevertheless, the same basic principles will apply in all cases. These start with the early identification of fluid flow into the well bore. Once the kick has been identified the steps listed below should be followed.

1. Pull the bit off bottom and close the BOP.
2. Pump 100 to 150 barrels of mud (at the density calculated to be slightly above that needed to balance the formation pressure) down the annulus and note the annular pressure.

3. If the pressure is still noted, pump another 50 to 100 barrels of mud into the annulus.
4. Repeat step 3 until no annular pressure is observed.
5. When no pressure registers on the annulus, open the choke line and check for flow.
6. If there is no flow, open the BOP, run to bottom and drill, circulating water down the drill-pipe at the calculated rate and adding mud continuously to the annulus.
7. If the well flows, close the BOP and pump a small quantity of mud into the annulus.
8. Repeat steps 5 and 6.

Annular Fill

Two types of annular fill methods have been used:

- The Batch Fill method
- The Continuous Fill method

The continuous fill method has generally been found to be most successful, as it provides for a more constant annular pressure. This constant pressure minimizes the amount of intruding fluid and reduces the effect of any intruding fluid migrating up the well bore, thus minimizing the blow out potential.

Once a stable mud level in the annulus has been achieved, batch treatment has been successfully used to keep the annulus full. The success of the technique depends on not allowing the annular mud level to drop too far before more mud is added. Therefore, the mud level and pressure in the annulus must be constantly monitored and kept within that required for well control.

Bottom Hole Assembly and Drill string

Consideration should be given to maximizing the effectiveness of hole-cleaning and the use of other measures for controlling lost circulation. Bit nozzles should be selected to maximize flow rate and provide the largest opening for coarse lost circulation material. If cement may be pumped, thought should be given to removing the nozzles from the bit.

MWD equipment and mud motors should be removed from the string as they will not allow many lost circulation materials to pass. They will also be damaged by cement and other thixotropic LCM slurries. For safety, install a float valve in the bit sub. Consider installing a second internal BOP. The additional BOP should be placed in every stand for additional safety, should a kick occur during the trip.

Rig

The drilling rig should be capable of mixing large quantities of mud and pumping controlled amounts of the mud into the annulus while drilling with water. This requires adequate mud storage capacity and good fluid transfer equipment.

Versatile choke and kill manifolds will plan for well control, easier.

Tripping

Tripping is the most hazardous, most time and mud consuming part of mud cap drilling. The number of trips attempted during mud cap drilling should be minimized. In addition, the makeup of the BHA should be as simple as possible to reduce the amount of time of the BHA in the BOP stack. Control of the fluid level in the annulus and the pressure in the drill-pipe is crucial for safe tripping. The exact procedure to use during tripping will depend on the conditions prevailing in the well, including the type of reservoir fluid, the flow capacity and the mud properties. Prior to tripping, a procedure like the following should be implemented.

1. Select a method to monitor the level of fluid in the annulus and measure the fluid level. Monitor the level carefully during the trip and maintain as constant a level as possible.
2. Monitor annular pressure to detect any intruding reservoir fluid. Bull head mud into the annulus as necessary to control kicks.
3. Displace the drill-pipe with kill weight mud. This relieves the dependence on the float valve as the primary pressure control device in the drill string.
4. Assess the effects of the swab and surge on the annular pressure. Either attempt to adjust the mud properties to minimize the

effects or trip at a speed that reduces these effects. Be aware that thinning the mud may result in a less stable mud column in the annulus.

5. Generally, the most difficulty is experienced moving the bottom hole assembly through the BOP. When tripping out, increase the annular mud addition rate to compensate for the BHA volume. Tripping in may cause larger than usual mud losses to the zone and annular mud additions may have to be increased.

Summary

All alternatives to drilling conventionally should be exhausted before committing to floating mud cap drilling, as it is a very high-risk technique. The well is on the verge of being uncontrolled at any time during the drilling operation. When mud cap drilling is anticipated, specific well control procedures should be in place prior to penetrating the loss zone. It should be matched to the characteristics of the zone. Successful use of a floating mud cap can result in the penetration of the loss zone without setting an extra string of casing.

CHAPTER 14

SUITABILITY OF RIG TO HANDLE OIL BASE MUD/SYNTHETIC OIL BASE MUD

Of late, use of SOBM has become a fashionable type of trend due to the convenience offered by SOBM to drilling community. However, the use of SOBM comes with many strings attached, when it comes to the selection of rig. Unfortunately, often these points do not attract as much attention as they deserve, causing grief to everyone during actual use. With the advent in technology and every minute of rig time attracting attention to save time, the trend is changing gradually and a lot of awareness has evolved over a decade to enrich a rig with features which helps handling SOBM much comfortable. An effort has been made to capture the salient features of each part of the rig and add value. The check list is generic and all the points need not hold true to all the rigs, nor is it exhaustive enough to not to consider any useful point which may come up during inspection. It serves a sort of ready-reckoner table which can be a minimum requirement for making the rig suitable to run SOBM.

When properly audited and promptly followed up, this list can be very useful to avoid last minute surprises in terms of ability to manage operations in SOBM environment without any environment / equipment related issues

A.	**RIG FLOOR & RIG FLOOR DRAINAGE**				
1.	Mud saver valve installed	☐	Yes	☐	No
2.	Mud bucket	☐	Yes	☐	No
3.	Condition of mud bucket seals	☐	Good	☐	Poor
4.	Spare mud bucket seals available (oil friendly)	☐	Yes	☐	No
5.	Condition of mud bucket drain line	☐	Good	☐	Poor
6.	Mud bucket drain line diameter	☐	4”	☐	6”
7.	Mud bucket drain line properly secured	☐	Yes	☐	No
8.	Individual shaker running lights in view of driller	☐	Yes	☐	No
9.	Drip tray under rig floor	☐	Yes	☐	No
10.	Drip tray drained to pits	☐	Yes	☐	No
11.	All drains from rig floor into collection tank	☐	Yes	☐	No
12.	Alarm on collection tank	☐	Yes	☐	No
13.	Facility to pump tank contents to pit	☐	Yes	☐	No

14.	Mouse hole piped to collection tank			☐	Yes	☐	No
15.	Rathole piped to collection tank			☐	Yes	☐	No
16.	Pipe wipers available for use when tripping			☐	Yes	☐	No
17.	Standpipe drain line			☐	Yes	☐	No
B.	**TRIP TANK**						
1.	Bund surrounding trip tank pump			☐	Yes	☐	No
2.	Type of gland packings			☐	Mechanical	☐	Pressurized
3.	Condition of gland packings			☐	Good	☐	Poor
4.	Overflow line fitted to trip tank			☐	Yes	☐	No
5.	Trip tank dump valve lockable			☐	Yes	☐	No
6.	Trip tank dump valve type	☐	Gate	☐	Knife	☐	Butterfly
7.	Drains with bunds			☐	Open	☐	Sealed
C.	**FLOWLINE**						
1.	Open sections			☐	Yes	☐	No
2.	Splash guards fitted to open sections			☐	Yes	☐	No

D. SOLIDS CONTROL PITS

1.	Dump valve lockable	☐ Yes	☐ No
2.	Common drain line	☐ Yes	☐ No
3.	Master dump valve in common drain line	☐ Yes	☐ No
4.	Suction lines recessed onto bottom of solids control pits	☐ Yes	☐ No
5.	Facility to circulate sand traps over shakers	☐ Yes	☐ No
6.	Solids control pits can be completely emptied without discharge to the sea	☐ Yes	☐ No

E. SHAKER HOUSE

3.	Header box configured to allow even flow over shakers	☐ Yes	☐ No
4.	Jetting facility in header box	☐ Yes	☐ No
5.	High pressure shaker wash-down system	☐ Yes	☐ No
6.	Cuttings retaining wall	☐ Yes	☐ No

7.	Shale Shaker house floor drains	☐	Open	☐	Sealed
8.	Shaker house floor drains to common collection tank	☐	Yes	☐	No
9.	Shaker house ventilation	☐	Adequate	☐	Inadequate
10.	Mud cleaner type and number				
11.	Bunds around centrifugal pumps for mud cleaners	☐	Yes	☐	No
13.	Pump gland packing condition	☐	Good	☐	Poor
14.	Centrifuge type and number				
15.	Bunds around centrifugal pumps for centrifuge	☐	Yes	☐	No
17.	Pump gland packing condition	☐	Good	☐	Poor
19.	Drains to common collection tank	☐	Yes	☐	No

F. MUD PITS

1. Number and capacity:	☐	ACTIVE PIT(S)	☐	m3	☐	bbl.
	☐	RESERVE PIT(S)	☐	m3	☐	bbl.

	☐ SURFACE PIT(S)	☐	m3	☐	bbl
2.	Pit level indicators fitted to active pit(s)	☐	Yes	☐	No
3.	Graduated volume indicators fitted in all pits	☐	Yes	☐	No
4.	Agitators *in all mud pits*	☐	Yes	☐	No
5.	Mud guns installed in all pits	☐	Yes	☐	No
6.	Lockable dump valves to all pits	☐	Yes	☐	No
7.	Condition of dump valves	☐	Good	☐	Poor
8.	Common drain from dump valves	☐	Yes	☐	No
9.	Master dump valve in common drain	☐	Yes	☐	No
10.	Suction lines recessed into bottom of pits	☐	Yes	☐	No
11.	Mud pits can be completely emptied on back-loading without discharge to the sea	☐	Yes	☐	No

G. PUMPS

1. Mud pump type and number ______________________

2.	Bunds surround all mixing and charge pumps	☐	Yes	☐	No
3.	Centrifugal pump gland packings	☐	Mechanical	☐	Pressurized
4.	Condition of pump gland packings	☐	Good	☐	Poor
5.	Drains in bunded areas	☐	Good	☐	Poor
6.	Drains in common collection tank	☐	Yes	☐	No
7.	Drains	☐	Open	☐	Sealed
8.	Condition of lines to cement unit	☐	Good	☐	Poor
H.	**MIXING AREA**				
1.	Hoppers surrounded by bunds	☐	Yes	☐	No
2.	Drains in bund area	☐	Yes	☐	No
3.	Drains	☐	Open	☐	Sealed
4.	Drains to collection tank	☐	Yes		No
I.	**BASE OIL**				
1.	Accurate base fluid storage volume indicator installed	☐	Yes	☐	No

2.	Metering system for base fluid movements	☐ Yes ☐ No
J.	**LOADING HOSES**	
1.	Dedicated hoses for loading/back-loading oil based mud	☐ Yes ☐ No
2.	Dedicated hoses for loading/back-loading or oil	☐ Yes ☐ No
3.	Date of last pressure test of hoses	
4.	Check valve at boat end of loading hose to prevent backflow of base oil (Avery Hardoll or similar)	☐ Yes ☐ No
5.	Book check valve at boat end of loading hose to prevent backflow of oil based mud	☐ Yes ☐ No
K.	**MUD VACUUM PUMPS**	
1.	Number and type of mud vacuum units on board	

2.	Condition of mud vacuum hose and couplings	☐ Good	☐ Poor
3.	Length of mud vacuum hoses sufficient to access all areas	☐ Yes	☐ No

L. PROCEDURES FOR HANDLING OIL BASE MUD

Are written procedures in place to cover…

1.	Integrity testing of tanks and valves prior to the loading of oil based mud	☐ Yes	☐ No
2.	Loading/backloading of oil based mud <u>and</u> base oil to & from a supply vessel	☐ Yes	☐ No
3.	Loading/backloading of either, oil base mud <u>and</u> base oil controlled by a permit to work (PTW) system	☐ Yes	☐ No
4.	Transfer of base oil or oil base mud from pontoon (where applicable)	☐ Yes	☐ No

5.	Cleaning of pits on completion of the use of oil base mud	☐ Yes	☐ No
6.	Use of mud vacuum units/air diaphragm pumps	☐ Yes	☐ No
7.	Displacement of oil base mud to/from the well	☐ Yes	☐ No
8.	Displacement of the riser on a semi-submersible from oil based mud to seawater (and vice versa)	☐ Yes	☐ No
9.	Pre-circulation checks	☐ Yes	☐ No
10.	Loss of oil base mud at the shakers whilst circulating	☐ Yes	☐ No
11.	Operation of dump valves controlled by permit to work (PTW)	☐ Yes	☐ No
12.	Displacement of oil base mud during cementing operations	☐ Yes	☐ No
13.	Handling of cement contaminated oil base mud	☐ Yes	☐ No

M. BASE OIL & ELASTOMER COMPATIBILITY

List rubber type for each component…, Use only those elastomers which are compatible with most of the types of base oils

1. BOP elements	
2. Flexible choke and kill line	
3. Flexible HP mud lines and Howson line to rig floor	
4. Flexible mud lines to top-drive	
5. Mud pump swabs	
6. Mud pump pulsation dampners	
7. Centrifugal pump pulsation dampeners	
8. Flowline pneumatic seals	
9. Other mud hoses (e.g. mud line to cement unit, casing fill line…)	

10. Seals and gaskets (e.g. flowline inspection hatches & flanges, vacuum degasser, mud bucket…)	
11. Plastic air, water and duct hoses in shaker area	
12. Plastic coated cables in shaker area	
13. Plastic cuttings discharge chute	
14. Shear unit	

CHAPTER 15

BASIC CALCULATIONS

Although in the ever-increasing dependence on software, we still need to be familiar with certain basic calculations. The list below is very basic and far advanced calculations are available. The intention here is to include the information only for beginners.

1. Increase Mud Density

Mud weight, ppg, increase with Barite (average specific gravity of Barite - 4.2)

$$\text{Barite, sk / 100 bbl} = \frac{1470\ (W_2 - W_1)}{35 - W_2}$$

Example: Determine the number of sacks of Barite required to increase the density of 100 bbl of 12.0 ppg (W_1) mud to 14.0 ppg (W_2):

$$\text{Barite sk / 100 bbl} = \frac{1470(14.0 - 12.0)}{35 - 14.0}$$

$$\text{Barite, sk / 100 bbl} = \frac{2940}{21.0}$$

$$\text{Barite} = 140 \text{ sk / 100bbl}$$

Volume increase, bbl, due to mud weight increase with Barite

$$\text{Volume increase, per 100 bbl} = \frac{100(W_2 - W_1)}{35 - W_2}$$

Example: Determine the volume increase when increasing the density from 12.0 ppg (W_1) to 14.0 ppg (W_2):

$$\text{Volume increase, per 100 bbl} = \frac{100(14.0 - 12.0)}{35 - 14.0}$$

$$\text{Volume increase, per 100 bbl} = \frac{200}{21}$$

$$\text{Volume increase} = 9.52 \text{ bbl per 100 bbl}$$

Starting volume, bbl, of original mud weight required to give a predetermined final volume of desired mud weight with Barite

$$\text{Starting volume, bbl} = \frac{V_F(35 - W_2)}{35 - W_1}$$

Example: Determine the starting volume, bbl, of 12.0 ppg (W_1) mud required to achieve

$$\text{Starting volume, bbl} = \frac{100(35 - 14.0)}{35 - 12.0}$$

$$\text{Starting volume, bbl} = \frac{2100}{23}$$

$$\text{Starting volume} = 91.3 \text{ bbl}$$

Mud weight increase with Calcium Carbonate (SG — 2.7)

NOTE: The maximum practical mud weight attainable with Calcium Carbonate is 14.0 ppg.

$$\text{Sacks / 100 bbl} = \frac{945(W_2 - W_1)}{22.5 - W_2}$$

Example: Determine the number of sacks of Calcium Carbonate/100 bbl required to increase the density from 12.0 ppg (W_1) to 13.0 ppg (W_2):

$$\text{Sacks / 100 bbl} = \frac{945\ (13.0 - 12.0)}{22.5 - 13.0}$$

$$\text{Sacks / 100 bbl} = \frac{945}{9.5}$$

$$\text{Sacks / 100 bbl} = 99.5$$

Volume increase, bbl, due to mud weight increase with Calcium Carbonate

$$\text{Volume increase, per 100 bbl} = \frac{100(W_2 - W_1)}{22.5 - W_2}$$

Example: Determine the volume increase, bbl/100 bbl, when increasing the density from 12.0 ppg (W_3) to 13.0 ppg (W_2):

$$\text{Volume increase, per 100 bbl} = \frac{100(13.0 - 12.0)}{22.5 - 13.0}$$

$$\text{Volume increase, per 100 bbl} = \frac{100}{9.5}$$

$$\text{Volume increase} = 10.53 \text{ bbl per 100 bbl}$$

Starting volume, bbl, of original mud weight required to give a predetermined final volume of desired mud weight with Calcium Carbonate

$$\text{Starting volume, bbl} = \frac{\text{VF}\ (22.5 - \text{W2})}{22.5 - W_1}$$

Example: Determine the starting volume, bbl, of 12.0 ppg (W_1) mud required to achieve

$$\text{Starting volume, bbl} = \frac{100(22.5-13.0)}{22.5-12.0}$$

$$\text{Starting volume, bbl} = \frac{950}{10.5}$$

$$\text{Starting volume} = 90.5 \text{ bbl}$$

Mud weight increase with hematite (SG — 4.8)

$$\text{Hematite, sk / 100 bbl} = \frac{1680(W_2 - W_{\text{-}})}{40 - W_2}$$

Example: Determine the hematite, sk/100 bbl, required to increase the density of 100 bbl of 12.0 ppg (W_1) to 14.0 ppg (W_2):

$$\text{Hematite, sk / 100 bbl} = \frac{1680(14.0-12.0)}{40-14.0}$$

$$\text{Hematite, sk / 100 bbl} = \frac{3360}{26}$$

$$\text{Hematite} = 129.2 \text{ sk / 100 bbl}$$

Volume increase, bbl, due to mud weight increase with hematite

$$\text{Volume increase, per 100 bbl} = \frac{100\ (W_2 - W_1)}{40 - W_2}$$

Example: Determine the volume increase, bbl/100 bbl, when increasing the density from

$$\text{Volume increase, per 100 bbl} = \frac{100(14.0-12.0)}{40-14.0}$$

$$\text{Volume increase, per 100 bbl} = \frac{200}{26}$$

$$\text{Volume increase} = 7.7 \text{ bbl per 100 bbl}$$

Starting volume, bbl, of original mud weight required to give a predetermined final volume of desired mud weight with hematite

$$\text{Starting volume, bbl} = \frac{VF(40.0 - W2)}{40 - W_1}$$

Example: Determine the starting volume, bbl, of 12.0 ppg (W_1) mud required to achieve 100 bbl (V_F) of 14.0 ppg (W_2) mud with hematite:

$$\text{Starting volume, bbl} = \frac{100(40 - 14.0)}{40 - 12.0}$$

$$\text{Starting volume, bbl} = \frac{2600}{28}$$

$$\text{Starting volume} = 92.9 \text{ bbl}$$

2. Dilution

Mud weight reduction with water

$$\text{Water, bbl} = \frac{V_1(W_1 - W_2)}{W_2 - Dw}$$

Example: Determine the number of barrels of water weighing 8.33 ppg (Dw) required to reduce 100 bbl (V_1) of 14.0 ppg (W_1) to 12.0 ppg (W_2):

$$\text{Water, bbl} = \frac{100\ (14.0 - 12.0)}{12.0 - 8.33}$$

$$\text{Water, bbl} = \frac{2000}{3.67}$$

$$\text{Water} = 54.5 \text{ bbl}$$

Mud weight reduction with diesel oil

$$\text{Diesel, bbl} = \frac{V_1(W_1 - W_2)}{W_2 - Dw}$$

Example: Determine the number of barrels of diesel weighing 7.0 ppg (Dw) required to reduce 100 bbl (V_1) of 14.0 ppg (W_1) to 12.0 ppg (W_2):

$$\text{Diesel, bbl} = \frac{100(14.0 - 12.0)}{12.0 - 7.0}$$

$$\text{Diesel, bbl} = \frac{200}{5.0}$$

$$\text{Diesel} = 40 \text{ bbl}$$

3. Mixing Fluids of Different Densities

Formula: $(V_1 D_1) + (V_2 D_2) = V_F D_F$

where V_1 = volume of fluid 1 (bbl, gal, etc.)
V_2 = volume of fluid 2 (bbl, gal, etc.)
V_F = volume of final fluid mix
D_1 = density of fluid 1 (ppg, lb/ft^3, etc.)
D_2 = density of fluid 2 (ppg, lb/ft^3, etc.)
D_F = density of final fluid mix

Example 1: A limit is placed on the desired volume:

Determine the volume of 11.0 ppg mud and 14.0 ppg mud required to build 300 bbl of 11.5 ppg mud:

Given: 400 bbl of 11.0 ppg mud on hand, and
400 bbl of 14.0 ppg mud on hand
Solution: let V_1 = bbl of 11.0 ppg mud
V_2 = bbl of 14.0 ppg mud

then a) $V_1 + V_2 = 300$ bbl
b) $(11.0)\,V_1 + (14.0)\,V_2 = (11.5)\,(300)$

Multiply Equation A by the density of the lowest mud weight (D_1 = 11.0 ppg) and subtract the result from Equation B:

$$\begin{aligned} \text{b) } (11.0)(V_1) + (14.0)(V)_2 &= 3450 \\ -\ \text{a) } (11.0)(V_1) + (11.0)(V_2) &= 3300 \\ \hline 0(3.0)\ (V_2) &= 150 \\ 3V_2 &= 150 \\ V_2 &= \frac{150}{3} \\ V_2 &= 50 \end{aligned}$$

Therefore: $V_2 = 50$ bbl of 14.0 ppg mud
$V_1 + V_2 = 300$ bbl
$V_1 = 300 - 50$
$V_1 = 250$ bbl of 11.0 ppg mud

Check: V_1 = 50 bbl D_1 = 14.0 ppg
V_2 = 150 bbl D_2 = 11.0 ppg
V_F = 300 bbl D_F = final density, ppg

$$(50)\ (14.0) + (250)\ (11.0) = 300\ D_F$$
$$700 + 2750 = 300\ D_F$$
$$3450 = 300\ D_F$$
$$3450 \div 300 = DF$$
$$11.5 \text{ ppg} = DF$$

Example 2: No limit is placed on volume:

Determine the density and volume when the two following muds are mixed together:

Given: 400 bbl of 11.0 ppg mud, and
400 bbl of 14.0 ppg mud

Solution: let V_1 = bbl of 11.0 ppg mud — D_1 = density of 11.0 ppg mud
V_2 = bbl of 14.0 ppg mud — D_2 = density of 14.0 ppg mud
V_F = final volume, bbl — D_F = final density, ppg

Formula: $(V_1\,D_1) + (V_2\,D_2) = V_F\ D_F$

$$(400)\ (11.0) + (400)\ (14.0) = 800\ D_F$$
$$4400 + 5600 = 800\ D_F$$
$$10{,}000 = 800\ D_F$$
$$10{,}000 \div 800 = D_F$$
$$12.5 \text{ ppg} = D_F$$

Therefore: final volume = 800 bbl
final density = 12.5 ppg

4. Oil based mud Calculations

Density of oil/water mixture being used

$$(V_1)(D,) + (V_2)(D_2) = (V\sim + V_2)D_F$$

Example: If the oil/water (o/w) ratio is 75/25 (75% oil, V_1, and 25% water V_2), the following material balance is set up:

NOTE: The weight of diesel oil, $D_1 = 7.0$ ppg
The weight of water, $D_2 = 8.33$ ppg

$$(0.75)\ (7.0) + (0.25)\ (8.33) = (0.75 + 0.25)\ D_F$$
$$5.25 + 2.0825 = 1.0\ D_F$$
$$7.33 = D_F$$

Therefore: The density of the oil/water mixture = 7.33 ppg

Starting volume of liquid (oil plus water) required to prepare a desired volume of mud

$$SV = \frac{35 - W_2 \times DV}{35 - W_1}$$

where SV = starting volume, bbl
W_1 = initial density of oil/water mixture, ppg
W_2 = desired density, ppg
Dv = desired volume, bbl

Example: $W_1 = 7.33$ ppg (o/w ratio — 75/25) $W_2 = 16.0$ ppg
Dv = 100 bbl

Solution:

$$SV = \frac{35 - 16}{35 - 7.33} \times 100$$

$$SV = \frac{19}{27.67} \times 100$$

$$SV = 0.68666 \times 100$$

$$SV = 68.7 \text{ bbl}$$

Oil/water ratio from retort data

Obtain the percent-by-volume oil and percent-by-volume water from retort analysis or mud still analysis. From the data obtained, the oil/water ratio is calculated as follows:

a) $\%\text{ oil in liquid phase} = \dfrac{\%\text{ by vol oil}}{\%\text{ by vol oil} + \%\text{ by vol water}} \times 100$

$$\text{b) \% water in liquid phase} = \frac{\text{\% by vol water}}{\text{\% by vol oil} + \text{\% by vol water}} \times 100$$

c) Result: The oil/water ratio is reported as the percent oil and the percent water.

Example: Retort analysis: % by volume oil = 51
% by volume water = 17
% by volume solids = 32

Solution:

$$\text{a) \% oil in liquid phase} = \frac{51}{51 \times 17} \times 100$$

$$\text{\% oil in liquid phase} = 75$$

$$\text{b) \% water in liquid phase} = \frac{17}{51 + 17} \times 100$$

$$\text{\% water in liquid phase} = 25$$

c) Result: Therefore, the oil/water ratio is reported as 75/25: 75% oil and 25% water.

Changing oil/water ratio

NOTE: If the oil/water ratio is to be increased, add oil; if it is to be decreased, add water.

Retort analysis: % by volume oil = 51
% by volume water = 17
% by volume solids = 32

The oil/water ratio is 75/25.

Example 1: Increase the oil/water ratio to 80/20:

In 100 bbl of this mud, there are 68 bbl of liquid (oil plus water). To increase the oil/water ratio, add oil. The total liquid volume will be increased by the volume of the oil added, but the water volume will not change. The 17 bbl of water now in the mud represents 25% of the liquid volume, but it will represent only 20% of the new liquid volume.

Therefore: let x = final liquid volume

then, 0.20x = 17
x = 17 : 0.20
x = 85 bbl

The new liquid volume = 85 bbl

Barrels of oil to be added:
Oil, bbl = new liquid vol − original liquid vol
Oil, bbl = 85 − 68
Oil = 17 bbl oil per 100 bbl of mud

Check the calculations. If the calculated amount of liquid is added, what will be the resulting oil/water ratio?

$$\%\text{ oil in liquid phase} = \frac{\text{original vol oil} + \text{new vol oil}}{\text{original liquid oil} + \text{new oil added}} \times 100$$

$$\%\text{ oil in liquid phase} = \frac{51+17}{68+17} \times 100$$

$$\%\text{ oil in liquid phase} = 80$$

$$\%\text{ water would then be} : 100 - 80 = 20$$

Therefore: The new oil/water ratio would be 80/20.

Example 2: Change the oil/water ratio to 70/30:

As in Example I, there are 68 bbl of liquid in 100 bbl of this mud. In this case, water will be added and the volume of oil will remain constant. The 51 bbl of oil represents 75% of the original liquid volume and 70% of the final volume:

Therefore: let x = final liquid volume

then, 0.70x = 51
x = 51 : 0.70
x = 73 bbl

Barrels of water to be added:

Water, bbl = new liquid vol — original liquid vol
Water, bbl = 73 — 68
Water = 5 bbl of water per 100 bbl of mud

Check the calculations. If the calculated amount of water is added, what will be the resulting oil/water ratio?

$$\% \text{ water in liquid phase} = \frac{17+5}{68+5} \times 100$$

$$\% \text{ water in liquid} = 30$$

$$\% \text{ oil in liquid phase} = 100 - 30 = 70$$

Therefore, the new oil/water ratio would be 70/30.

Dilemma of calculations: In a rarest of the rarest case a very peculiar and uncharacteristic observation came to light while playing with OWR and mud weight changes in OBM system. A mud of 1.28 SG with 90/10 OWR was required to be converted to 97/3 OWR of 1.40 SG weight using Calcium Carbonate as weighing material. Since both OWR and mud weight were changing together, no simple calculation method could correctly provide the answer. The material balance equation and the software calculations showed that the actual volume of Oil required in 1.40 SG at 97/3 OWR is in fact less than the volume of Oil required in 90/10 OWR of 1.30 SG. While his could have been easy, if the amount of water and oil remained in same ration of 90/10 and weight obtained as 1.40 SG from 90/10 OWR of 1.28 SG. The fact that the required volume of oil in 97/3 OWR is less than 90/10 OWR is a riddle, but all available options provide same results, including the retort observations as well as material balance equation! This is an interesting feature and it opens a new arena for an in-depth investigation! This again proves- we must know how to calculate, but more importantly we MUST know how to judiciously use the information in real case. In real life, if there is a need to prepare 97/3 OWR mud from 90/10 OWR and simultaneously raise the weight using Calcium Carbonate, then it is not recommended to convert the 90/10 mud to 97/3 mud, but prepare fresh 97/3 OWR mud. It reminds a proverb, knowledge is to know that Tomato is a fruit, but wisdom says not to use tomato in fruit salad preparation!!

CHAPTER 16

COMPLETION FLUID

Completion fluids can be defined as, any fluid pumped down-hole to conduct operations after the initial drilling of a well. Work-over fluids are those used during remedial operations after a well has been completed and produced oil and/or gas. Clear, solids-free brine completion/work-over fluids serve to control down-hole formation pressures while reducing the risk of permanent formation damage (permeability damage) resulting from solids invasion or some form of incompatibility between the completion fluid and the in-situ matrix. The clear brines used for completion and work-over applications are pure solutions of dissolved salt in water and must be stable at the surface and down-hole conditions. Depending on the application, other completion/work-over fluid types are sometimes used, including solids-laden, oil based and emulsions. For this book, no distinction is made between completion and work-over fluids and the terms are used interchangeably throughout. Packer fluids are those that fill the annular volume above a production packer. The term drill-in fluid is recognized as referring to a drilling fluid designed specifically for the productive interval. Drill-in fluids are designed to minimize damage to the interval, typically by eliminating insoluble solids such as Barite, minimizing the total solids content and formulating such that a thin, resilient, removable, non-damaging filter cake is placed.

Among the typical operations in which clear brines are applied are well kills, fishing, perforating, washing, drilling and gravel packing and as packer fluids. To perform the desired function, completion fluids must control formation pressures, circulate and transport solids, protect the productive zone, be stable under the surface and down-hole conditions, be safely handled, be environmentally friendly or be used with controlled exposure, and be cost effective. Completion fluids have no defined purpose within the formation and may in fact reduce the permeability.

Clear brine completion fluids are formulated and applied in the field, according to performance specifications that ensure well control with minimal permeability reduction. These specifications are not always expressly identified, but should always be understood and assigned. Density and solids content (expressed as clarity - NTU) are typical performance specifications for clear brine, although the selection of completion fluid according to these alone can be dangerous to the productivity of a well. Proper density is necessary for pressure control. Clarity is necessary to eliminate formation plugging by solids. In addition to these, the all-encompassing term "formation compatible" is also a requirement and must not be overlooked. To select a completion or work-over fluid appropriate for the application, one must understand the various types and properties of clear brine fluids.

The Planning Process

The planning process steps are organized in such a way as to assist in using basic pieces of information to develop a plan that encompasses all aspects of selecting the correct clear brine fluid, additives, and associated equipment.

The planning process steps are arranged to enable to:

- Determine appropriate fluid density using true vertical depth (TVD), bottom-hole pressure (BHP), and bottom-hole temperature (BHT);
- Select the correct true, crystallization temperature (TCT);
- Estimate the volume of clear brine fluid for the job;
- Select the proper clear brine fluid family (single, two, or three salt); or

- Where compatibility issues, corrosion concerns, or sensitive formations exist, select an engineered fluid system or a specialty fluid.

Fluid Categories

Planning any completion is an iterative process and will most likely require more than one pass as we gather more information and refine a selection. Using basic design information, true vertical depth, bottom-hole pressure, and environmental temperature considerations as outlined in the following sections, we can determine which clear brine fluids are a good match for the conditions.

Low density systems usually consist of single salt fluids, which can range in density from slightly above the density of water, such as 3% potassium chloride (KCl), to as high as 13.1 lb/gal Potassium Formate., and still higher on few cases. Unique formation properties or concerns about the compatibility of conventional brines with formation water may suggest the use of Sodium bromide (NaBr), Calcium bromide (CaBr2), Sodium Formate (NaO2CH), potassium Formate (KO2CH), or cesium Formate (CsO2CH) —the latter three of which are halide free, containing no chloride or bromide.

Midrange density fluids, 11.7 lb/gal to 15.1 lb/gal, are typically two salt mixtures of Calcium chloride (CaCl2) and Calcium bromide (CaBr2). The boundary between two and three salt fluids in Figure 2 is influenced by the lower of the expected atmospheric temperature or mudline temperature. In many cases, the lowest temperature in the entire fluid column is at the ocean floor (mudline) where temperatures can routinely be less than 40°F. This temperature will often dictate the CBF category.

Fluid Density

Expected bottom-hole conditions, is the basic criteria that influence the selection of a clear brine completion fluid density. The fluid density required for a job is largely determined by the true vertical depth (TVD) planned for the well and the expected bottom-hole pressure (BHP). True vertical depth is normally given in feet (ft), and bottom-hole pressure is given in pounds per square inch (psi or lb/in^2). These two values are used to determine the pressure gradient in pounds per square inch per foot of depth (psi/ft). An additional margin of safety should be added to the BHP to ensure that control of the well is achieved, usually 100 psi.

General Fluid Density Ranges

Table 1 below provides an extensive list of conventional and specialty clear brine fluids and their working density ranges.

Density Ranges

There are many fluid options at the lower ranges of density, up to about 10.0 lb/gal. The choice of one brine over another may be based on unique formation properties. Bromide-chloride, two salt fluids and formates reach densities up to 13.0 lb/gal. When the density requirement is more than 14.0 lb/gal the selection of brine is limited to two and three salt halides, zinc bromide ($ZnBr_2$), and cesium Formate (CsO_2CH).

TABLE 1. General Density Ranges for Clear Brine Fluids

Completion brine	Max. density achievable, without considering temperature effect – in ppg
Ammonium chloride	8.9
Potassium chloride	9.7
Sodium chloride	10
Sodium formate	11
Calcium chloride	11.6
Sodium Bromide	12.7
Potassium Formate	13.1
Calcium bromide	15.1
Cesium Formate	19.2
Zinc Bromide	20.5

In addition to it, various other desnsity ranges are achievable by combination of two orthree different brine types, which are compatible with each pther.

True crystallization Temperature:

The true-crystallization temperature of a brine is the temperature at which a solid phase begins to form from with-in a clear solution of brine, resulting in a mixture of solid particles and solution. These solids may be salt crystals or water crystals (ice).

The crystallization point is determined by cooling a brine until salt crystals form, and recording temperatures at various times during the process. *First Crystal to Appear* (FCTA) is the point at which salt crystals first form. The formation of salt crystals generates a small amount of heat, which causes a slight rise in the solution's temperature. This higher temperature corresponds to the *true crystallization temperature* (TCT) of the brine.

Once the crystals have formed, the brine can then be heated until all the crystals are re-dissolved. The temperature which corresponds to the temperature at which the salt goes back into solution is labeled *Last Crystal to Dissolve* (LCTD). Generally, the TCT is the most commonly reported crystallization point. In practice, the TCT is extremely valuable since it is a strong reflection of the composition of the brine. It is, in fact, the most reliable and reproducible measure of the safe working limits of heavy brines. The measurement of TCT is governed by an API protocol.

In the case of multi-salt brines, the least soluble component will crystallize at the TCT. Thus, if heavy brine is contaminated with minor amounts of NaCl or KCl from formation brine or seawater, the TCT may be shifted to a much higher temperature. This is due to the limited solubility of NaCl and KCl in heavy brines. Although the brine at the altered TCT may appear cloudy, it can be cooled to the original TCT with no further crystallization occurring.

Temperature Considerations

Temperature conditions that are likely to be encountered over the length of the fluid column may cause heating or cooling of the brine. Rapid or unanticipated changes in weather conditions may also cause cooling of a fluid as it travels through surface piping and equipment. It is important to anticipate, as closely as possible, the weather conditions that may occur during the entire course of the completion project.

Critical points in the flow path are:

1. Ocean water surface temperature,
2. Water temperature at the ocean floor (mud line),
3. Atmospheric conditions—temperature changes in surface tankage and distribution piping due to weather,
4. Filtration equipment, and
5. Pill tanks and storage/transfer tanks.

If a completion fluid cools below its stated TCT, solid salt(s) will begin to separate out from it. This will greatly increase the demands placed on pumping equipment due to increased resistance to flow. The solids may impede filtration, two ways—through a cake buildup in the plate and frame diatomaceous earth (DE) filters and/or by plugging cartridges.

The loss of soluble salts, either by settling out or filtration, will drastically reduce the density of the completion fluid. Loss of density could result in a dangerous underbalanced situation.

It is vital to make a temperature profile for the entire flow system expected for the completion fluids. The lowest temperature likely to be encountered will determine the safe crystallization temperature.

To provide an adequate safety margin, the TCT for the fluid should be set 10°F (5.5°C) below the lowest temperature expected to be encountered at any point along the flow path.

Pressure Considerations—Pressurized Crystallization Temperature

Deep-water and subsea completions require a greater attention to detail, especially in terms of TCT. At ocean water-depth greater than approximately 1,500 feet, additional adjustment must be made to the fluid formulation. Experience has shown that at low temperatures and high pressure likely to occur in deep-water wells, both can cause changes in measured TCT. At pressures, likely to be attained during the testing of a blowout preventor (BOP) for example—a fluid which functions correctly under normal hydrostatic pressure, may begin to crystallize with the increased testing pressure. Similarly, at a low sea bed temperature of sea bed a brine solution of concentration may separate out in liquid and solid form which otherwise is a solution at ambient temperature conditions.

TCT is a measure of crystallization temperature under atmospheric pressure. For most applications, the effect of pressure on solubility is slight. The effect of pressure can be significant under conditions involving a combination of high pressure and low temperature such as in deep-water applications at seabed or when pressure testing in colder climates. Under these conditions, some solutions of brine may crystallize at a temperature higher than the expected TCT, possibly varying by as much as 20°F. Therefore, the pressurized crystallization temperature (PCT) behavior of the brine should be known prior to usage. Unexpected crystallization could have disastrous results. At the mud line, where temperatures are lower, choke or kill lines can plug and valves can seize, jeopardizing operations. Removal of crystals deposited during pressure testing in the BOP stack or in other locations where circulation is poor can be extremely difficult.

It is strongly recommended that the PCT be determined for fluids where low temperature and high-pressure conditions may coexist.

Temperature and pressure Effect

When brine is put into use, the down-hole temperature profile will cause the brine to expand, lowering the average density of the fluid column. Pressure has the opposite effect and causes an increase in density. Adjustments need to be made to the fluid density to compensate for the combination of bottom-hole pressure and bottom-hole temperature.

For fluids with densities less than approximately 12.0 lb/gal, thermal expansion will typically be in the range of 0.26 lb/gal to 0.38 lb/gal per 100°F (lb/gal/100°F) increase in temperature. From 12.0 lb/gal to 19.0 lb/gal, the expansion ranges from 0.33 lb/gal to 0.53 lb/gal per 100°F increase. Typically, the density correction is made for the average temperature of the fluid column. Pressure effects are much smaller and range from 0.019 lb/gal per thousand psi to 0.024 lb/gal per thousand psi.

Before the advent of deep-water drilling, pressure related contraction of brine was never taken seriously as now. In normal shallow water and land wells the impact of increase in density is minimal and after applying temperature correction, the usual practice was to either live with that number or occasionally add 50 psi overbalance.

Following formula provides temperature corrected brine densities. For a more accurate selection of densities, especially in deep-water environment, it is strongly recommended that a more complicated formula be used, which is available in all big company's completion manuals.

Temperature Correction

EQUATION.

Temp. corrected weight = A (BHT-surf) / 200
Where
A is thermal expansion factor ppg/100 deg F
Surf is surface temperature
BHT is bottom hole temperature

Estimating Required Fluid Volume

Objectives

- Maintain well control—ensure a full column of clear brine fluid of an adequate density
- Respond to pressure changes
- Plan for fluid contingency needs

Factors Affecting

- Well design and surface equipment
- Formation permeability
- Distance to the supply point

Carefully estimating the required fluid volume will allow you to maintain an adequate volume of completion fluid to ensure smooth, uninterrupted completion operations.

Determination of the appropriate quantity of completion fluid should be based primarily on the capacity of the casing and tubing used during completion operations. The quantity of fluid circulating at any time is the total of the well volume, less the tubing displacement, plus all surface equipment, piping, pumps, tanks, and filtration equipment. Contingency planning for additional fluid needs will include potential fluid loss and density control. Finally, the distance to the supply point may suggest additional volume to ensure a timely response. Generally, the initial fluid order should be at least two to three times the circulating volume of the well.

Calculating Volume Requirements

A volume calculation worksheet should include the following:

1. Circulating volume
2. Holding tanks
3. Filtration equipment
4. Surface piping
5. Contingency needs and pill demands

Circulating Volume

Determining the volume of the CBF required to fill the hole and maintain the required hydrostatic pressure is a matter of adding up the casing, liner, and openhole volumes and then subtracting the volume displaced by drill-pipe or tubing.

Cased Hole with Tubing.

Holding Tanks

The tank capacity necessary for a CBF job is often substantially greater than that required for circulating a drilling fluid. Since brines are continuously filtered, two holding tanks are required, one for returning fluid that may be carrying solids and another of equal volume of filtered fluid. Holding tank volume may also be limited by rig space.

At least one complete hole volume should be available in surface holding tanks to allow filtration operations to keep pace with circulating requirements.

Surface Piping

Any unusual requirements for positioning equipment can result in additional volumes in hoses, pumps, and piping. An allowance of 10 barrels is a reasonable recommendation.

While planning for volume we also need to remember that the volume should be that of a filtered brine to chosen NTU and not the

brine shipped from supply vessel or prepared on rig, which is not yet filtered!

Contingency Planning and Pill Demands

Fluid Loss Pills. On occasion, it may become necessary to pump a viscous pill into the producing zone to slow fluid loss. The volume of the pill will be equal to at least the combined annular and tubing volume through the perforated zone plus some additional footage for safety. As a rule of thumb, about 1.5 times the volume of the perforated zone can be used.

Spike material, or spike fluid, is the high-density fluid that is transported to and stored on location in case it is necessary to raise fluid density to control pressure or respond to a kick. The volume usually ranges between 200 and 300 bbl of a selected high density blending stock. The volume of spike material that is held in reserve should be based on many factors, including:

- Uncertainty regarding bottom-hole pressure,
- Treating dilutions of working fluid,
- Available storage space on the location or rig,
- Density difference between the working fluid and the spike fluid,
- Environmental discharge/spill limitations, and
- Cost considerations.

A detail that is often overlooked when determining the density and volume of spike material is the relative amount of spike fluid needed to raise the density of the working fluid by an increment. Often, it is more economical to use a much heavier spike fluid, even if its unit cost is higher. The reason for this is that it may take substantially less of the heavier spike material to obtain the same density increase.

For example, it can take twice as much 16.2 ppg heavy spike fluid to raise the density of a 14.85 lb/gal working fluid by 0.2 lb/gal than it would if an 18.5 lb/gal spike fluid was used. The numbers are illustrative and not real. The intention is to show the impact of the weigh tof spike fluid. In addition to the smaller storage needs of the higher density spike fluid, there is the added benefit that, when it is used to achieve a given density adjustment, it will create a smaller volume increase in the working fluid.

Permeability and Pressure Conditions in a Producing Zone

Formation characteristics will play a large role in determining the amount of fluid that is held in reserve. Large quantities of fluid may be lost to highly permeable formations or formations that contain fracture permeability. Experience in a particular producing horizon may dictate carrying extra fluid inventory to allow for seepage into the formation.

Distance to the Supply Point

The distance to the nearest supply point, the uncertainty about bottom-hole conditions, and seasonal factors such as temperature changes should be considered in determining the volume delivered at the beginning of the job. Deep-water offshore platforms will probably have longer supply lead times than shallow water or onshore projects. In cases where substantial delays could impact operations, additional volume should be purchased to ensure that volume losses can be made upon a safe and timely basis, in order to avoid delays.

Volume Calculation Worksheet

According to the general rule, the initial fluid order should be two to three times the circulating volume of the well. Another method for determining the initial fluid quantity is to use a tool similar to the volume calculation table below.

Volume calculation table

Equipment	Volume
Circulating Volume	Deduct tubing volume
Holding Tanks	Leave dead volume
Filtration Equipment capacity	Expect waste during cartridge change to some extent
Surface pipe	Expect interface volume contamination
LCM Pills	
Conssideration for losses	One and half times. This can vary depending upon severity of the case
Total	****

Fluid Compatibility

Mineralogy

Reservoir mineralogy, especially the percentage of active clay, may influence the decision as to the type of CBF best suited to a particular formation. The dominant Cation in the brine, for example, ammonium (NH_4^+), Sodium (Na^+), potassium (K^+), Calcium (Ca^{+2}), or zinc (Zn^{+2}), can react with clay minerals to promote / demote stability. Compatibility testing of core samples from the reservoir is the most reliable means of assessing the response of clay minerals to a brine. Experience in offset wells should also be considered if existing data indicate the sensitivity of clay minerals.

Reservoir Fluid Chemistry

Reservoir fluids are in a state of chemical equilibrium with the reservoir minerals. This state of equilibrium will be disturbed once a formation is penetrated and production activities begin. Prior to producing the well, the potential for formation damage resulting from reactions between formation fluids and drilling or completion fluids will exist. The chemical composition of formation waters should be evaluated for compatibility, paying attention to the degree of saturation with salt (NaCl) and any bicarbonate and sulphate ion concentrations. Incompatibility between formation minerals and brine composition can result in precipitation insidethe reservoir, seriously impairing the productivity of the well.

Metallurgy and Elastomers

Clear brine fluids must also be compatible with the metallic components used in down-hole equipment as well as with any tools, with which they will come into contact. Temperature, pressure, and stresses can result in corrosion induced by the interaction between clear brine fluids and various types of metals. The increase in HPHT drilling has led to greater use of corrosion resistant alloys (CRAs) in the production tubing.

Acid Corrosion

Corrosion of metallic surfaces that encounter brines is strongly accelerated by the presence of the hydrogen ion (H^+). The hydrogen ion can be essentially eliminated by raising the pH of the brine. The pH of fluids containing Sodium, potassium, or Calcium can be raised into a range where only negligible concentrations of hydrogen ions are present. Adjusting the pH of fluids containing ammonium or zinc ions is not recommended, as those ions are not stable at the pH levels that can be attained in other CBFs, leading to the formation of precipitate of their respective hydroxides.

Carbonate

Certain calcareous reservoirs with a high partial pressure of Carbon dioxide may be incompatible with fluids that contain the Calcium ion. Mixing formation water and Calcium containing CBFs may result in the precipitation of Calcium Carbonate at the point of contact between the two fluids. The formation of Calcium Carbonate can result in permeability reduction. If formation water analysis indicates high levels of the bicarbonate ion (HCO_3^{-1}), fluids containing Calcium should be avoided.

CHAPTER 17

FIELD TIPS WHICH NO CLASSROOM EVER TEACHES

This chapter has been included to guide new and aspiring entrants only. Those who have spent enough time on rig as a mud engineer, may be already conversant.

The information shared here is more a personal perception and less of a scientific truth, but mud engineering is never 100 % science. These traits are like a mother being able to pre-empt upcoming trouble to her child, which no one else can predict or feel before the occurrence. Similarly, the great mud engineer only will know exactly what the returning mud has in store, for the hole in short future, and how to mitigate any possible threat. There are occasions where one is left of its own to find a solution on the rig and at that time, your confidence as well as the acquired knowledge are the only two friends who stay with you.

- While mixing SOBM, always invariably emphasize on adequate mechanical agitation for brine and base oil combination. The finer the droplets of brine phase interspersed, the stronger is the emulsion. Use all available sources of mechanically agitating the emulsion before mixing emulsifier.

- While mixing water base mud, one should first begin with clay and then add enough of the F/L control agent as well as weighing agent. Salt/s can be added at a later stage, but do remember that add Barite only to the tune of 3/4th of the calculated amount, which should also have a provision of rise in mud weight due to salt addition. This skeleton, when checked for rheology, shall give an indication whether or not as well as how much low-end rheology modifier is required. Split your treatment in two phases with an intermittent partial mud check to work upon for final tweaking.
- Before leaving everything on mixing hands, make sure that they differentiate, distinguish and segregate different sacks and exactly understand which additive to be mixed for how many sacks. It may sound silly, but unknowingly they may end up mixing wrong additive in the wrong number, leaving you baffled with mud check results.
- Shale inhibitors can be added in last, but you will often find that they do contribute their two cents in controlling F/L also, either by additional solids or by a suitable chemical mechanism depending upon the type of shale inhibitor.
- Unless drilling through Chert, there is most likely no need to have additional lubricant in WBM and any treatment suggested in mud program can be simply an effort by the program writer to be politically correct in his planning. Therefore, save on precious rig time and change over without adding lubricant, which can anyway be added once LOT recording is over and drilling resumed.
- Changing over to treated mud before resuming drilling of next section also takes an hour or two. Be a smart planner to use that time in mixing shale inhibitors, lubricant, etc. Your efforts to save costly rig time will always be appreciated.
- The mud testing equipment needs thorough cleaning to minimize the creeping of error in final readings reflected by them. If the equipment is cleaned no sooner than the tests are over, it is very easy to clean, whereas dry mud sticking equipment pose real challenge and yet leave room for causing error in readings.

- While dealing with cement contamination eradication treatment, show patience and do not over-treat the mud, as it can be counterproductive.
- The addition of Glycol, PHPA, and other lubricants can result in drop of stand pipe pressure due to increased lubricity of mud. Therefore, as a measure or precaution, this treatment should be communicated to driller and other relevant persons on board to let them understand that drop in stand pipe pressure should not be mis-read by them as any sign of unproductive incident. As a matter of policy all treatments and actions in the pit room must be made known to the relevant person/s on the rig. Everything and anything we do on rig to perform our duty is linked to other important operations and hence it helps to be aware of what is happening on the rig and let the rig staff also be aware of what is happening in your domain area. a **communication saves lot of misunderstanding and lack of it gives rise to as many misunderstandings.**
- While checking turbidity of brine, if you come across consistently high value of NTU readings of turbidity despite two cycles of online filtration, add a drop or two of 0.1 N H Cl in your sample used for turbidity measurement so that if there is any Iron particle causing undesirable reading, it gets dissolved. The filtration unit cartridges may not be able to handle Iron.
- Monitoring active pit volume is a very important and necessary part of mud engineering, but alas, this often does not get as much attention as required. By the time we realize this, the damage is already done. Thanks to sensors and mud loggers, the situation is not that bad as it was almost 30 years ago. Yet knowing a change and responding to change are two different aspects of any operation. Any change coming to notice should be probed and reason known, as well as acted upon, as deemed fit.
- While on duty, make it a point to spend time in shaker area. Closely look for wear and tear in screens of shaker as well as shake the content of that area of shaker which first receives cuttings laden mud before the same comes on screens. At times, some cuttings never reach to shaker screen and sink in underneath area of shale shaker. This can give the false impression that the mud is not bringing cuttings out of the hole. Ideally stir

the entire area during connection time and confirm presence/absence of cuttings. If the area is full of cuttings, then it pays to sacrifice the small volume of mud from the area holding these cuttings and throw the stuff out of the circulatory system.

- The sand trap needs to be inspected without opening it every time for establishing as to when is it necessary to dump it and get rid of cuttings settled underneath. This can be easily done with the help of a long bar by inserting the same up to the bottom of the sand trap and "feel" the condition at the bottom. If the bar faces resistance in touching the bottom of the sand trap, then it is probably the time to get rid of settled solids. Much will depend upon how high is the rod while facing the resistance. Sand trap can't be dumped too often, nor can it be left un-dumped for too long. The periodicity of dumping of sand trap differs from well to well and interval to interval. There can't be any rule to it except the fact that one has to keep a check and use wisdom on when to dump.
- During milling operations while trying to catch the metal shavings and prevent them from entering in main active tank, the magnet bars are placed in the flow channel to catch the metal contained in the mud coming out of the hole and entering the channel after flowing through shakers. The pattern of placement of these magnets and inspecting them at every available opportunity to clear the magnet goes a long way in improving the efficiency of the system for preventing entry of metal shavings in active pit.
- While storing Polymeric hi-vis pill in the tank, it is advisable to turn OFF the agitator, as the continuous mechanical shearing will favor degradation of pill, making it less efficient. However, if it is Bentonite hi-vis pill, keep the agitator ON as Bentonite yields more with time and mechanical agitation.
- While preparing hi-vis pill with Bentonite, do not fill the tank more than 75% of its capacity to begin with. The pill yields with time and require additional space. The water used for such pill preparation should be treated with just enough amount of Soda ash to bring the Total hardness below 250 ppm. This should be the first step which can then be followed by the addition of Bentonite. Treating the pill with Caustic should be the last step. As Bentonite is being mixed and there is no alkalinity

in the system so far, one can add extra amount of Bentonite which then will easily convert into a thick pill as soon as Caustic is added to it.

- While preparing conventional LCM pill with a calculated amount of LCM, ensure that the final viscosity of the pill is exceedingly high even if it calls for treating the pill with additional viscosifier. On almost all occasions, it has been noticed that such really thick pills have a higher chance of controlling / curing losses. The simple thumb rule is – make it so thick that it does not flow through the marsh funnel.
- The active pit level should be around 70 % of its total capacity to account for rise in volume in the pit during pump closure and having an allowance for treatment, which also will add up to total volume apart from the fact that yet, there is enough volume in pit to monitor losses for a while without transferring.
- It helps a lot if two pits can be hooked up to active system, one hooked up with hopper, receiving return mud and other hooked up to pump for suction. This helps in the immediate treatment on return mud, which is the one that requires immediate attention.
- In a deep - water environment prior to conducting LOT after first casing is lowered to the desired depth, it pays dividends to keep the area around shoe, surrounded with mud treated with bridging and sealing agent prior to conducting LOT. Ideally speaking after drilling out shoe and rat hole and prior to conducting LOT, this must be planned in conjunction with the rig team that the fluid entering the hole (However small it may be) to facilitate recording of LOT is the fluid treated with excess doses of bridging and sealing agent but of the same weight as active mud. Thus, whatever fluid enters the wellbore as a result of leak-off, has a property of sealing off the crack and mitigate down-hole losses in the vicinity of shoe and rat hole. It might sound unconventional to do so, but in practical cases, it does save the well from a possible situation of lost return. Simultaneously, it also strengthens the rock artificially. The same pill is then absorbed in the system itself.
- In a HTHP well while tripping out of hole and using SOBM, it is likely that the hole may start flowing / trickling – giving an indication of activity from the hole. While that possibility

cannot be ruled out, but what one should also remember is the static mud in the hole starts equilibrating with BHST which is invariably higher than BHCT. This increase in temperature of mud in annulus results in expansion of mud despite no influx entering the wellbore! However, the best way to distinguish it from influx from the hole is, not to close BOP immediately and keenly observe the trend of mud flowing. If it is continuous bleeding with gradually decreasing volume coming out, most probably it is expansion of SOBM causing initiation of flow. If without close monitoring, the BOP is closed and well-control procedure initiated, it will easily result in couple of hours of NPT which was avoidable by simple common sense.

- While drilling through a potentially known gas bearing zone, it is at times mandatory to not to offer any treatment to mud (including raising the mud weight) in order to strictly monitor pit volume and not to miss a kick of even less than 3 bbl pit gain. Any gain during mixing due to increase in volume can otherwise be misjudged as an influx from the hole. If hole conditions warrant for any treatment of mud, then drilling needs to be suspended and resumed only after the treatment or conditioning is over.
- Likewise, when the mud circulation is started after a prolonged period of stagnation the mud in the sea bed area is in a contracted state (due to low temp. contraction of SOBM) giving a false impression that mud is getting lost in the formation. Again – take some time and observe the trend before acting on loss control/loss curing exercise
- Deep-water, ultra-deep-water HTHP wells are tricky and they require experienced as well as professionals with calm composure. Keep the thinking cap "ON" and act only after gathering sufficient confirmation of the real situation.
- In a situation of string getting stuck instantaneously due to collapse of hole the first thing which most of us think of is to attempt freeing the string by working on string and raising the mud weight. It is not wrong to do so, but often we shall notice that despite raising the mud weight the well-bore collapses again. A simple thought process on the probable cause can avert a repetition of this situation. If we are able to clear the pack-off and circulate conventionally, then the mud weight

increase should not be attempted again. What has happened is, the matrix near the well-bore is probably excessively stressed and irrespective of any mud weight, when a hole is made in such vulnerable rock, the nearby area is bound to fall, but soon stabilize. This stabilization will invariably raise the hole diameter by an unknown number, but now since the stress is released, the hole will remain stable. The point then is, why raise the mud weight in a stable hole?

- If it is known before resuming drilling that we are likely to encounter such stressed rock, the consideration must be given to start drilling with higher mud weight in first place, provided it is possible. If not, then just buy peace with the hole and do not attempt to strangulate it by higher mud weight and call for other problem associated with excess mud weight.
- Do not be scared of taking unconventional decisions, but one MUST provide that comfort and confidence to the team before execution. It is often easier said than done. The ultimate decision on many such matters is beyond the responsibility window of mud engineers. Therefore, your charisma, your approach towards the team and your ability to infuse faith in the team really makes a big difference in such circumstances.
- While looking after the mud on a deep-water / ultradeep-water well, always think large. Plan for everything in large volume and expect the things to go wrong despite best intentions. A simple increase in the dose of any additive in a circulating volume of 6000 bbls by one ppb, one needs 6000 lbs. That is 120 sx of 50 lbs each, spread over a cycle time close to two hours! Therefore, ordering of chemicals should always be planned such that there is a provision for surviving even if supply from shore is cutoff due to weather playing a sport-spoil without any warning, throwing supply boat out of action! It helps to conduct more than one operation simultaneously and yet ensure that you never miss on conducting the mud check by yourself at-least for rheology, HTHP F/L, Retort, ideally everything.
- Calibrate mud balance every week, both on lower and higher sides with the use of water and high-density brine. The higher side weight calibration should be carried out to a number around 2 ppg higher than running mud weight, if

possible. If non-availability of heavy fluid is an issue at least saturated $CaCl_2$ should always be handy to calibrate.

- Do remember to clean the cones of De-silter and De-sander after switching them off. This will ensure that they are available to serve every time and any time you need at their optimum efficiency. The same also holds good for centrifuges. Check for holes in shaker screens periodically and ensure that screens are always in tip-top condition to expel the cuttings at the first place itself.

Planning for mixing initial mud and treatment selection

Before resuming mixing of mud in the pits, take some time to plan the sequence and timing of each chemical. By the time you do this, let your team get engaged in stacking the "first to be mixed" chemicals near hopper and silos be full of Barite or any other bulk powder you may need.

Depending upon the type of mud to be mixed, the sequences and timing can and will vary, but invariably the pit will need some liquid in it to resume circulation through the hopper. By the time you plan, let that fluid flow in the pit which is connected to the hopper. One may often get tempted to mix faster, but polymers often do not permit this and are unforgiving if we do not respect their choice. Unless the powder is thoroughly dispersed in the liquid it neither yields nor will it allow other chemicals to get mixed. Use of gun line and high pressure shearing line (wherever available) is recommended to expedite but that also should abide by the rule that additives must go into liquid homogeneously. Depending upon the content already present in the hole, it makes sense to deliberate which additive to be used at what time. Although a carefully written mud program may often include these finer points, but non-inclusion cannot be an excuse for not doing so. A simple example is the inclusion of PHPA in the mud while mixing. No one undermines the importance of PHPA, but if we should first drill cement, shoe-track and rat-hole, then PHPA treatment can wait because cement drilling is going to degenerate PHPA. Although the first changeover and drilling of cement, shoe track and rat hole can be carried out with mud, but during this process it is not necessary to have PHPA in mud, rather it is detrimental. Again, common sense and presence of mind must prevail at all the times.

Such examples can be never-ending, but the objective is not to spoon-feed. It is intended to spread awareness and make field engineers understand the root cause of every step so that there is a purposeful and mindful activity. As the popular saying goes- Know the rules thoroughly, so that you know when you can break them!

Inventory and logistics management

Once a mud engineer is on-board, these two factors can make him Hero or Zero, depending upon his skill of managing inventory and planning logistics. The material-man is your first line of all sorts of help in these two aspects. Plan for the entire section and stagger your requirement/s such that the supply chain never gets disrupted. Think about delays due to unpredictable weather playing against you, sudden breakage of the supply chain from warehouse due to unforeseen reasons. If you can stock the stuff for an entire section and additional 2000 bbls for well-bore instability cause, then you are pretty much covered.

Logistics management is something which comes naturally to some while the rest struggle a lot to cope with. This depends much upon individual traits than anything else. Nevertheless, working in remote locations, offshore environment puts logistics management on high priority. Knowing by the overall standards of performance of the team, one should align and amalgamate the requirement in such a way that the available rig space is optimally used not only for activities of mud engineering, but also for other affiliated activities. At times, we need to compromise on our own requirements to accommodate the other department's requirement depending upon the ongoing activities. Having said that, the requirement must be chalked out and shared with all responsible persons with risk assessment. Never make a prestige issue if you should surrender your comfort zone to accommodate others' necessary inclusions.

Planning is a very crucial part of logistics, so is the fact that do not keep your plans close with yourself. Let the team know about it, especially the company man and the material man. Always remember Murphy's law – If things should go wrong, they will go at a time when you need them the most! Creativity and innovation always find extra room to squeeze in additional requirement or manage without what we consider as essential.

SPACE MANAGEMENT

It might sound awkward, but at times a mud engineer will find himself in a situation where there is not enough pit space or storage space available to execute his plan and / or to have enough stock of mud chemicals due to shortage of storage place in sack room. Storing chemicals in open deck always have an additional factor of availability of crane or other moving equipment to shift them in mixing room, whenever required.

A unique situation came up on an old generation deep-water floater due to inadequate pit space. The entire capacity of the pit storage was less than the volume required to change over to treated mud after lowering the BOP above risers to resume drilling. Those who have worked on such rigs are pretty much aware and can correlate themselves with this, often. An out of box situation turned this crisis into an opportunity to innovate a novel method, where one of the mud pumps was put to sea-chest to directly suck seawater from the ocean. The available pit space was utilized in preparing mud which had enough additives mixed in lesser volume of water to match with inadequate availability of water. The pump strokes on both the pumps were so adjusted on two pumps that the resultant volume of mixed fluid (sea water from sea-chest and over-treated mud contained in pits) came up with final mud parameters/properties close to what was planned. As the mud kept going in the hole, the pit space got free to facilitate additional preparation of a buffer volume of mud. Thus, what appeared an almost impossible situation eventually turned out as a simple revised procedure for changing over.

Similarly stacking of mud chemicals has a lot to do with common sense than anything else. A mud engineer always knows what are the chemicals going to be consumed faster and shall always invariably stack them closer to hopper and slowest moving chemicals always got stacked behind. No rocket science, but such simple actions make life easy for the entire crew. The list can be still longer, but the essence is to think and continue to analyze the dynamics of the system. Be open to adapt to changed circumstances and always keep the interest of the well above your own personal requirements pertaining to mud engineering. Despite not being wrong, taking part of the blame on you for the larger goal of team spirit, brings you closer to the entire team!

It might sound philosophical, but it is true and it invariably works in your favor as guys on-board will always reciprocate and stand by you, should something go wrong in your area of responsibility. During formative days as a mud engineer, many, often rub staff on the wrong side of the shoulders for reasons as varied as one can imagine. Author himself being not an exception to it, has learnt it the hard way that no matter how intelligent, knowledgeable, competent and wise you may be, if the team does not accept you as one among them, your performance can never be as good as you really are, or you think you are!

UNDERSTANDING MUD PARAMETERS

We all know mud parameters by our acquired knowledge, but the same cannot be said about understanding mud parameters! In earlier chapters, enough has already been written about the science and knowledge related facts about the parameters of mud and following discussion shall focus on what to derive out of the numbers and how to best use it with the goal of helping the drilling community to bring hydrocarbons to surface by delivering a healthy well without jeopardizing its ability to flow to the surface.

Knowing comes with knowledge, but understanding comes from your wisdom and your ability to learn from your failures. Everybody has been taught that undesirable solids are bad for mud, but how many of us understand that even inert solids are alert! Yes, their alertness is demonstrated in the form of their role of forming the skeleton, around which a body of mud then develops. In scientific terms, the so called inert solids helps in filling the micro-voids on a plaster of well-bore to help slow down the process of filtrate invasion. This process is of great importance in drilling through soft and unconsolidated rock matrix in a deep-water environment. The cobwebs made by inert particles are more effective in controlling/preventing seepage losses. Since these solids silently perform their role without appearing on the right slot of parameters, they often fail to get the recognition due to them. This is not to be confused that inert solids should be left in mud un-attended! We need to stretch ourselves in controlling inert solids only up to a level that the natural process of plastering the well-bore to partly block the cracks is not adversely affected.

We often get frightened by the way the adverse effects of fine particles are narrated, but the situation does not always turn that bad overnight. Imagine drilling a well by using a fluid which is completely devoid of solids and ensuring that drilled-solids do not enter the mud in the form of drilled cuttings. It will be interesting to visualize such a situation! This precisely is the meaning of understanding the parameters. Every parameter and property has a finite and useful range with-in which it is an asset to the system and is a liability if it is outside the range. The range varies from type of mud to type of well and a specific situation. Thus, a great mud engineer wisely treads through and a good mud engineer will turn a slave to what has been taught in mud school.

Never be a slave to the numbers, but draw meaningful inference from them and use them wisely. Different temperature, pressure and geological variations in a hole can have mutually opposing expectation from mud in terms of keeping the well-bore healthy and safe. A wise engineer will prioritize and have that formulation running in his pits which caters to all the situations in an integrated matter. A long open hole can have a huge difference in temperature, but the mud circulating in the hole should be compatible over the entire range. This is especially true of deep-water environment where reversal of temperature cycle poses challenges unheard of. It is not intended to preach and condition the mind of the reader in a specific direction and hence more examples are not being quoted. The intention here is to convey the message with few illustrative examples to explain the relevance and then leave it to the individual to build his own set of norms, with-in the framework of acceptability.

To give a comprehensive view of the entire planning process a generic program of a deep-water well drilled with SOBM is being shared. This is included with the intention of sharing what all goes as an input from the support office for the field engineers on the rig. This is only a generic program and anything as well as everything stated here is for educational purpose only. It is strongly recommended not to use the program as such, but can be viewed as a support document, based upon which one can design his own program which shall be specific to the well under question.

CHAPTER 18

GENERIC DRILLING FLUIDS PROGRAM WELL ABC

This is an example of a deep-water well, where pump and dump mode of drilling was adopted to push the landing depth of the first casing.

The XYZ well will be the 1st well to be drilled by ABC Company in Offshore. The closest offset well DEF is the exploratory well. The XYZ well will be located in *** m of water depth. The temperature at the mud line is expected to be * °C with a thermal gradient of *** deg C / 100m.

The 36" hole shall be jetted to a depth of **** m and a 26" hole will be drilled from **** m to **** m with returns to seabed. The last > 100+ m of this interval shall be drilled with mud of 10.5 ppg PAD mud (Pump and dump) to counter the perceived pore pressure rise as indicated in pre-drill geo-pressure plot.

After setting 20" casing at **** a 17 ½" *20" hole shall be drilled to **** m and 16" liner lowered. Below this interval a 14 ¾"*17" hole shall be drilled to **** m, and 13 3/8*13 5/8" casing lowered. The 12-1/4" hole will then be drilled to **** m and logged. Afterwards, it will be cased and cemented.

The last part of the hole shall be completed in 8 ½" size for the depth of **** m.

Following general discussion shall facilitate planning for a trouble-free mud work.

If possible prior to the start of drilling, all Barite and Bentonite silos on the rig are to be filled. Planning for re-stocking of these items must consider that 50 kg sacks will have to be cut in town to fill silos for loading on the boat (Over 2,000 sacks for 100 MT). To be able to use Bentonite throughout the top-hole sections for sweeps and large displacement volumes, as much drill water as possible is to be kept on board the drill-ship always. Delivery of water at the dockside is about ** MT per hour. It is important to ensure that boats carry supplies of Barite, Bentonite and drill-water whenever possible.

Approximately 1000 bbls of 45 ppb pre-hydrated bentonite mud should be made up prior to jetting. More volume can be prepared as and when required while assuring that we do not end up with excess volume at the conclusion of the interval. To begin with, attempt should be made to have as much 17.5 ppg WBM ready as possible because the last 150 m of the hole is to be drilled with 10.5 ppg mud in PAD (pump and dump) mode and the 26" open hole shall be displaced with 13 ppg mud. This calls for consumption of ~ 5500 bbls 17.5 ppg WBM to be able to generate ~ 15000 bbls 10.5 ppg WBM and ~ 5000 bbls of 13 ppg mud. Advanced resumption of preparation of heavy mud is the key to successful completion of drilling of 26" hole in PAD mode. If there is any perceived shortage of heavy mud, immediate preparation of more volume of heavy mud can begin as soon as pit space is free during the consumption of heavy mud. Adequate volume of Calcium chloride brine and ~ 1000 MT Barite should be facilitated on rig / boats along with PAC-UL, Flo-Vis-L, Duo-vis, Guar gum to ensure that the Barite doesn't settle out when the heavy mud is being cut back to 10.5 ppg mud with the help of sea water through continuous blending and pumping unit.

A continuous blending and pumping manifold is available on the rig and the same should be function tested in advance to ensure proper functioning.

General

The drilling fluid program has been designed to combat and control the possible problems associated with drilling in Deep-water. They are

- o Low fracture gradient with resultant possible mud losses.
- o Swelling and dispersion of recent soft and sticky marine formation.
- o Possibility of kicks and the formation of gas hydrates.
- o Hole-cleaning.
- o Ballooning of plastic clay-stone formations.
- o Tectonic stresses, micro-fractures and hole stability.
- o Minimize filtrate invasion, size the particles minimize formation damage, particularly to the reservoir section to be cored.

Operational Guidelines

A continuous blending and pumping manifold will be rigged up on the rig prior to jetting. An engineer, especially dedicated for this job, will be available to use this equipment.

Volume management will be very important to this well to prevent the rig from shutting down for waiting on mud related matters. A mud management plan should be drawn up on the rig, and this should be continuously monitored. Boat movements will also need to be carefully planned so the supply boats can offload the base oil when needed.

All mud additives that are to be added to the drilling fluid must be added from the rig itself, not from the supply boats or somewhere else. This will ensure that the chemicals are added at the correct concentrations as formulated, without mistakes being made with the amounts added.

H_2S is not regarded as a serious issue on this well, although Zinc Carbonate H_2S scavenger may be kept on board the rig if possible to mitigate against any possible problem caused by this gas. CO_2 is not present in large amounts and is not anticipated to be a problem.

Lost circulation material should be stored on the rig, always in case a loss situation occurs. No LCM with known damage potential to reservoir shall be used. Only acid soluble CaC03 (like Gel Flake or any other form of Calcium Carbonate) and non-damaging type of cross linking polymers shall be used to control down hole losses. During the entire course of drilling with returns to surface the cross linking polymeric LCM including accelerator and retarder should also be available on the rig for immediate use, along with various standard forms of sized Calcium Carbonate. Details of the formulation will be sent to the rig just prior to drilling into the reservoir, once temperatures are known.

The SBM products to be used should be first tested for their environmental acceptance approved by a recognized laboratory. The rig should be audited for use of SOBM and shortcomings, if any, be removed.

Escaid 110 base oil will be used as it has environmental clearance and is a thin base fluid, which is the preferred type on deep-water wells. Enough 5 Darcy PPT ceramic discs (at least 15 should be stored on a rig to measure the PPT as frequently as required for optimizing the bridging / sealing material concentration. This is to try and achieve the minimum possible value of filtrate from SBM at bottom hole dynamic conditions.

Utmost care is needed when handling SBM. The lead mud engineer should be actively involved in SBM transfers in conjunction with the rig crew, as care should be taken to work in a zero-discharge environment. Detailed instructions relating to SBM movements should be made available on the rig to ensure compliance.

Additions of fluids to the active system should be notified to all concerned, and the transfer pit included in the total pit system, so that any influx can be rapidly seen even when transferring fluid. When using SBM the shale shakers should be always monitored closely, including during tripping, to ensure that surface losses do not occur. The shale shakers should also have the small drying screens fitted to the front of them. Every effort should be made to ensure that the screen size is optimized to retain essential bridging and sealing material.

The centrifuge should be used with utmost care as it may strip out the Calcium Carbonate, which is required in the mud. As such its use should be restricted to a minimum to ensure that we do not end up throwing away essential Calcium Carbonate. Screen selection for shakers should be driven by the fact that we do not end up throwing out essential Ca CO_3 and yet are able to discard the drilled cuttings as much as possible. Priority should be given to discarding as much drilled-solidsas possible and any loss of Ca CO_3 should be replenished by continuous addition of fresh Ca CO_3 of the correct size as recommended in the fluids program.

SBM ordinarily takes more time to show a pit gain than WBM due to dissolution of gas, which eventually liberates out while traversing up the annulus. This feature should be considered while monitoring changes in pit level during well kill operations. The exact depth at which the gas comes out of the mud will depend upon the temperature and pressure to which the fluid is subjected, along with the type of base oil, the concentration and the composition of the gas intruded.

As a result, it is not possible to accurately predict the depth at which the dissolved gas will break out. This makes 'kick' detection more difficult and hence a constant check of pit volume changes becomes essential. While using SBM it should also be remembered that it takes a long time for the flow to stop after the pumps have been switched off. As a result, adequate time must be given to flow check the well to confirm the presence or absence of flow, with 30 minutes being the optimum time.

A close watch has to be kept on possible 'ballooning' to stop any unnecessary mud weight increases being made and any false 'kicks' being taken. High gas levels should not in any way be the only criteria for increasing the mud weight, with these increases in mud weight based on a number of factors such as cavings, D-Exponent and drill breaks.

During the drilling of zones where shows may be expected from the formation, mud treatment should be kept at a minimum as there needs to be a strict control of volume changes enabling the drillers to pick up the slightest of influxes.

Any transfer of fluid from one pit to another should be carried out with utmost care and communicated to all the concerned persons so that even the slightest pit gain can be noted immediately.

While circulating a 'kick' out of the hole the mud weight at the shale shakers should be monitored as frequently as practically possible. As a general practice, the mud weight should be taken every 30 minutes, while drilling or circulating, and the mud weight announced on the rig loud-speaker system. All mud pumped down the hole should be checked constantly whilst circulating out a 'kick' to ensure that only the correct weight fluid is being pumped back down the hole.

Running Calcimetry on the SBM will help in having an idea of the total concentration of $Ca\ CO_3$ in mud and accordingly the treatment of fresh $Ca\ CO_3$ can be planned.

While treating the mud for raising weight, the strategy should be to first add the suggested dose of Calcium Carbonate (both 500 microns and sized) and then add Barite, if required. The mud weight should not be raised while continuing to drill. This is to ensure that even a slight gain of influx from the formation is not misinterpreted as volume increase due to Barite addition.

A pressurized mud balance is to be used at all times to record the mud weights on this well. The mud weight will be announced every 30 minutes during drilling / circulation and as frequently as possible when circulating out a kick. It is the mud engineers' responsibility to ensure that the Derrick man is performing his duties with respect to these instructions. Any deviation must be immediately reported to the on-board company man. It is essential to calibrate the pressurized mud balance at least once a day, with high weight as well as low weight fluids, to ensure accuracy of readings.

36" Conductor Jetting to m TVD BRT

The 36' conductor will be jetted in with seawater, along with 60 bbl un-weighted, pre-hydrated bentonite sweeps pumped every 10 meters, to a depth of meters TVD BRT. PHB Sweeps clean hole better than guar gum and hence PHB sweep should be used in preference to guar Gum whenever possible. Although Guar gum sweeps can be substituted for pre-hydrated bentonite if logistics or time constraints dictate, but all efforts should be made to avoid this from happening, while ensuring that we never have to spot the guar Gum treated mud in open hole and use this only as a sweep. Larger volume sweeps can be pumped if the hole conditions demand while jetting and 12.0 ppg heavy mud can also be pumped if problems persist. Approximately 500 bbls of pre-hydrated bentonite will be needed to complete this interval.

This surface section presents a major challenge in hole-cleaning with every meter of jetting resulting into the generation of 3.47 bbls of cuttings. This material must be effectively cleaned from the well bore, and failure to do so would possibly result in the 36" casing being unable to be jetted to bottom. Effective removal of cuttings can be achieved by maximizing flow rates and ensuring that an adequate volume of un-weighted, high viscosity pre-hydrated bentonite is pumped regularly.

Sweeps should be pumped regularly at an interval of every 10 meters while jetting and shall comprise of at least 60 bbls of ~ 30 ppb un-weighted (45 ppb - diluted with sea water), pre-hydrated bentonite, representing approximately one and a half times the volume of cuttings drilled. If problems with hole-cleaning occur, then the sweep volume should be increased and the duration between sweeps decreased. If problems still occur, then neat, un-weighted, pre-hydrated bentonite sweeps should be pumped that have not been cut back with seawater.

26" Hole Interval m TVD BRT

The 26" hole will be drilled to a depth of ***meters TVD BRT, using seawater and high viscosity, un-weighted, pre-hydrated bentonite sweeps or Guar Gum sweeps if logistics or availability of drill-water becomes a limiting factor. This should only be done in an emergency as hole conditions and POOH are much easier and less time consuming when the hole has been swept with pre-hydrated bentonite rather than Guar Gum or XC Polymer sweeps. Pre-hydrated bentonite sweeps should be pumped due to the fact that large sand bodies are likely to be encountered, and PHB is a better fluid than Guar Gum. Sweeps of 50 bbls should be pumped at regular intervals (+/- 15 meters). If hole conditions deteriorate the frequency and size of these sweeps should be increased. The open-hole should then be displaced with weighted 13 ppg mud prior to pulling out of the hole for a wiper trip, and displaced again to weighted 13 ppg mud prior to running the 20" casing, which will be set at meters MD BRT. If the casing needs to be washed to bottom and the pre-hydrated bentonite mud runs out, then seawater should be used in its place. The un-used 13.0 ppg kill mud should be retained in a tank (if logistics do not pose any problem) till the 20" casing is landed to bottom as it can be used to contain any influx, if noticed during casing lowering. Up to this stage all the returns are being dumped to seabed and hence enough volume should be built well in advance to avoid any shortage of any type of fluid. The centrifuge should be function-tested during this period to ensure that the same is available for use in 17-1/2*20 "section.

This section presents some challenge in hole-cleaning with every meter drilled representing 2.15 bbls of cuttings. This material must be effectively cleaned from the well bore, and failure to do so can possibly result in a build-up of hydrostatic head in the annulus, lost circulation and subsequent 'pack off'. As a result of the possibility of losses and to aid with hole-cleaning, the ROP may have to be controlled. Effective removal of cuttings can be achieved by maximizing flow rates and ensuring that an adequate volume of un-weighted, pre-hydrated bentonite mud is regularly pumped.

Un-weighted pre-hydrated Bentonite sweeps should be pumped every 15 meters, with the volume pumped being 50 bbls, representing approximately 1.5 times the volume of cuttings drilled. If the driller reports excess drag on connections, or increased torque while drilling, the hole should be swept with 60 bbls of un-weighted, pre-hydrated bentonite mud and the hole circulated clean. If problems with hole-cleaning persist, then the sweep volume should be increased and the duration between sweeps decreased. From *** m onwards (the depth from where the pump and dump drilling resumes with 10.5 ppg brine based WBM) the hi-vis sweep should be made up of Duo-vis and 10.5 ppg brine only. Once we resume the PAD operations, the use of PHB should be discontinued completely. At section TD, a 200 bbl 10.5 ppg hi-vis sweep will be pumped followed by displacing the same with 13 ppg mud heavy mud.

The 36" casing needs to be thoroughly cleaned out on the wiper trip, by back-reaming, circulating as fast as possible and pumping sweeps, to ensure the 20" casing string can be run easily and all cuttings are removed from the large size 36" casing prior to the 20" casing being run.

At section TD of meters TVD BRT, the open hole will be displaced to 2400 bbls weighted 13 ppg mud and a wiper trip performed. Once back on the bottom, again displace the hole again with 2400 bbls of 13 ppg mud prior to running the 20" casing. Enough mud should be pumped at this stage to consider any hole washout with a minimum mud volume of 125% of gauge hole being pumped. This should allow for some mud to be pumped whilst pulling out of the hole if required. Although the requirement for displacement is 2400*2 bbls only, preparation should be made for 6000 bbls as the excess can be used in case required.

17-1/2"x 20" Hole Interval m TVD BRT

***The 17 ½*20" hole will be drilled to *** m with a 67/33 OWR, synthetic based mud (SBM) with an initial mud weight of 9.1 ppg at surface to give an equivalent mud weight of 9.2 ppg at bottom hole and the mud weight will be increased to 9.2 / 9.3 ppg after LOT. A 16" liner will be lowered to drilled depth. Mud weight will be raised, if required.

The leftover mud at shore base should be used and treated such that it contains ~ 25 ppb of 500 microns Ca Co3, ~ 15 ppb of sized Calcium Carbonate and 7 to 8 ppb Graphite in the entire system. SBM is being used to minimize borehole instability problems by not allowing any water to come in contact with the shale and thereby not allowing the formation to swell or disperse. Better ROP, less corrosion, more resistant to solids, better control of torque and drag and eliminating the risk of the formation of hydrates in the mud or BOP stack during a 'kick' are other advantages of SBM. The fluid has been selected after obtaining environmental clearance for the use of Escaid 110 base oil. The mud weight has been selected after looking at the pore pressure data available and keeping in mind that SBM tends to balloon more than WBM. With a YP of +/-24and 6 RPM value of +/- 18 at ambient temperature, the ballooning feature can be kept under control, and the mud weight will maintain just enough overbalance to hopefully eliminate enlargement of the well-bore due to mechanical stresses. The mud weight should be maintained at 9.2 ppg (effective 9.3 in bottom hole conditions) and a constant check on this has to be maintained. If possible, the town should be contacted about mud weight changes, but if the town cannot be contacted for any reason the company man has the authority to alter the mud weight if he thinks the situation so demands. Sweeps (high viscosity) should be pumped preferably after every three stands drilled. However, while doing so a very close monitoring of YP is required. If the YP reading increases beyond 30, then pumping of sweeps may be stopped temporarily and should be continued only as a response action to tackle any unstable bore hole condition due to insufficient hole-cleaning. This can be better handled by closely monitoring the volume of cuttings returning at the shakers. At section TD before the wiper trip for logging / casing, tandem sweeps of Low and High viscosity may be pumped to ensure cuttings-free hole (100bbls). These sweeps should be pumped irrespective of whether they bring back additional cuttings or not. The weight of the hi-weighted viscous mud should be planned in such a way that it counters for the loss of hydrostatic head due to pumping of Lo-vi's (Base-oil) and the combined minimum volume of both the pills should be such that it covers the entire length of BHA at a given time while the pills travel up in the annulus. The screen selection for shakers should be optimized to retain the bridging material and at the same time to minimize the use of the centrifuge and thereby avoid discarding any sized Calcium Carbonate, which has to be an integral part of mud composition, in the process. The 20" shoe track, rat hole and three meters of 17 1/2-20" formation will be drilled with 9.3 ppg SBM (effective 9.4 at bottom) which is treated and thoroughly conditioned to suggested parameters and the LOT carried out with same SBM containing 15 ppb Sized Ca Co3, 25 ppb 500 microns CaC03 and 7 to 8 ppb Graphite.

There is no need to pump any spacer between SBM and Sea water as both the fluids are distinctly distinguishable from each other and a clean sharp interface will be seen.

As soon as the 26" hole section is completed and 20" casing run, cemented and the plug bumped, all the available WBM should be dumped, tanks, cleaned to receive / blend / treat the brought forward and fresh mud (if required) to have +/- 6500 SBM and shear the entire volume to finally have the SBM of suggested parameters along with 25 ppb of Oil wet 500 microns CaC03, 15 ppb Sized Ca Co3 and 7 to 8 ppb Graphite. The back-loaded mud from mud plant is available on the supply vessels as well as at the shore base mud plant. The same needs to be re-conditioned to the requirement of present well drilling.

The SBM should be treated on the rig and sheared through the high - pressure - shearing devise to give the mud the desired rheology and properties with the optimum dose of additives. Drilling ahead with SBM of parameters widely different than programmed should not be attempted.

Principle properties of 17-1/2" x 20" phase mud:

Property	**Value**	**Remarks**
Mud Weight ppg	9.1 to 9.3 in pits	
OWR	67/33	
Electrical Stability	> 300 V	
HTHP – 500 psi	< 3.5 ml	All Oil
PV	ALAP	
YP lbs / 100 ft2	+/- 24	
6 RPM reading	+/- 18	
Excess Lime	7 ppb	
AW	0.76	+/- 0.01

Once the SBM is mixed it should be sheared, not only to give properties close to those of the specifications in this drilling fluids program, but also for the Calcium Carbonate to get properly oil wet. This may take considerable time and swift action will be required by all concerned to ensure that the rig loses no time waiting on SBM. Mud engineers will have to closely monitor this activity to ensure that the drilling fluid is prepared to the correct specifications and is ready for use whenever required.

Due to longer trip time from rig to shore-base, planning is essential for ensuring that the required additives and fluids are in place at the correct time. Boat movements need to be monitored to ensure the left over SBM, base oil and Calcium chloride brine from the previous well are brought onto the rig in a timely manner, and close co-ordination has to be done between Shore base, the rig and the boats. The Calcium chloride brine stored at the shore-base mud plant for use on this well is 11.6 ppg brine and will need to be cut back to give the required salt concentration before adding to the SBM.

There is no need to have kill mud for this interval.

Stringent control of mud weight will be required whilst drilling this well, as there are low fracture gradients, and the anticipated LOT below the 20" casing shoe is only very less (9.7 ppg) . Maintaining a good rheological profile is also essential, to ensure that good transport of cuttings is maintained in the riser and the ECD is not too high. Ensure circulation is broken slowly, or the pipe is staged in the hole circulating occasionally, after trips or periods of non-circulation to ensure pump pressures do not become too high and break down the formation. The circulation rate may need to be maximized and the ROP controlled to prevent cuttings loading, but this will be seen from the APWD tool on the LWD/MWD, and accordingly a decision can be made. Another equally important part is to have 25 ppb 500 Microns CaC03, 15 ppb Sized CaC03 and 7 to 8 ppb Graphite in the system to minimize / eliminate seepage losses as well as to minimize filtrate invasion.

Every time the mud weight is changed the choke and kill lines should also be displaced with the same type and weight of fluid, although this is not anticipated in this hole section.

The pumping of sweeps should be closely monitored with the ECD value obtained from APWD tool and the packing-off tendency of the string if any, during connections. Hence a real time-based decision can be considered. However, if ECD values do not show any spike during hi-vis sweep, then the default decision should be to pump sweeps as long as the YP does not exceed 30 . As far as possible, High-vis sweeps should be constituted with either low end rheology booster or Low-end rheology modifier.

At TD prior to tripping out for logging and / or wiper trip, the rat hole above regular 17-1/2" x 20" hole must be thoroughly cleaned of cuttings by pumping a tandem low viscosity / high weight viscosity pill when the bit is just below the rat hole.

Drilling fluid formulation - Synthetic Based Mud (67/33 Oil/ Water Ratio):

Escaid 110	:	0.62 bbl.
CaCl2 brine	:	0.30 bbl (118.5 ppb CaCl2).
Emulsifier	:	16.0 ppb.
ORGANO-PHILLIC CLAY	:	8.0 ppb.
Fluid loss control agent	:	3.5 ppb.
Lime	:	10.0 PPb.
Filtratin control ahent	:	1.5 ppb.
Calcium Carbonate	:	30.0 ppb (500 microns).
Sized Ca Co3	:	10.0 ppb
Graphite	:	6.0 ppb
Barite	:	As required for suggested ppg

The suggested formulation is a guideline only and the emphasis should be on achieving the properties of the final fluid as suggested. The concentration of Calcium Carbonate should be maintained as proposed, due to the fact that it has been carefully selected to minimize filtrate loss, after carrying out pilot studies on a core sample from a well in same field drilled in the past.

SBM requires thorough shearing and adequate mixing time to exhibit the properties expected of it, and hence mud preparation should proceed well in advance to ensure that the fluid is ready when required. Insufficiently sheared mud can result in overflowing of mud from shakers and hence it is essential to prepare the mud well in advance and shear the same on the surface with the high - pressure shear device as much as possible to prevent this from happening.

Addition of fluids to the active system should be notified to all concerned and the transfer pit included in the total pit system so that any influx can be rapidly seen even when transferring fluid. When using SBM the shale shakers should be monitored at all times even when tripping to ensure no losses occur. The active system should be as full as possible when starting the interval to minimize the amount of transfers needed during drilling. Proper reporting of cuttings size, shape and texture needs recoding as does hole-cleaning efficiency, and these should be mentioned daily in the drilling fluids report.

The 17-1/2" x 20"section presents comparatively less challenge for hole-cleaning than the last section, with every meter drilled representing 1.27 bbl of cuttings. Maximizing the flow rate and ensuring that the booster pump is run as frequently as possible can achieve effective removal of drilled cuttings. The rat hole has posed problems in the past, with the logging tool standing up, and not reaching the drilled depths on several occasions. Drilling out the shoe area carefully and ensuring the BHA is clear of the area before maximum drilling parameters are started, should help in reducing wash out in the shoe area, thereby minimizing the probability of this happening. Hi-vi's pills should be pumped, preferably after every three stands drilled as well as at section TD prior to wiper trip and again when back on the bottom after the wiper trip. The pills can be composed with low end rheology booster or Low-end rheology modifier, depending upon then existing rheological profile of SBM. These sweeps can be incorporated into the active system when back at the surface. The hole should be circulated until the shakers are clean, and cuttings size and quantity should be monitored and reported on the daily mud report.

Another important aspect of this phase shall be the use of a booster pump for cleaning any cuttings in the riser, where there will be a drop in annular velocity. If hole-cleaning poses problems, consideration should be given to pumping low end rheology booster / Low end rheology modifier (depending upon then existing conditions) treated high viscosity pills, in addition to the programmed sweep every five stands drilled.

The use of centrifuge on the active system should be restricted to a minimum for the following reasons:-

- **To have a better control of pit volume changes.**
- **To retain Calcium Carbonate in the system, which otherwise gets discarded and need replenishment with fresh addition. Nevertheless, if the mud weight control is not possible without running centrifuge, it must be run and loss of sized Calcium Carbonate should be replenished as soon as possible.**

A close watch has to be kept on possible 'ballooning' to stop any unnecessary mud weight increase being made and any false 'kicks' being taken. High gas levels also have to be monitored and understood to prevent unnecessary mud weight increases being made. High gas levels should not in any way be the only criteria for increasing the mud weight, with these increases in mud weight based on a number of factors such as cavings, D-Exponent and drill breaks. Fingerprinting of all connections is essential to ensure ballooning is not mistaken for mud loss / gain.

As we shall be drilling some of the reservoir sections in this interval, cross-linking polymer lost circulation material (LCM), including the accelerator and the retarder should be available on the rig for immediate use, along with Calcium Carbonate. It is imperative that lost circulation material be stored on the rig at all times during this section, but only acid soluble material should be pumped only when specific instructions are received from the town. Details of the cross-linking polymer make up will be sent to the rig just prior to drilling into the reservoir, once BHST is known, although procedures for the pumping and spotting of these fluids is available in the procedures part of the program. If losses occur, then Calcium Carbonate 500 Microns and sized (90 ppb) or cross-linking polymer, or a combination of both can be pumped down-hole without causing problems with the LWD/MWD tools.

It should be noted that if large-scale losses occur during this interval, then the hole should not be continually filled with synthetic-based mud. The hole should be filled from the top with water until equilibrium is reached and the equivalent mud weight can be worked out. Once this has been done the mud system can be reduced in weight until the losses are cured. This procedure will prevent unnecessary mud being pumped and lost down-hole.

Prior to the cementing of the 16 "liner, thin SBM (400 bbls) should be pumped to aid in better removal of mud cake, thereby improving the probability of a better cement job. This mud should have ALAP PV and YP of ~ 12. Suitable thinner should be available for thinning back this mud.

14-3/4" x 17" Hole Interval 2510 m TVD BRT

The 14-3/4" x 17" section will be drilled to m MD BRT with a 67/33 to 70:30 OWR, synthetic based mud similar to that used in the previous section, but with a mud weight of 9.4 ppg (effective 9.5 at bottom) up to *** m. The mud weight, then needs to be raised to 10 ppg in pits and maintained throughout the interval. The screen selection for shakers should be optimized to retain the bridging material and at the same time to minimize the use of the centrifuge and thereby avoid discarding any sized Calcium Carbonate, which has to be an integral part of mud composition, in the process. Pumping of high viscosity pills must be carried out, preferably after every three stands drilled. The pumping of sweeps should be closely monitored with the ECD value obtained from APWD tool and, the packing-off tendency of the string if any, during connections. Thus, a real time-based decision can be taken. However, if ECD values do not show any spike while hi-vis sweep is passing through the annulus, then the default decision should be to pump sweeps as long as YP does not exceed 30. The frequent use of the booster pump becomes more important in this section as the pump discharge might not provide satisfactory hole-cleaning in the riser. Keeping the hole free of cuttings is of utmost importance to eliminate the possibility of any pack-off, leading to lost-circulation / stuck-string.

There is only a small requirement to prepare fresh mud for this section as almost enough mud will be available on board after 16 " liner cement job. Preparation of fresh mud should be minimized as much as possible by proper planning, bearing in mind that at the end of the section, additional volume will be available and ever-increasing volume causes logistics related problems. Having said this, enough products should be kept on-board the rig to build at least 2000 plus mud, should the situation demand. As the mud weight is being changed in this interval the choke and kill lines should also be displaced with the same type and weight of the fluid. The brought forward mud of previous section should be weighed up in such a way that the FIT for the current section is recorded with 9.4 ppg (effective 9.5 ppg at bottom)mud, which is the chosen one for recording LOT in the current section.

Stringent control of mud weight will be required whilst drilling this well, as there are low fracture gradients, and the anticipated FIT at the 16" liner shoe is moderate (10.3 ppg). Maintaining a good rheological profile is also essential, to ensure that good transport of cuttings is maintained in the riser and the ECD is not too high. Ensure circulation is broken slowly after trips or periods of non-circulation to ensure pump pressures do not become too high and break down the formation. The circulation rate may need to be maximized and the ROP controlled to prevent cuttings loading. This can be also seen from the APWD tool, and a decision can be made accordingly. Another equally important part is to have enough sized bridging material in the system to minimize / eliminate seepage losses as well as to minimize filtrate invasion, so ensure 25 ppb of 500 Microns Calcium Carbonate, 15 ppb Sized CaC03 and 6 to 8 ppb of Graphite is maintained in the active system.

Principle properties of 14-3/4" x 17" phase mud:

Property	Value	Remarks
Mud Weight	9.4 / 10 ppg	
OWR	70/30	
Electrical Stability	> 300 v	
HTHP – 500 psi	3 - 2.5 ml	All Oil
PV	ALAP	
YP lbs / 100 ft2	+/- 24	
6 RPM reading	16	
Excess Lime	5 ppb	
AW	0.76	+/- 0.01

The 14-3/4" x 17" section presents less challenge for hole-cleaning than the last section, with every meter drilled representing 0.98 bbl of cuttings. Maximizing the flow rates and ensuring that the booster pump is run as frequently as possible, can achieve effective removal of drilled cuttings. High viscosity sweeps should be pumped every three stands drilled, as well as at section TD prior to a wiper trip and again when back on the bottom after the wiper trip. The pumping of sweeps should be closely monitored with the ECD value obtained from APWD tool and, the packing off tendency of the string if any, during connections. Thus, a real time-based decision can be taken. However, if ECD values do not show any spike while hi-vis sweep is passing through the annulus, then the default decision should be to pump sweeps as long as the YP does not exceed 30. At section TD sweeps should be pumped both prior to and after the wiper trip and the hole circulated until the shakers are clean even if this is longer than a complete circulation.

It should be noted that if large-scale losses occur during this interval, then the hole should not be continually filled with synthetic-based mud. The hole should be filled from the top with water until equilibrium is reached and the equivalent mud weight can be worked out. Once this has been done the mud system can be reduced in weight until the losses are cured. This procedure will prevent unnecessary mud being pumped and lost down-hole.

Drilling fluid formulation - Synthetic Based Mud (70/30 Oil/ Water Ratio):

CaCl2 brine	:	0.30 bbl (118.5 ppb CaCl2).
Emulsifier	:	16.0 ppb.
ORGANO-PHILLIC CLAY	:	8.0 ppb.
Fluid loss control agent	:	3.5 ppb.
Lime	:	10.0 ppb.
Filtratin control ahent	:	1.5 ppb.
Calcium Carbonate	:	10.0 ppb (sized).
500 microns Calcium Carbonate	:	30 ppb
Graphite	:	10.0 ppb
Barite	:	As required for desired ppg in pit

The formulation is only suggestive and the emphasis should be on achieving the parameters as mentioned in the program.

Prior to the cementing of the 13-5/8 " casing, thin SBM (300 bbls) should be pumped to aid in better removal of mud cake, thereby improving the probability of a better cement job. This mud should have ALAP PV and YP of ~ 12.

12-1/4" Hole Interval 5040 m TVD BRT

The 12-1/4" section will be drilled to m MD BRT with a 70/30 OWR, synthetic based mud similar to that used in the previous section. The mud weight at the end of the previous section should be the mud weight for recording LOT. The mud weight required for drilling he section shall be 10.2 to 10.5 ppg subject to obtaining a LOT of 10.9 ppg. If the LOT figure is different than 10.9 ppg, then the actual mud weight shall be accordingly adjusted and maintained throughout the interval. The screen selection for shakers should be optimized to retain the bridging material and at the same time to minimize the use of the centrifuge and thereby avoid discarding any sized Calcium Carbonate, which has to be an integral part of mud composition, in the process. Pumping of high viscosity pills must be carried out, preferably after every three stands drilled. The pumping of sweeps should be closely monitored with the ECD value obtained from APWD tool and, the packing-off tendency of the string if any, during connections. Thus, a real-time based decision can be taken. However, if ECD values do not show any spike while hi-this sweep is passing through the annulus, then the default decision should be to pump sweeps as long as the YP does not exceed 30. The frequent use of the booster pump becomes more important in this section as the pump discharge might not provide satisfactory hole-cleaning in the riser. Keeping the hole free of cuttings is of utmost importance to eliminate the possibility of any pack-off, leading to lost-circulation / stuck-string.

There is only a small requirement to prepare fresh mud for this section as almost enough mud will be available on board after 13-5/8 " casing cement job. Preparation of fresh mud should be minimized as much as possible by proper planning, bearing in mind that at the end of the section, additional volume will be available and ever-increasing volume causes logistics related problems. Having said this, enough products should be kept on-board the rig to build at least 2000 plus mud, should the situation demand. As the mud weight is being changed in this interval the choke and kill lines should also be displaced with the same type and weight of the fluid. The brought forward mud of previous section should be weighed up in such a way that the LOT for the current section is recorded with 10.0 ppg mud.

Stringent control of mud weight will be required whilst drilling this well, as there are low fracture gradients, and the anticipated FIT at the 13-5/8 "casing shoe is moderate (10.9 ppg). Maintaining a good rheological profile is also essential, to ensure that good transport of cuttings is maintained in the riser and the ECD is not too high. Ensure circulation is broken slowly after trips or periods of non-circulation to ensure pump pressures do not become too high and break down the formation. The circulation rate may need to be maximized and the ROP controlled to prevent cuttings loading. This can be also seen from the APWD tool, and a decision can be made accordingly. Another equally important part is to have enough sized bridging material in the system to minimize / eliminate seepage losses as well as to minimize filtrate invasion, so ensure 25 ppb of 500 Microns Calcium Carbonate, 15 ppb Sized CaC03 and 6 to 7 ppb of Graphite is maintained in the active system.

Principle properties of 12-1/4" phase mud:

Property	Value	Remarks
Mud Weight	10.0 to 10.5 ppg	
OWR	70/30	
Electrical Stability	> 300 v	
HTHP – 500 psi	2 to 2.5 ml	All Oil
PV	ALAP	
YP lbs / 100 ft2	+/- 24	
6 RPM reading	14	
Excess Lime	5 ppb	
AW	0.76	+/- 0.01

The 12-1/4" section presents less challenge for hole-cleaning than the last section, with every meter drilled representing 0.47 bbl of cuttings. Maximizing the flow rates and ensuring that the booster pump is run as frequently as possible can achieve effective removal of drilled cuttings. High viscosity sweeps should be pumped every three stands drilled, as well as at section TD prior to a wiper trip and again when back on the bottom after the wiper trip. The pumping of sweeps should be closely monitored with the ECD value obtained from APWD tool and, the packing-off tendency of the string if any, during connections. Thus, a real time-based decision can be taken. However, if ECD values do not show any spike while hi-vis sweep is passing through the annulus, then the default decision should be to pump sweeps as long as the YP does not exceed 28 respectively. At section TD sweeps should be pumped both prior to and after the wiper trip and the hole circulated until the shakers are clean even if this is longer than a complete circulation.

A close watch has to again be kept on possible 'ballooning' – see previous section for details.

The rat hole needs cleaning properly on the way out of the hole – see previous section for details.

Acid soluble lost circulation material of various types should be kept on the rig at all times – see previous interval for details.

It should be noted that if large-scale losses occur during this interval, then the hole should not be continually filled with synthetic-based mud. The hole should be filled from the top with water until equilibrium is reached and the equivalent mud weight can be worked out. Once this has been done the mud system can be reduced in weight until the losses are cured. This procedure will prevent unnecessary mud being pumped and lost down-hole.

Drilling fluid formulation - Synthetic Based Mud (67/33 Oil/ Water Ratio):

Escaid 110	:	0.60 bbl.
CaCl2 brine	:	0.30 bbl (118.5 ppb CaCl2).
Emulsifier	:	16.0 ppb.
ORGANO-PHILLIC CLAY	:	8.0 ppb.
Fluid loss control agent	:	3.5 ppb.
Lime	:	10.0 ppb.
Filtratin control ahent	:	1.5 ppb.
Calcium Carbonate	:	15.0 ppb (sized).
500 microns Calcium Carbonate	:	25 ppb
Graphite	:	6.0 ppb
Barite	:	As required for 10.0 ppg

The formulation is only suggestive and the emphasis should be on achieving the parameters as mentioned in the program.

Prior to the cementing of the 10-3/4" casing, thin SBM (300 bbls) should be pumped to aid in better removal of mud cake, thereby improving the probability of a better cement job. This mud should have ALAP PV and YP of ~ 12. SOBM thinner should be available for thinning back this mud. Pilot tests should be done on the active mud with Versthin as it is a powerful thinner and over-treatment can cause serious problems.

8-1/2" Hole Interval 4585 m TVD BRT

The 8-1/2" section will be drilled to m MD BRT with a 70/30 OWR, synthetic based mud similar to that used in the previous section but with a mud weight of 10.7 ppg. The same mud weight should be maintained throughout the interval. The screen selection for shakers should be optimized to retain the bridging material and at the same time to minimize the use of the centrifuge and thereby avoid discarding any sized Calcium Carbonate, which has to be an integral part of mud composition, in the process. Pumping of high viscosity pills must be carried out, preferably after every three stands drilled. The pumping of sweeps should be closely monitored with the ECD value obtained from APWD tool and, the packing off tendency of the string if any, during connections. Thus, a real time-based decision can be taken. However, if ECD values do not show any spike while hi-vis sweep is passing through the annulus, then the default decision should be to pump sweeps as long as the YP does not exceed 27. The frequent use of the booster pump becomes more important in this section as the pump discharge might not provide satisfactory hole-cleaning in the riser. Keeping the hole free of cuttings is of utmost importance to eliminate the possibility of any pack-off, leading to lost-circulation / stuck-string.

There is only a small requirement to prepare fresh mud for this section as almost enough mud will be available on board after 10-3/4 " casing cement job and from the increased volume due to the base oil addition for lowering the mud weight to 10.7 ppg. Preparation of fresh mud should be minimized as much as possible by proper planning, bearing in mind that at the end of the section, additional volume will be available and ever-increasing volume causes logistics related problems. Having said this, enough products should be kept on board the rig to build at least 2000 bbls mud, should the situation demand. As the mud weight is being changed in this interval the choke and kill lines should also be displaced with the same type and weight of the fluid.

Stringent control of mud weight will be required whilst drilling this well, as there are low fracture gradients, and the anticipated FIT at the 10-3/4 " casing shoe is 11.5 ppg. Maintaining a good rheological profile is also essential, to ensure that good transport of cuttings is maintained in the riser and the ECD is not too high. Ensure circulation is broken slowly after trips or periods of non-circulation to ensure pump pressures do not become too high and break down the formation. The circulation rate may need to be maximized and the ROP controlled to prevent cuttings loading. This can be also seen from the APWD tool, and a decision can be made accordingly. Another equally important part is to have enough sized bridging material in the system to minimize / eliminate seepage losses as well as to minimize filtrate invasion, so ensure 25 ppb of 500 Microns Calcium Carbonate, 15 ppb Sized CaC03 and 6 to 7 ppb of Graphite is maintained in the active system.

Principle properties of 8-1/2" phase mud:

Property	**Value**	**Remarks**
Mud Weight	10.7 ppg	
OWR	70/30	
Electrical Stability	> 300 v	
HTHP – 500 psi	2 to 2.5 ml	All Oil
PV	ALAP	
YP lbs / 100 ft2	+/- 22	
6 RPM reading	11	
Excess Lime	5 ppb	
AW	0.76	+/- 0.01

The 8-1/2" section presents less challenge for hole-cleaning than the last section, with every meter drilled representing 0.23 bbl of cuttings. Maximizing the flow rates and ensuring that the booster pump is run as frequently as possible can achieve effective removal of drilled cuttings. High viscosity sweeps should be pumped every three stands drilled, as well as at section TD prior to a wiper trip and again when back on the bottom after the wiper trip. The pumping of sweeps should be closely monitored with the ECD value obtained from APWD tool and, the packing-off tendency of the string if any, during connections. Thus, a real time-based decision can be taken. However, if ECD values do not show any spike while hi-vis sweep is passing through the annulus, then the default decision should be to pump sweeps as long as the YP does not exceed 27. At section TD sweeps should be pumped both prior to and after the wiper trip and the hole circulated until the shakers are clean even if this is longer than a complete circulation.

A close watch has to again be kept on possible 'ballooning' – see previous section for details.

The rat hole needs cleaning properly on the way out of the hole – see previous section for details.

Acid soluble lost circulation material of various types should be kept on the rig at all times – see previous interval for details.

It should be noted that if large-scale losses occur during this interval, then the hole should not be continually filled with synthetic-based mud. The hole should be filled from the top with water until equilibrium is reached and the equivalent mud weight can be worked out. Once this has been done the mud system can be reduced in weight until the losses are cured. This procedure will prevent unnecessary mud being pumped and lost down-hole.

Drilling fluid formulation - Synthetic Based Mud (70/30 Oil/ Water Ratio):

Escaid 110	:	0.60 bbl.
CaCl2 brine	:	0.30 bbl (118.5 ppb CaCl2).
Emulsifier	:	16.0 ppb.
ORGANO-PHILLIC CLAY	:	8.0 ppb.
Fluid loss control agent	:	3.5 ppb.
Lime	:	10.0 ppb.
Filtratin control ahent	:	1.5 ppb.
Calcium Carbonate	:	15.0 ppb (sized).
500 microns Calcium Carbonate	:	25 ppb
Graphite	:	6.0 ppb
Barite	:	As required for desired ppg

The formulation is only suggestive and the emphasis should be on achieving the parameters as mentioned in the program.

Prior to the cementing of the 7-5/8" liner, thin SBM (250 bbls) should be pumped to aid in better removal of mud cake, thereby improving the probability of a better cement job. This mud should have ALAP PV and YP of ~ 12. SOBM thinner should be available for thinning back this mud. Pilot tests should be done on the active mud with Versthin as it is a powerful thinner and over-treatment can cause serious problems.

Ballooning

Ballooning occurs with the opening and closing of induced (surge and swabbing) or in situ micro fractures, in a claystone, often caused during the drilling process. These fractures are elastic enough to allow the filling up with fluid under a certain pressure, and then have the ability to allow back flow of this fluid when the pressure has been relaxed, and the fractures close. Historically, this phenomenon occurs in the smaller diameter holes such as 8.5" and 6" although there have been some instances in 12.25" sections. This problem is more likely to arise, when using oil or mineral based mud such as those planned for this well.

The stable pressure environment for shale bodies is called the confining pressure, and if the hydrostatic pressure caused by the mud column is too low, the hole will tend to come in, causing sloughing shales or tight hole. If the mud weight is too high for the confining pressure, a loss of fluid to the formation is caused, which sometimes never comes back to the well-bore. In ballooning, most of the lost fluid comes back to the well-bore. Preventative measures such as adding bridging agents such as Calcium Carbonates have been tried with little effect on stemming the loss to formation, although reduction in pump rates reduces losses very quickly. Other parameters affected by ballooning are increased penetration rates, which, if left unchecked will raise the ECD's further. Severely fluctuating pump pressures, foaming, false kicks with increased return flows and high levels of gas cut mud can also be seen in a ballooning incident. With the pressure increase caused by the whole mud invasion in the claystone, adjacent sand formations are influenced, with resultant pressured fluids and gas entering the well bore creating SICP & SIDPP if the well is shut in. If the well is allowed time to flow back its gained contents, then shut in pressures are zero on both sides SIDPP & SICP.

Well-bore Instability

Well-bore instability is caused by a radical change in both the mechanical stress and the chemical and physical environments when a hole is drilled, exposing the formation to the drilling fluid. Well-bore instability is mostly seen as swelling and sticky clays, cavings, hole enlargement, bridges and hole fill. The most common consequences are stuck pipe, sidetracks, poor cement jobs as well as difficulty running drill-pipe, casing and logging tools.

If well-bore instability does occur, it is very important to work out which one of the below mechanisms is causing the problems. Only after concluding this, will it be possible to work out what is the way forward.

Unconsolidated Formations

Unconsolidated formation can fall into the well-bore if it is loosely packed with little or no bonding between particles, pebbles or boulders. The collapse of the formation is caused by the removal of the supporting rock as the well is drilled and generally the faster the ROP the faster the hole collapses. It mainly happens in a well-bore when there is little or no filter cake is present. The un-bonded formation (sand, gravel, boulders, etc.) cannot be supported by a hydrostatic overbalance as the fluid simply flows into the formation and then sand or gravel can fall into the hole and pack off the drill string. The effect can be a gradual increase in drag over a number of meters, or can be very sudden, with the mechanism normally associated with shallow formations. These formations can be controlled, with a good filter cake which stabilizes the formation, in conjunction with fine lost circulation material to control seepage loss. If possible, prevent excessive circulating time with the BHA opposite unconsolidated formations to reduce hydraulic erosion and a gel pill should be spotted before POOH. Tripping speeds should be slowed down when the BHA is opposite unconsolidated formations to avoid mechanical damage. A successful method to prevent this is to drill 10 meters, pull back to the top of the section and wait 10 minutes.

Note: Any fill on bottom when returning to drill ahead and if the fill is significant, then ensure the process is repeated every 10 meters. This way it is possible to prevent the hole collapsing entirely and allow it to stabilize by itself, with the BHA up out of danger area.

Mobile Formations

Mobile formations can squeeze into the well-bore due to being compressed by the overburden forces, and behave in a plastic manner, deforming under pressure. The deformation results in a decrease in the well bore size, causing problems running bottom hole assemblies, logging tools and casing. A deformation occurs because the mud weight is not sufficient to prevent the formation squeezing into the well bore. To prevent this from happening, a sufficient mud weight and appropriate mud system have to be chosen that will not aggravate the mobile formation. Frequent reaming/wiper trips over the troublesome section of the hole and reducing the open-hole exposure time can help minimize these problems. If the mobile formation is a salt, then consideration should be given to using a slightly under-saturated mud system to allow a controlled washout.

Fractured & Faulted Formations

A natural fracture system in the rock can often be found near faults where the rock can be broken into large or small pieces. If they are loose, they can fall into the well bore and jam the string in the hole. Even if the pieces are bonded together, the impacts from the BHA due to drill string vibration can cause the formation to fall into the well bore. This type of situation can also occur due to tectonically stressed formations being drilled and is particularly unusual in that the well-bore can become unstable and stuck pipe can occur whilst drilling ahead. When this has happened in the past, the first sign of a problem has been the string torquing up and sticking. There is a risk of well-bore instability in fractured / faulted formations, particularly when drilling through a fault and when drilling through fractured limestone formations. It is an advantage to minimize drill string vibration if at all possible and this can often be done by choosing an alternative RPM or changing the BHA configuration if high shock vibrations are observed. Slowing down the trip speed before the BHA enters a suspected fractured/faulted area and reaming this area carefully as careful movement o the the drill string to avoid pressure surges to the well bore, helps to stabilize. Generally, fractured formations require time to stabilize.

Naturally Overpressure Shale Collapse

A naturally over-pressured shale, is one with a natural pore pressure greater than the normal hydrostatic pressure gradient. Naturally over-pressured shales are most commonly caused by geological phenomena such as under-compaction, naturally removed overburden (*i.e. weathering*) and uplift. Using insufficient mud weight in these formations will cause the hole to become unstable and collapse. These shales normally occur in rapid depositional areas. To help alleviate these problems it is essential to plan a mud weight that is adequate and once the shale has been exposed the mud weight should not be lowered. It is also essential to minimize hole exposure time in these over-pressured shale sequences. Rigorous use of gas levels and the use of other information to predict pore pressure trends, for example D exponent, cavings at the shale shakers, increasing ROP etc should be monitored carefully.

Induced Overpressure Shale Collapse

Induced over-pressure shale occurs when the shale assumes the hydrostatic pressure of the well bore fluids after a number of days exposure to that pressure. When this is followed by no increase or a reduction in hydrostatic pressure in the well bore, the shale, which now has a higher internal pressure than the well bore, collapses in a similar manner to naturally over-pressured shale. This generally occurs in water based mud after a reduction in mud weight has taken place, and is most often worse in the casing rat hole. This situation is easily avoidable if non-water based mud is used and it is not planned to reduce the mud weight after exposing large sequences of shale. Indications of the problem include splintery cavings showing no sign of hydration, tight hole in the rat hole area and hole fill on bottom.

Reactive Formations

This is when a water sensitive shale is drilled with less inhibition than is required. The shale absorbs the water and swells into the well bore with the reaction being 'time dependent', as the chemical reaction takes time to occur. This time dependency can range from hours to days and normally occurs when using water based mud in shales and clays in young formations. It occurs when drilling with an incorrect mud specification, particularly, an insufficient concentration of inhibition additives such as salts (potassium chloride and Sodium chloride), glycol and polymer. The preventative action for this type of hole instability is to use an inhibited mud system and maintain the mud properties as planned. The addition of various salts (*potassium, Sodium,* Calcium, etc.), in particular, potassium chloride, will reduce the chemical attraction of the water to the shale and various encapsulating (coating) polymers, particularly PHPA, can be added to WBM mud to reduce water contact with the shale. Regular wiper trips or reaming trips may help if shale begins to swell. Indications of this happening include, hydrated or mushy cavings at the shale shakers, increase in LGS, filter cake thickness, PV, YP, MBT and an increase or fluctuations in pump pressure. It is advisable to POOH to prevent swabbing if this is thought to be occurring.

Tectonically Stressed Formations

Well bore instability is caused when highly stressed formations are drilled and there exists a significant difference between the near well bore stress and the restraining pressure provided by the drilling fluid density. Tectonic stresses build up in areas where rock is being compressed or stretched due to movement of the earth's crust. The rock in these areas is being buckled by the pressure of moving tectonic plates. When a hole is drilled in an area of high tectonic stresses the rock around the well bore will collapse into the well bore and produce splintery cavings similar to those produced by over-pressured shale as well as large blocky cavings. In the tectonic stress case, the hydrostatic pressure required to stabilize the well bore may be much higher than the fracture pressure of the other exposed formations. Preventative measures, include casing-off these formations as quickly as possible, with increasing the mud weight often the only other way to control the well bore instability. Signs of this type of well bore instability include, pack-offs and bridges, splintery and blocky cavings at the shakers, increase in torque and drag and an increase in volume of returns at the shakers relative to the hole volume drilled.

Wetting, Fluid Invasion & Pressure Transmission

Wetting along pre-existing shale micro-fractures, fluid invasion and equalizing the overbalance pressure all compromise well-bore stability. Water wetting along shale micro-fractures, is the principle cause of shale instability, with the drilling fluid filtrate or whole mud invasion, creating weaknesses along bedding planes and causing splitting and caving of the shales. Fluid invasion equalizes the overbalance pressure away from the well-bore and tends to destabilize the mechanical integrity of the well-bore, which is the case for both permeable (sandstone) and impermeable (claystone) formations. Chemical inhibition and mud weight increases can help alleviate these problems, but the best way is to seal the micro-fractures with a plugging or plastering agent such as bentonite, sulphonated asphalt or Calcium Carbonate. These chemicals form a good filter cake and help plug pores, thus preventing fluid invasion into the micro-fractures.

Recommended Fluid Properties and well / hole configuration:

Hole Sizes	**Units**	**36" Jetting**	**26" hole**	**12-1/4" investigation hole & 17-1/2" x 20" Hole**	**14-3/4"x 17" hole**	**12-1/4" x 13-1/2" hole**	**9-1/2" hole**
Casing Size	Inches			**16**	**13-5/8**	**10-3/4**	**7-5/8"**
Drilled depth	mts						
Casing Depth	mts						
Max Pore Pressure With gas effect	ppg	**N/A**	**8.9**	**9.7**	**9.7**	**10.1**	**9.8**
Planned FIT	ppg			**10.2**	**11.5**	**12.5**	**13.7**
Planned M W	ppg.	8.55	8.55	**9.8**	**9.8**	**10.2**	**10.0**
Oil/Water Ratio	%	-	-	67/33	67/33	67/33	67/33
PV	cp.	ALAP	ALAP	ALAP	ALAP	ALAP	ALAP
Yield Point	lbs/100 ft^2	>40	>40	+/- 24	+/- 24	+/- 24	+/- 22

Continues

Hole Sizes	Units	36" Jetting	26" hole	12-1/4" investigation hole & 17-1/2" x 20" Hole	14-3/4"x 17" hole	12-1/4" x 13-1/2" hole	9-1/2" hole
6 rpm Reading	Rotation meter reading	>15	>15	15-16	15-16	15-16	10-11
PPT Value	ml			+/-2	+/-2	+/-2	+/-2
HTHP Fluid Loss	cc/30 min	-	-	< 3 (all oil)	< 3 (all oil)	< 3 (all oil)	< 3 (all oil)
pH		9.5	9.5	-	-	-	-
Excess Lime	ppb	-	-	5	5	5	5
MBT	ppb.	>25	>25	-	-	-	-
AW		-	-	+/- 0.78	+/- 0.78	+/- 0.78	+/- 0.78
ES	Volts	-	-	> 300	> 300	> 300	> 300
Type of mud		WBM	WBM	SBM	SBM	SBM	SBM

The mud weight is suggested only for planning purposes and the actual number could depend upon then existing hole conditions and FIT values obtained in respective section.

CHAPTER 19

LESSER KNOWN / LESSER USED TYPES OF MUD

There are certain types of drilling related operations which are not so common as the rest, which have their own specific requirement related to mud system management. However, small it may seem, but when properly executed, they too add value to the **operations.**

1: COIL TUBING DRILLING:

The coil tubing is a continuous string of pipe without connections over most of its length. Its inherent properties provide strength, toughness and flexibility that allows the pipe to be spooled onto a drum, straightened as it is run into a well, perform under a variety of harsh conditions, and then be retrieved and re-spooled numerous times without failing.

The fact to remember is, mud will always perform the same role / function as expected, but due to change in extraneous as well as rock-matrix related features, the composition changes to suit to the changed conditions. The diameter of the hole in this case is very small, which leads to the even smaller annulus. Therefore, all precautions / measures that are required to be taken for drilling through a restricted annular passage apply here with the fact that the annulus is not restricted due to deposition of particles of mud or any external source. Since the original diameter planned for the hole is also small, we need to ensure that the content of the mud should cater for this change. Under such circumstances we need reasonably low lifting/carrying capacity and minimum solids in mud, which are just enough to provide, the body to mud. Often, the amount of solids which gets picked up during drilling itself are adequate and one can start drilling straight away with the brine of appropriate weight and F/L control as well as rheology imparting additives to begin with. The discharge of the pump should be kept in the range which does not generate additional ECD and yet has lifting/carrying capacity just sufficient to flush the cuttings out. Being almost solids-free, the ECD will anyway be lower. Efforts also to be focused on boosting low end rheology as YP values cannot be kept too high. Some bullet points below should suffice to explain and help manage drilling fluids to suit these conditions.

(1) Drilling fluid design to focus on imparting Low shear rate viscosity by using visco-elastic surfactants and solids free heavy brine or normal brine with ultra-fine, but heavy weighing agents whichever is more cost effective. The final composition often comes from the mud weight requirement and cost considerations.
(2) There may not be any real need to use SOBM. However, if at all, it is an option for any specific reason, Pre-shear the SBM in mud plant and obtain desired Emulsion strength in the first place, because unlike larger rigs CT rigs cannot provide rigorous shearing to SBM during circulation.
(3) This also calls for very careful selection of emulsifier to rely more on the chemical mechanism of emulsion strength rather than mechanical shearing.
(4) Lubricants may be required if the long horizontal drain hole is to be drilled. Consider using Glycol in WBM to facilitate lubricity as well as shale inhibition. SBM need not have any lubricant.
(5) Solids control equipment is not a real necessity. Finest screen size can often suffice to expel the cuttings in first place.

(6) Shaker screens as fine as > 300 mesh can be installed to enable discarding of entire drilled solid and yet enable retain the essential weighing agent, which is so small in size that it can pass through such fine screens and stay in the system.

2: HORIZONTAL DIRECTIONAL DRILLING (HDD)

This type of drilling is mostly carried out at a very shallow depth, but the drain hole length is often long, as the aim of such drilling is to facilitate laying of pipeline for carrying oil/gas through underground.

Unlike normal drilling, formation damage is not an issue here, nor do we need higher mud weight to counter any Geo-pressured rock with trapped hydrocarbon/s. Although the hole size is usually large (30 – 40 inches) since there is not enough depth, the hole-cleaning can be managed by simple suspension of Bentonite without worrying about any additional F/L control. The challenges posed by such drilling are often of mechanical nature due to hard rock which is not weathered enough to lose its strength, but collapsing near the well-bore due to the stresses being released. Often, such holes find connectivity to large water bodies at shallow depth, leading to complete loss of return. Of late it is becoming fashionable to use polymers, which is nothing else but a thoughtfully planned marketing trap by manufacturers of such chemicals. Bentonite serves the purpose better than polymers and by selecting the right pH and dose of Bentonite, one can play around the desired lifting capacity to lift even the large boulders. The abnormally high drag observed in a long drain hole and inability to lift he boulders from the hole, coupled with loss of circulation are the main challenges and following steps should eliminate/minimize them

Use a lubricant with Bentonite. There are plenty of lubricants available in the market and anyone should suffice. Since there is no high temperature, high pressure condition encountered here, simple, cheap lubricant should do the job.

One can also think of drilling with water and sweep high-vis pills of Bentonite to lift cuttings or push them in large thief zone, should there be complete loss of circulation.

Strong doses of PHPA in the form of slug can also be pumped to maximize lubrication if no other lubricant performs to satisfaction.

Ideally, Bentonite suspension should be adequate for drilling and PHPA sweeps should do the job of the hi-vis pill as well as the lubricant.

The role of the mud engineer in such operations is still not widespread. Often operators manage with a guy who can simply throw some bags of Bentonite in water and a sack of caustic here and there. The sector is still unorganized as far as a mud engineering concept is concerned. The encouraging thing is, it is slowly catching up the attention of some operators and simultaneously the discouraging fact is – they are falling a prey to fancy chemicals being pushed upon them by smart marketing people.

3: COAL BED METHAND DRILLING (CBM)

This is a newly developed technology to exploit cleanest form of energy trapped between coal cleats which occur in conjunction with other lithological sequences. Being clean and completely combustible gas, Methane ($C\ H_4$) has caught the attention of oil/gas companies equal as that of government agencies. The drilling fluid, in this case should be formation non-damaging due to the inherent nature of such rocks getting seriously impaired even with slight damage caused.

There are various types of fluids which has attracted attention and under-balanced drilling is one potent type gaining attention as there is almost zero chance of rock damage because no fluid can ever enter the rock in UBD technique.

Plain water, air, foam and mist are some of the types of fluids which are used in abundance. Various types of surfactants are used to create foam and cuttings are expelled with pneumatic pressure applied in the annulus by external sources. In between Bentonite sweeps have been also pumped sporadically to clean the hole. At times safety features negate the benefits accrued by UBD techniques and a brine based inhibited fluid often comes at rescue. Many companies have come up with their own specialized systems and the final choice is often a suitable compromise between safety and formation damage.

CHAPTER 20

INTERDISCIPLINARY INTERACTIONS ON RIG

Since mud has got several roles to perform, it is quite natural that the quality of mud will have an impact on the performance of other associated disciplines. Converse is also true. Consider this as a part of the body which affects and gets affected by the health of other parts of the body. The dynamics of this interaction impacts the outcome and one can't stay aloof to this. To make the discussion appear more friendly and realistic, let us pick up the interdependence with each discipline.

Driller / drilling shift crew and mud engineer.

This is the most frequently interacted discipline and action, lack of action or inaction of one affects the other almost always. Very popularly and colloquially mud is also referred as the blood of the well. Whether a mud engineer likes it or not, his role is to support the process of drilling, which means in a way, a mud engineer should take a step back when it comes to final decision-making process which can affect his work! The situation sometimes gets sensitive when one of the two personalities gets the feeling that they are trying to play mind game with each other! A driller will always invariably want that the mud should facilitate faster drilling and he is very well with-in his right to think so. The trouble arises when there is very little a mud engineer can technically actually do to facilitate and driller is not prepared to accept that. Similarly, driller invariably has an opinion that he should have first access to the equipment available on the rig, which are also required by a mud engineer to carry out his job. However genuine both may be in terms of priority of deployment of shared resources, preferably one over the other, both invariably feel that they are at receiving end when the other guy gets it preferentially over him. Driller is the in-charge of the entire crew of his shift and mud engineer always depends upon many crew, including but not restricted to Derrick man. Basically, it is a conflict of interest, which can be easily overcome if both look at the situation from each other's perspective instead of their own. It is easier said than done, as at times both lock horns on resource sharing. The easier way of handling this is to build a relationship of faith and respect for each other's requirements, while remembering that nothing personally belongs to either of them. Some of the best friends and worst enemies can be seen in this love-hate relationship where emotions sometimes run high.

Some tricky practical situations, which trigger conflict is

Driller and mud engineer both requiring crane at the same time

Driller and mud engineer requiring derrick-man at the same time

Driller and mud engineer requiring mixing hands at the same time

Driller feeling frustrated due to splash of mud while POOH, whereas the mud engineer not being able to do anything about it.

Any anomaly in the well, whether well collapsing due to mechanical stresses or excess cavings returning on shakers – mud engineer is a soft target. The list could be endless, but the point to be highlighted here is not the length of the list, but the interaction with other discipline.

Geologist and mud engineer

This one is not that intense and is manageable, fairly- easy. There are some specific cases where a geologist might suggest a mud engineer to avoid treating the mud with any additive which can interfere his efforts to pick up an oil show by checking the fluorescence of the cuttings. Most of the times, this does not turn into a conflict. Some demanding types of geologists or mud engineers sometimes do rub each other on the wrong side of shoulders, but that seldom gets serious to the extent that their job gets affected adversely by each other.

Mud logger and mud engineer

They do get well with each other because of interdependence on each other to perform their respective roles in a professional manner.

Material-man and mud engineer

This interaction is often situation based and never permanently good or bad. Mud engineer feels left alone when his timely requests to get his stuff from the shore / warehouse does not get attention even in comparison to last minute request of drilling discipline. Without getting into the debate of who was right and wrong, often the company-man comes to each other's rescue and they then live happily together till recurrence of another similar situation.

Logger and mud engineer

They do not have much common in terms of interaction except the fact that logger requests for filtrate sample prior to start of open hole logging, which is shared by the mud engineer without any qualm. At times logger may also seek information on parameters and some of the properties of mud, which are happily shared. The only time when they seem to be frustrated at each other is the logging tool, not reaching to the bottom and logger expressing his opinion about the condition of mud without knowing much about it.

Company-man and mud engineer

Perhaps this is the most interesting interactions among all. A company-man is the ultimate authority on all matters of rig and to be fair to him, he needs to be in that position to deliver desired results. While everyone agrees that he has more knowledge than the rest on the rig, but equally true is the fact that no one knows the mud better than a mud engineer on-board. Most of the times company-man is wise enough to realize (but never admit) and gives the freedom to mud engineer for running the mud as he thinks is the best. Few headstrong mud engineers can upset company-man by behaving like an autocrat and always boasting of his independence in their body language / gestures which can give a testing time for patience of company-man. Some not so wise company-men may sometimes carry this false impression despite mud engineer being nice – due to their experience. When two strong personalities interact, conflict is inevitable and it can sometimes motivate both to give their best, which is a healthy output. The real world However, is not ideal always and when such personalities misunderstand each other, the ultimate sufferer is either the hole or the mud engineer. A situation of well-control often requires meticulous coordination between company man and mud engineer as being an emergency, company-man takes over the reins in his control.

Cementer and Mud engineer

There is a situation based interaction with cementer, when a cement slurry pumping and displacement job required to be carried out for various reasons, most often, to place cement behind the casing lowered. The placement of cement slurry behind casing in an open hole may at first glance appear like simply pumping and displacing a fluid, but it has huge impact on then overall health and longevity of the hole and should never be taken lightly. Specialized cement slurry design is invariably required meeting specific requirement/s depending upon the conditions of the hole.

Cementer often requires to know the salinity of the water with which the slurry is going to be prepared and mud engineer comes at rescue. Apart from it keeping a close track of volume changes during all stages of cement job is perhaps the most important job of a mud engineer in this operation. Pre-planning is required to have enough volume available in the active pit to preferably accommodate the entire return mud in the active tank without the need to make any changes once the job execution has started. A clear split of volume changes during each job is also a requirement. Although now a-days a lot of automation has taken place to monitor these changes, but often one needs to verify these automated figures with actual figures as noted by mud engineer. Therefore, precision and accuracy in tracking these changes are always a preferred option. This clearly indicates whether the well-bore was completely stable and healthy during the entire job. Often, loss of mud is experienced during placement of the slurry in annulus due to restricted passage and a nicely conditioned mud to the lowest possible and realistic rheology, coupled with completely removed filter-cake goes a long way in achieving this goal. Despite automating Densitometers constantly measuring the density of slurry going in the hole, a physical check in between with the help of a pressurized mud balance is always a welcome step. During pumping of slurry, collection of representative samples of slurry is another responsibility of the mud engineer which attracts attention, especially if the flaws in job execution, if any needs to be scrutinized later to find the underlying root cause. These samples are then kept at the surface as well as in a pre-set constant temperature oven at BHST to get an idea of the time a slurry shall take to harden enough for taking a known decision in terms of resuming the operations which require to allow the slurry to completely set.

Although pit management is often a joint responsibility of cementer and the mud engineer, it is the mud engineer who plays an active role by this part of operation falling in his domain area. At times, cementer requires preparation of mix water to prepare the slurry in his unit and this requires adequate cleaning of the pit/s and transfer lines to ensure that the mix water does not pick up any contamination of extraneous matter, as that can be detrimental to slurry properties, leading to deterioration in the ability of slurry to perform the role/s expected of it. Pumping a slurry requires handling lots of chemicals, for which a cementer is often required to share the sack-room space, otherwise being singularly used by mud engineer. A healthy relationship between a mud engineer and a cementer often makes the entire process conducive to better achievements of results.

Radio operator and mud engineer

This too is a smooth relationship and it hardly reaches a stage where it attracts attention publicly for the wrong reasons.

There are many more disciplines and personalities and one which deserves special mention is the Offshore installation manager (OIM) on offshore and the rig - manager / client manager on land location. This gentleman looks after the rig as his top priority and a mud engineer seldom reacts directly with him, unless they are personally good to each other. The interaction of mud engineer and OIM are invariably through company-man

HSE representative, Electrician, Barge engineer and Mechanic are some other important inter-disciplinary interaction nodes. In an unlikely situation of any rift between any of them, company-man invariably mediates.

Conflict resolution.

The oil/gas filed professionals are by and large very confident type of personalities and almost all of them suffer from their own share of frustration/stress of staying away from home for longer periods. This is reflected in their behavior which can become explosive sometimes. When two such personalities interact, conflict can erupt. To eliminate such situation, a sense of humor at the right time in right proportion and a right spirit, can diffuse this otherwise tough situation. Therefore, it helps to share a coffee with them sometimes and indulge in fun activities as and when possible. These small moments are real stress busters and it not only adds to the productivity, but also keeps everyone healthy by providing them some light moments which help forget / minimize the agony of staying away from loved ones. Another equally important factor which adds to the unrest among the people is the pressure of delivering and dealing with unknown always all the time. Although the technology has advanced to a stage where a lot of information is available on screen, but nature has its own ways of playing with them and still pose challenges, not known before the occurrence. It is like a clock ticking at the back of your mind and sometimes it can take on your nerves. Stay calm, composed and try not to be aloof on the rig. This simple feature can go a long way in eliminating situations of conflict. Should you still get into such unpleasant situation? Think of it this way – the other guy deserves your appreciation and sympathy as he is unable to find a vent valve to off-load his frustration. Walk away calmly from there and speak to the same person tomorrow morning, as if nothing untoward happened between you and him. You will see he is completely different and better person by that time!

Expanded version of the abbreviations used	
ppg	Pounds per gallon
bph	Barrels per hour
Bbl/hr	Barrels per hour

Continues

S G	Specific gravity
Lbs/100 ft2	Pounds per 100 square feet
Cp	Centi poise
Yp	Yield point
AV	Apparent viscosity
PV	Plastic viscosity
Gels 10 s	Gel strength after 10 seconds static
Gels 10 m	Gel strength after 10 min static
Temp.	Temperature
Deg F	Degree/s farenheit
Deg C	Degreed Celsius or centigrade
psi	Pounds per square inch
Kg/m3	Kg per cubic metre
bbl	Barrel
Sce/qt	Seconds per quart
Mud wt.	Mud weight
M W	Mud weight
g/l	Grams per litre
gms	Grams
cc	Cubic centimeter
M^3	Cubic meter
Ft^3	Cubic feet
Lb /Lbs	Pound / Pound/
Cm^2	Centimeter square
F I T	Formation integrity test
L O T	Leak-off test
A LA P	As low as possible
N/A	Not applicable
O W R	Oil water ratio
HTHP	High-temperature high pressure
O I M	Off-shore installation manager
ml	Milli-liter

Continues

E S	Electrical stability
ppb	Pounds per barrel
Cc/30 min	Cubic centimeter per 30 minutes
PPT	Permeability plug tester
SIDPP	Shut-in drill-pipe pressure
SICP	Shut-in casing pressure
RPM	Revolutions per minute
SOBM	Synthetic oil base mud
ECD	Equivalent circulating density
WBM	Water base mud
OBM	Oil base mud
BHCT	Bottom hole circulating temperature
BHST	Bottom hole static temperature
C T	Coil tubing
SBM	Synthetic base mud
DIF	Dril-in fluid
CBM	Coal bed methane
HDD	Horizontal directional drilling
TVD	True vertical depth
ERD	Extended reach drilling
TD	Target depth
UBD	Under-balanced drilling
MPD	Managed pressure drilling
F M C	Fine Medium Coarse
NIF	Non-invasive fluid
APWD	Annular pressure while drilling
MWD	Measurement hile drilling
LWD	Logging while drilling
TLC	Tough logging conditions
CBL	Cement bond log
VDL	Variable density log
SP	Standard potential

Continues

GR	Gamma ray
PHPA	Partially hydrated poly-acryl amide
PHB	Pre-hydrated bentonite
CMC	Carboxy-methyl cellulose
PAC	Poly-aninic cellulose
MBT	Methylene blue test
CEC	Cation exchange capacity
HEC	Hydroxy-ethyl cellulose
MEG	Mono-ethylene glycol
MEA	Mono-ethanol amine
IPA	Iso-propyl alcohol
LGS	Low gravity solids
HGS	High gravity solids
EDTA	Ethylene di-amine tetra acetic acid
BRT	Below rotary table
MSL	Mean sea level
ART	Above rotary table
API	Americal petroleum indstitute
IBP	Initial boiling point
FBP	Final boiling point
F/L	Fluid loss
POOH	Pull out of hole
RIH	Run in hole
D/P	Drill-pipe
D/C	Drill-collar
DST	Drill stem test
ESD	Equivalent static density
F L	Flow-line
FLT	Flow-line temperature
ROP	Rate of prevention
BOP	Blow out preventor
Cv	Criticle velocity

Continues

Av	Annular velocity
IADC	International association of drilling contractors
HPWBM	High performance water base mud
BHA	Bottom hole assembly
CTU	Coil-tubing unit
MD BRT	Measured depth below rotary table
TVD BRT	True vertical depth below rotary table
PAD	Pump and dump
XC	Xanthomnas compestris
LCM	Loss ontrol material
MT	Metric Tonne
m	Metre
NPT	Non-productive time
Ppm	Parts per million
mg/l	Mill grams per litre
NTU	Nephelo-turbiditic unit
CBF	Clear brine fluid
CRA	Corosion resistant alloy
BHT	Bottom hole temperature
SSC	Sulphie tress cracking
Gal/bbl	Gallons per barrel
Lbs/bbl	Pounds per barrel
SAPP	Sodium Ammonium Pyro-phosphate
Pf	Filtrate phenolphthaleine alkalinity
mf	Filtrate methyl orange alkalinity
pm	Mud alkalinity
LO	Low
HI	High
VIS	Viscous
OEDP	Open End Drill-Pipe
MDT	Modular formation dynamic tester

Made in the USA
Las Vegas, NV
29 June 2026

50063225R00239